APPLIED CHEMISTRY AND CHEMICAL ENGINEERING

Volume 5

Research Methodologies in
Modern Chemistry and Applied Science

APPLIED CHEMISTRY AND CHEMICAL ENGINEERING

Volume 5

Research Methodologies in
Modern Chemistry and Applied Science

Edited by

A. K. Haghi, PhD
Ana Cristina Faria Ribeiro, PhD
Lionello Pogliani, PhD
Devrim Balköse, PhD
Francisco Torrens, PhD
Omari V. Mukbaniani, PhD

AAP APPLE ACADEMIC PRESS

Apple Academic Press Inc.
3333 Mistwell Crescent
Oakville, ON L6L 0A2 Canada

Apple Academic Press Inc.
9 Spinnaker Way
Waretown, NJ 08758 USA

Library and Archives Canada Cataloguing in Publication

Applied chemistry and chemical engineering / edited by A.K. Haghi, PhD, Devrim Balköse, PhD, Omari V. Mukbaniani, DSc, Andrew G. Mercader, PhD.
Includes bibliographical references and indexes.
Contents: Volume 1. Mathematical and analytical techniques --Volume 2. Principles, methodology, and evaluation methods --Volume 3. Interdisciplinary approaches to theory and modeling with applications --Volume 4. Experimental techniques and methodical developments --Volume 5. Research methodologies in modern chemistry and applied science.
Issued in print and electronic formats.
ISBN 978-1-77188-515-7 (v. 1 : hardcover).--ISBN 978-1-77188-558-4 (v. 2 : hardcover).--ISBN 978-1-77188-566-9 (v. 3 : hardcover).--ISBN 978-1-77188-587-4 (v. 4 : hardcover).--ISBN 978-1-77188-593-5 (v. 5 : hardcover).--ISBN 978-1-77188-594-2 (set : hardcover).
ISBN 978-1-315-36562-6 (v. 1 : PDF).--ISBN 978-1-315-20736-0 (v. 2 : PDF).-- ISBN 978-1-315-20734-6 (v. 3 : PDF).--ISBN 978-1-315-20763-6 (v. 4 : PDF).-- ISBN 978-1-315-19761-6 (v. 5 : PDF)
1. Chemistry, Technical. 2. Chemical engineering. I. Haghi, A. K., editor

| TP145.A67 2017 | 660 | C2017-906062-7 | C2017-906063-5 |

Library of Congress Cataloging-in-Publication Data

Names: Haghi, A. K., editor.
Title: Applied chemistry and chemical engineering / editors, A.K. Haghi, PhD [and 3 others].
Description: Toronto ; New Jersey : Apple Academic Press, 2018- | Includes bibliographical references and index.
Identifiers: LCCN 2017041946 (print) | LCCN 2017042598 (ebook) | ISBN 9781315365626 (ebook) | ISBN 9781771885157 (hardcover : v. 1 : alk. paper)
Subjects: LCSH: Chemical engineering. | Chemistry, Technical.
Classification: LCC TP155 (ebook) | LCC TP155 .A67 2018 (print) | DDC 660--dc23
LC record available at https://lccn.loc.gov/2017041946

Apple Academic Press also publishes its books in a variety of electronic formats. Some content that appears in print may not be available in electronic format. For information about Apple Academic Press products, visit our website at **www.appleacademicpress.com** and the CRC Press website at **www.crcpress.com**

ABOUT THE EDITORS

A. K. Haghi, PhD

A. K. Haghi, PhD, holds a BSc in Urban and Environmental Engineering from the University of North Carolina (USA), an MSc in Mechanical Engineering from North Carolina A&T State University (USA), a DEA in applied mechanics, acoustics and materials from the Université de Technologie de Compiègne (France), and a PhD in engineering sciences from the Université de Franche-Comté (France). He is the author and editor of 165 books, as well as of 1000 published papers in various journals and conference proceedings. Dr. Haghi has received several grants, consulted for a number of major corporations, and is a frequent speaker to national and international audiences. Since 1983, he served as professor at several universities. He is currently Editor-in-Chief of the *International Journal of Chemoinformatics and Chemical Engineering* and the *Polymers Research Journal* and on the editorial boards of many international journals. He is also a member of the Canadian Research and Development Center of Sciences and Cultures (CRDCSC), Montreal, Quebec, Canada.

Ana Cristina Faria Ribeiro, PhD

Ana C. F. Ribeiro, PhD, is a researcher in the Department of Chemistry at the University of Coimbra, Portugal. Her area of scientific activity is physical chemistry and electrochemistry. Her main areas of research interest are transport properties of ionic and non-ionic components in aqueous solutions. She has experience as a scientific adviser and teacher of different practical courses. Dr. Ribeiro has supervised master degree theses as well as some PhD theses, and has been a theses jury member. She has been referee for various journals as well an expert evaluator of some of the research programs funded by the Romanian government through the National Council for Scientific Research. She has been a member of the organizing committee of scientific conferences, and she is an editorial member of several journals. She has received several grants, consulted for a number of major corporations, and is a frequent speaker to national and international audiences. She is a member of the Research Chemistry Centre, Coimbra, Portugal.

Lionello Pogliani, PhD

Lionello Pogliani, PhD, was Professor of Physical Chemistry at the University of Calabria, Italy. He studied Chemistry at Firenze University, Italy, and received his postdoctoral training at the Department of Molecular Biology of the C. E. A. (Centre d'Etudes Atomiques) of Saclay, France, the Physical Chemistry Institute of the Technical and Free University of Berlin, and the Pharmaceutical Department of the University of California, San Francisco, CA. Dr. Pogliani has coauthored an experimental work that was awarded the GM Neural Trauma Research Award. He spent his sabbatical years at the Centro de Química-Física Molecular of the Technical University of Lisbon, Portugal, and at the Department of Physical Chemistry of the Faculty of Pharmacy of the University of Valencia-Burjassot, Spain. He has contributed nearly 200 papers in the experimental, theoretical, and didactical fields of physical chemistry, including chapters in specialized books. He has also presented at more than 40 symposiums. He also published a book on the numbers 0, 1, 2, and 3. He is a member of the International Academy of Mathematical Chemistry. He retired in 2011 and is part-time teammate at the University of Valencia-Burjassot, Spain.

Devrim Balköse, PhD

Devrim Balköse, PhD, is currently a faculty member in the Chemical Engineering Department at the Izmir Institute of Technology, Izmir, Turkey. She graduated from the Middle East Technical University in Ankara, Turkey, with a degree in Chemical Engineering. She received her MS and PhD degrees from Ege University, Izmir, Turkey, in 1974 and 1977, respectively. She became Associate Professor in Macromolecular Chemistry in 1983 and Professor in process and reactor engineering in 1990. She worked as Research Assistant, Assistant Professor, Associate Professor, and Professor between 1970 and 2000 at Ege University. She was the Head of the Chemical Engineering Department at the Izmir Institute of Technology, Izmir, Turkey, between 2000 and 2009. Her research interests are in polymer reaction engineering, polymer foams and films, adsorbent development, and moisture sorption. Her research projects are on nanosized zinc borate production, ZnO polymer composites, zinc borate lubricants, antistatic additives, and metal soaps.

Francisco Torrens, PhD

Francisco Torrens, PhD, is lecturer in physical chemistry at the Universitat de València in Spain. His scientific accomplishments include the first implementation at a Spanish university of a program for the elucidation of crystallographic structures and the construction of the first computational chemistry program adapted to a vector facility supercomputer. He has written many articles published in professional journals and has acted as a reviewer as well. He has handled 26 research projects, has published two books and over 350 articles, and has made numerous presentations.

Omari V. Mukbaniani, D.Sc.

Omari Vasilii Mukbaniani, DSc, is Professor and Head of the Macromolecular Chemistry Department of Iv. Javakhishvili Tbilisi State University, Tbilisi, Georgia. He is also the Director of the Institute of Macromolecular Chemistry and Polymeric Materials. He is a member of the Academy of Natural Sciences of the Georgian Republic. For several years he was a member of the advisory board of the *Journal Proceedings of Iv. Javakhishvili Tbilisi State University* (Chemical Series) and contributing editor of the journal *Polymer News* and the *Polymers Research Journal*. He is a member of editorial board of the *Journal of Chemistry and Chemical Technology.* His research interests include polymer chemistry, polymeric materials, and chemistry of organosilicon compounds. He is an author more than 420 publications, 13 books, four monographs, and 10 inventions. He created in the 2007s the "International Caucasian Symposium on Polymers & Advanced Materials," ICSP, which takes place every other two years in Georgia.

Applied Chemistry and Chemical Engineering, 5 Volumes

Applied Chemistry and Chemical Engineering,
Volume 1: Mathematical and Analytical Techniques
Editors: A. K. Haghi, PhD, Devrim Balköse, PhD, Omari V. Mukbaniani, DSc, and Andrew G. Mercader, PhD

Applied Chemistry and Chemical Engineering,
Volume 2: Principles, Methodology, and Evaluation Methods
Editors: A. K. Haghi, PhD, Lionello Pogliani, PhD, Devrim Balköse, PhD, Omari V. Mukbaniani, DSc, and Andrew G. Mercader, PhD

Applied Chemistry and Chemical Engineering,
Volume 3: Interdisciplinary Approaches to Theory and Modeling with Applications
Editors: A. K. Haghi, PhD, Lionello Pogliani, PhD, Francisco Torrens, PhD, Devrim Balköse, PhD, Omari V. Mukbaniani, DSc, and Andrew G. Mercader, PhD

Applied Chemistry and Chemical Engineering,
Volume 4: Experimental Techniques and Methical Developments
Editors: A. K. Haghi, PhD, Lionello Pogliani, PhD, Eduardo A. Castro, PhD, Devrim Balköse, PhD, Omari V. Mukbaniani, PhD, and Chin Hua Chia, PhD

Applied Chemistry and Chemical Engineering,
Volume 5: Research Methodologies in Modern Chemistry and Applied Science
Editors: A. K. Haghi, PhD, Ana Cristina Faria Ribeiro, PhD, Lionello Pogliani, PhD, Devrim Balköse, PhD, Francisco Torrens, PhD, and Omari V. Mukbaniani, PhD

CONTENTS

LIST OF CONTRIBUTORS

Miguel A. Aguilar
Centro de Investigación y de Estudios Avanzados del Instituto Politécnico Nacional (Unidad Saltillo), Av. Industria Metalúrgica 1062, Parque Industrial Saltillo-Ramos Arizpe, 25900 Ramos Arizpe, Coahuila, México

Cristóbal N. Aguilar
Autonomous University of Coahuila. School of Chemistry, Department of Food Science and Technology, 25280 Saltillo, Coahuila, México. E-mail: cristobal.aguilar@uadec.edu.mx

Miguel A. Aguilar-González
CINVESTAV, Center for Research and Advanced Studies, IPN Unit Ramos Arizpe, Coahuila, Mexico

Olga B. Alvarez-Pérez
Department of Food Research, School of Chemistry, Universidad Autónoma de Coahuila, Saltillo 25280, Coahuila, Mexico

Ayten Baghiyeva
Institute of Catalysis and Inorganic Chemistry, Azerbaijan National Academy of Sciences, H.Javid Ave. 113, AZ 1143, Baku, Republic of Azerbaijan

Devrim Balköse
Department of Chemical Engineering, Izmir Institute of Technology, Izmir, Turkey. E-mail: devrimbalkose@gmail.com

Teuku Beuna Bardant
Indonesian Institute of Science, Jakarta, Indonesia

E. Besalú
Institut de Química Computacional i Catàlisi, Universitat de Girona, Girona 17003, Spain

Daniel Boone-Villa
Autonomous University of Coahuila. School of Medicine, 26090 Piedras Negras, Coahuila, México

Cecilia Castro-López
Autonomous University of Coahuila. School of Chemistry, Department of Food Science and Technology, 25280 Saltillo, Coahuila, México

Juan C. Contreras
Food Research Department, School of Chemistry, University Autonomous of Coahuila, Saltillo CP25280, Coahuila, Mexico

Juan C. Contreras-Esquivel
Autonomous University of Coahuila. School of Chemistry, Department of Food Science and Technology, 25280 Saltillo, Coahuila, México

Pınar Tüzüm Demir
Department of Chemical Engineering, Faculty of Engineering, Usak University, Usak TR64200, Turkey

E. García-España
Institut de Ciència Molecular, Universitat de València, Paterna, Valencia, Spain

Anna Ilyina
Facultad de Ciencias Químicas, Universidad Autónoma de Coahuila, Blvd. V. Carranza e Ing. José Cárdenas Valdés, 25280 Saltillo, Coahuila, México. E-mail: anna_ilina@hotmail.com

Güneş Boru Izmirli
Department of Chemical Engineering, Faculty of Engineering, Ege University, Bornova TR35100, Izmir, Turkey

J. V. de Julián-Ortiz
Departament de Química Física, Facultat de Farmàcia, Universitat de València, Av V. Andrés Estellés 0, Burjassot 46100, València, Spain. E-mail: jejuor@uv.es

Handan Kaplan
Hawle Armaturen GmbH, Izmir, Turkey

Guillermo C. G. Martinez-Avila
Autonomous University of Nuevo Leon School of Agronomy, Laboratory of Chemistry and Biochemistry, 66050 General Escobedo, Nuevo León, México

José L. Martínez-Hernández
Facultad de Ciencias Químicas, Universidad Autónoma de Coahuila, Blvd. V. Carranza e Ing. José Cárdenas Valdés, 25280 Saltillo, Coahuila, México.

Matanat Magerramova
Institute of Catalysis and Inorganic Chemistry, Azerbaijan National Academy of Sciences, H.Javid Ave. 113, AZ 1143, Baku, Republic of Azerbaijan

José L. Martínez
Food Research Department, Chemistry School, Coahuila Autonomous University, Saltillo Unit 25280, Coahuila, México

Gloria Alicia Martínez-Medina
Food Research Department, Chemistry School, Coahuila Autonomous University, Saltillo Unit 25280, Coahuila, México

Diana B. Muñiz Márquez
Engineering Department, Technological Institute of Ciudad Valles, National Technological of Mexico, Ciudad Valles 79010, San Luis Potosí, Mexico

Sukanchan Palit
Department of Chemical Engineering, University of Petroleum and Energy Studies, Energy Acres, Post-Office-Bidholi via Premnagar 248007, Dehradun, India E-mail: sukanchan68@gmail.com; sukanchan92@gmail.com
43, Judges Bagan, Post-Office-Haridevpur 700082, Kolkata, India

Jorge E. Wong Paz
Engineering Department, Technological Institute of Ciudad Valles, National Technological of Mexico, Ciudad Valles 79010, San Luis Potosí, Mexico

Arely Prado-Barragán
Biotechnology Department, Biological and Health Sciences Division, Metropolitan Autonomous University, Iztapalapa Unit, Ciudad de México 09340, Mexico

Sócrates Palacios-Ponce
Facultad de Ciencias Químicas, Universidad Autónoma de Coahuila, Blvd. V. Carranza e Ing. José Cárdenas Valdés, 25280 Saltillo, Coahuila, México

Rodolfo Ramos-González
CONACYT–Universidad Autónoma de Coahuila, Blvd. V. Carranza e Ing. José Cárdenas Valdés, 25280 Saltillo, Coahuila, México.

Rosa M. Rodríguez-Jasso
Department of Food Research, School of Chemistry, Universidad Autónoma de Coahuila, Saltillo 25280, Coahuila, Mexico

Rosa Ma. Rodríguez
Food Research Department, Chemistry School, Coahuila Autonomous University, Saltillo Unit 25280, Coahuila, México

Raúl Rodríguez-Herrera
Department of Food Research, School of Chemistry, Universidad Autónoma de Coahuila, Saltillo 25280, Coahuila, Mexico

Raúl Rodríguez
Food Research Department, School of Chemistry, University Autonomous of Coahuila, Saltillo CP25280, Coahuila, Mexico

Romeo Rojas
Autonomous University of Nuevo Leon School of Agronomy, Laboratory of Chemistry and Biochemistry, 66050 General Escobedo, Nuevo León, México

Orlando de la Rosa
Food Research Department, School of Chemistry, University Autonomous of Coahuila, Saltillo CP25280, Coahuila, Mexico

Héctor A. Ruiz
Facultad de Ciencias Químicas, Universidad Autónoma de Coahuila, Blvd. V. Carranza e Ing. José Cárdenas Valdés, 25280 Saltillo, Coahuila, México

Nazilya Salmanova
Azerbaijan State University of Oil and Industry, Ministry of Education, Azadlig Ave. 20, AZ 1010, Baku, Republic of Azerbaijan

Olga Sánchez
Facultad de Ingeniería Química, Instituto Superior Politécnico José Antonio Echeverría, 127 Marianao, La Havana, Cuba

R. Sathishkumar
Department of Botany, PSG College of Arts and Science, Coimbatore, India

Sevdiye Atakul Savrık
Akzo Nobel Boya A. Ş, Izmir, Turkey

Şefika Çağla Sayılgan
Chemical Engineering Department, Izmir Institute of Technology, Urla 35430, Izmir, Turkey

Elda P. Segura-Ceniceros
Facultad de Ciencias Químicas, Universidad Autónoma de Coahuila, Blvd. V. Carranza e Ing. José Cárdenas Valdés, 25280 Saltillo, Coahuila, México

Heru Susanto
Indonesian Institute of Science, Jakarta, Indonesia. E-mail: heru.susanto@lipi.go.id
Department of Information Management, College of Management, Tunghai University, Taichung, Taiwan

Ayben Top
Department of Chemical Engineering, Izmir Institute of Technology, Izmir, Turkey

Semra Ülkü*
Chemical Engineering Department, Izmir Institute of Technology, Urla 35430, Izmir, Turkey.
E-mail: semraulku35@gmail.com

Sevgi Ulutan
Department of Chemical Engineering, Faculty of Engineering, Ege University, Bornova TR35100,
Izmir, Turkey. E-mail: sevgi.ulutan@gmail.com

B. Verdejo
Institut de Ciència Molecular, Universitat de València, Paterna, Valencia, Spain

Janet M. Ventura-Sobrevilla
Autonomous University of Coahuila. School of Chemistry, Department of Food Science and
Technology, 25280 Saltillo, Coahuila, México

Ina Winarni
Forest Products Research and Development Center

Eldar Zeynalov
Institute of Catalysis and Inorganic Chemistry, Azerbaijan National Academy of Sciences, H. Javid Ave.
113, AZ 1143, Baku, Republic of Azerbaijan. E-mail: zeynalov_2000@yahoo.com

LIST OF ABBREVIATIONS

AB	amido black
ACE	angiotensin converting enzyme
ALL	acute lymphoblastic leukemias
AlO_3	aluminum oxide
AOPs	advanced oxidation processes
AR	androgen receptor
ATP	adenosine triphosphate
ATR	attenuated total reflection
AW	artificial weathering
BCC	basal cell carcinoma
BOD	biological oxygen demand
BPH	benign prostatic hyperplasis
Caco	colorectal adenocarcinoma cells
CAGR	compound annual growth rate
CB	carbon black
CCRD	central composite rotatable design
CdS	cadmium sulfide
CFMF	cross flow microfiltration
COD	chemical oxygen demand
CR	Congo red
CVD	cardiovascular disease
DBD	DNA-binding domain
DC	degree of crystallinity
DD	disc diffusion
DEFT	direct epifluorescent fiber technique
DFT	density functional theory
DHFR	dihydrofolate reductase
DSC	differential scanning calorimetry
E.U.	enzyme units
EBV	Epstein–Barr virus
EDS	energy dispersive spectroscopy
EFB	empty fruit bunch
EGFR	epidermal growth factor receptor
ER	estrogen receptors
FAO	food and agriculture organization
FDA	food and drug administration
FO	forward osmosis
FOS	fructooligosaccharides

FTIR	Fourier transform infrared
GALT	gut-associated lymphoid tissue
GG	guar gum
GI	glucose isomerase
GOS	galactooligosaccharides
GR	glucocorticoid receptor
HALS	hindered amine light stabilizer
HAP	hydroxyapatite
HDPE	high density polyethylene
HHP	high hydrostatic pressure
HLB	*hydrophile-lipophile balance*
HPRT	hydrostatic pressure resistance test
HRT	hydraulic residence time
IR	infrared spectroscopy
ISBP	in situ bioprecipitation process
LAB	lactic acid bacteria
LBD	ligand-binding domain
LDA	local density approximation
LDPE	low density polyethylene
LLDPE	linear low density polyethylene
MB	methylene blue
MCMM	Monte Carlo multiple minimum
MD	molecular dynamics
MF	microfiltration
MG	malachite green
MIC	minimum inhibitory concentration
MMFF	molecular mechanics force fields
MO	methyl orange
MOFs	metal-organic frameworks
mTOR	mammalian target of rapamycin
MTR	methyltransferase
MTX	methotrexate
MV	crystal violet
MW	microwave
NaCl	sodium chloride
NDOs	non-digestible oligosaccharides
NF	nanofiltration
NOM	natural organic matter
NPs	nanoparticles
NSC	non-structural carbohydrates
NW	natural weathering
OH	hydroxyl radical
OI	oxidation index
OIT	oxidation induction time

OIT	oxidation induction temperature
OSCC	oral squamous cell carcinomas
PAMP	pathogen-associated molecular patterns
PCNSL	primary central nervous system lymphoma
PDB	protein data bank
PDMS	polydimethylsiloxane
PE	polyethylene
PFO	pseudo first order
PHA	polyhydroxyalkanoates
PHGG	partially hydrolyzed GG
PI	isoelectric point
PLA	polylactic acid
PMMAs	polymethylmethacrylates
PRB	permeable reactive barriers
PS	polystyrene
PSA	prostate-specific antigen
PSMA	prostate specific membrane antigen
PSO	pseudo second order
PTFE	polytetrafluoroethylene
RB	rhodamine B
RCC	renal cell carcinoma
RO	reverse osmosis
RSM	response surface methodology
SARM	selective androgen receptor modulator
SASA	solvent accessible surface area
SBU	secondary building units
SCC	squamous cell carcinoma
SCFA	short-chain fatty acid
SEM	scanning electron microscopy
SF	submerged fermentation
SiO_2	silicon oxide
SmF	submerged fermentation
SPF	super critical-fluids hydrolysis
SPS	steam pretreated spruce
SSF	solid-state fermentation
SSG	sage seed gum
TGA	thermal gravimetric analysis
TLR	toll-like receptor
TMR	transparency market research
TNM	tetranitromethane
TRAMP	transgenic adenocarcinoma of the mouse prostate
TrxR	thioredoxin reductase
TS	thymidylate synthase
UF	ultrafiltration

US	ultrasound
UV	ultraviolet
VEGF	vascular endothelial growth factor
VF	virgin film
VOCs	volatile organic compounds
WAO	wet air oxidation
WAXS	wide angle X-ray scattering
WPC	whey protein concentrates
WPI	whey protein isolates
XRD	X-ray diffraction

PREFACE

Research methodologies in modern chemistry and applied sciences is an interdisciplinary and collaborative field. Most research methodologies in modern chemistry (analytical, inorganic, organic, physical, or theoretical) and applied sciences are closely related to other disciplines, including physics, biology, materials science, engineering, and medicine.

This volume, the last in the **Applied Chemistry and Chemical Engineering 5-volume set,** is designed to fulfill the requirements of scientists and engineers who wish to be able to carry out experimental research in chemistry and applied science using modern methods. Each chapter describes the principle of the respective method as well as the detailed procedures of experiments with examples of actual applications. Thus, readers will be able to apply the concepts as described in the book to their own experiments.

This volume:

- Addresses a selection of key issues in chemical technology
- Presents several new practical techniques for experimental research in the growing field of modern chemistry and applied science
- Provides well-documented presentations of the experimental methods
- Covers principles, practical techniques, and actual examples
- Presents ideas and methods from international researchers

This book traces the progress made in this field and its sub-fields and also highlights some of the key theories and their applications.

Applied Chemistry and Chemical Engineering, Volume 5: Research Methodologies in Modern Chemistry and Applied Science provides valuable information for chemical engineers and industrial researchers as well as for graduate students.

Applied Chemistry and Chemical Engineering, 5-Volume Set includes the following volumes:

- Applied Chemistry and Chemical Engineering,
  Volume 1: Mathematical and Analytical Techniques
- Applied Chemistry and Chemical Engineering,
  Volume 2: Principles, Methodology, and Evaluation Methods

- Applied Chemistry and Chemical Engineering,
  Volume 3: Interdisciplinary Approaches to Theory and Modeling
  with Applications
- Applied Chemistry and Chemical Engineering,
  Volume 4: Experimental Techniques and Methodical Developments
- Applied Chemistry and Chemical Engineering,
  Volume 5: Research Methodologies in Modern Chemistry and
  Applied Science.

PART I
Key Issues in Chemical Technology

CHAPTER 1

ADSORPTION OF MALACHITE GREEN TO SILICA HYDROGEL

AYBEN TOP[1], HANDAN KAPLAN[2], SEVDIYE ATAKUL SAVRIK[3], and DEVRIM BALKÖSE[1,*]

1Department of Chemical Engineering, Izmir Institute of Technology, Izmir, Turkey
2Hawle Armaturen GmbH, Izmir, Turkey
3Akzo Nobel Boya A. Ş., Izmir, Turkey
*Corresponding author. E-mail: devrimbalkose@gmail.com

CONTENTS

ABSTRACT

Malachite green (MG) is a poisonous dye which must be removed from wastewater before disposal. In the present study silicahydrogel which had a porous structure and might be formed in any shape by gelation of silica hydro sol was employed for the adsorption of MG from water. A transparent silica hydrogel in slab form was obtained from aqueous sodium silicate and sulfuric acid and purified by water. The equilibrium and kinetics of MG adsorption was investigated by in situ measurements with a conventional and fiberoptic spectrophotometer respectively. Adsorption equilibrium and kinetic models in both linear and nonlinear forms were applied. It was shown that nonlinear forms of these models fitted to the experimental data more thoroughly than their linear forms by giving evenly distributed errors.

1.1 INTRODUCTION

Malachite green (MG) is a triarylmethane dye with applications in aquaculture, textile, and medical industries. It is a quite effective biocide to combat protozoal and fungal infections of fish and other aquatic organisms. However, its therapeutic effects have been overshadowed by the toxicological concerns as it may enter into the mammalian bodies through food chain, resulting in carcinogenic, mutagenic, and teratogenic effects.[1–5] In textile industries, it is commonly used to dye cotton, silk, wool, jute, and leather, generating aesthetically unpleasant discharge. More importantly, the stability of this kind of dyes, coupled with their toxicity, necessitates efficient dye removal methods from the textile effluents.[2,5,6]

Flocculation, coagulation, ultrafiltration, electrochemical destruction, microbial degradation, precipitation, ion exchange, and adsorption are the proposed process alternatives to treat dye-containing wastewater. Of these processes, adsorption has been studied extensively due to its ease of operation as well as its cost-effective performance.[5–9] A wide variety of adsorbents including natural zeolites,[10,11] clays,[7,12,13] activated carbon,[14,15] fly ash,[16] fibers,[17,18] sawdust,[19] and silica gel[20] have been tested for MG adsorption.

Silica hydrogel, which is a water-containing cylindrical fiber network of silica[21,22] with chief applications in oil purification and wine and beer clarification, has also been tested for MG adsorption and diffusion. Adsorption equilibrium data were fitted to Freundlich isotherm, and diffusion coefficients of MG in silica hydrogels prepared at pH values of 4.5, 7.0, and 9.0

were measured. BET surface areas of these silica samples were found to increase with decreasing preparation pH.[23] We prepared a silica hydrogel at considerably lower pH, 1.0, to have different surface characteristics. In the scope of this study, it was aimed to investigate adsorption of MG onto this hydrogel with high surface area. Linear and nonlinear forms of equilibrium and kinetic models were tested to provide some insights about the fitting of the experimental adsorption data. Finally, we estimated apparent diffusion coefficient of MG in the hydrogel by using a simple diffusion equation and discussed the potential applications of these kinds of hydrogels.

1.2 MATERIALS AND METHODS

1.2.1 PREPARATION OF SILICA HYDROGELS

Silica hydrogel was prepared by simplifying the previously used method.[24] Basically, 50 mL of sodium silicate solution (Aldrich, $d = 1.390$ g/mL with 14% (w/w) NaOH and 27% (w/w) SiO_2) was mixed with equal volume of deionized water. While being stirred with a magnetic stirrer, diluted sodium silicate solution was added drop wise into 1.5 M of H_2SO_4 (Merck, 98%) until pH was equal to 1.0 to prepare sol. Aliquots of 20 cm³ of the sol in Schott bottles were allowed to gel approximately for a day. Silica hydrogel obtained was washed with equal volume of water ten times for 30 min in each washing. The pH of the washing water was recorded as around 1.6 after the washing process.

1.2.2 ADSORPTION EXPERIMENTS

In the adsorption experiments, oxalate salt of MG (Merck) was used as a dye source. Chemical structure of MG is given in Figure 1.1. Silica hydrogel was crushed and ground to particles with 1–3 mm size. An amount of 5 g of the hydrogel particles was contacted with 50 mL of freshly prepared MG solutions at 2, 4, 6, 10, and 15 ppm initial concentrations (C_i). The solutions were shaken for 1 week at 25°C in order to reach the equilibrium. After that, the solutions were centrifuged to separate solid and liquid phases. Absorbance values of the supernatants were recorded at 617 nm using Perkin Elmer Lambda 45 model UV/Vis spectrophotometer. These absorbance values were converted to equilibrium concentrations of MG in solution phase, C, by using the calibration curve based on the absorbance values

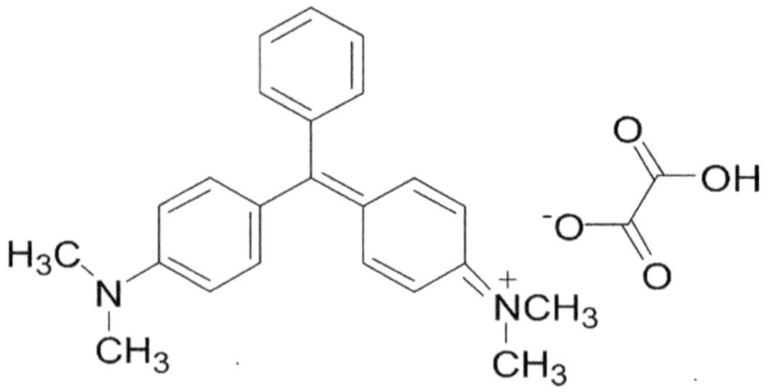

FIGURE 1.1 Chemical structure of oxalate salt of malachite green.

of known concentrations. Corresponding equilibrium concentrations of MG in hydrogel phase (q) were calculated using eq 1.1:

$$q\left[\frac{\text{mg MG}}{\text{g hydrogel}}\right] = \frac{(C_i - C)\left[\dfrac{\text{mg MG}}{l}\right]}{\text{amount of hydrogel(g)}} \times \text{solution volume}(l) \tag{1.1}$$

Adsorption kinetic measurements were carried out by contacting 70 mL of the dye solution (C_i = 1.8 or 6.1 ppm) with 20 mL of the silica hydrogel casted in a Schott bottle. Absorbance values of solutions at 617 nm were measured in situ using Aventes-2048 model fiber optic spectrophotometer equipped with a circulation pump (Masterflex C/L) to ensure detection of an average concentration rather than a local concentration value. The data points were collected at 1 min intervals up to 25 min followed by 5 min intervals up to 125 min.

Fitting of nonlinear kinetic and equilibrium models to the experimental data points was conducted using SigmaPlot software, which provides the values of model fitting parameters and regression coefficients (R^2). Root mean squared error, RMSE, and normalized standard deviation, Δq (%) values were determined by the following equations, respectively:[25,26]

$$\text{RMSE} = \sqrt{\frac{\sum(q_{\text{exp}} - q_{\text{cal}})^2}{k}} \tag{1.2}$$

$$\Delta q(\%) = 100 \times \sqrt{\frac{\sum\left[(q_{\text{exp}} - q_{\text{cal}})/q_{\text{exp}}\right]^2}{(k-1)}}, \tag{1.3}$$

where q_{exp} and q_{cal} are experimental and predicted values of amount of the dye in the silica hydrogel phase and k is the number of experimental data points. The course of adsorption process was monitored by taking the pictures showing the color changes in both MG solution and the silica hydrogel occasionally up to 2 weeks by using a Nikon, Coolpix 995 model digital camera.

1.3 RESULTS AND DISCUSSION

1.3.1 ADSORPTION EQUILIBRIUM

Equilibrium isotherm of MG–silica hydrogel pair was fitted to both linear and nonlinear forms of Langmuir, Freundlich, and Temkin models described by the equations given in Table 1.1. In these models, K_L, K_F, and K_T are respective Langmuir, Freundlich, and Temkin constants. q_m is the maximum adsorption capacity of the hydrogel, n is the other Freundlich constant, which determines the extent of surface heterogeneity, and B_T is the other Temkin constant related to heat of adsorption. Model parameters obtained are summarized in Table 1.2. Linearized plots of these models are presented in Figure 1.2. Comparisons of the experimental data points with the model predictions are shown in Figure 1.3a and b for linear and nonlinear models, respectively.

TABLE 1.1   Linear and Nonlinear Forms of Adsorption Equilibrium Models.

Isotherm	Linear model	Nonlinear model
Langmuir	$\dfrac{C}{q} = \dfrac{1}{q_m K_L} + \dfrac{C}{q_m}$	$q = \dfrac{q_m K_L C}{1 + K_L C}$
Freundlich	$\ln q = \ln K_F + n \ln C$	$q = K_F C^n$
Temkin	$q = B_T \ln K_T + B_T \ln C$	$q = B_T \ln (K_T C)$

As indicated in Table 1.2, different model parameters were obtained for linear and nonlinear forms of Langmuir and Freundlich models, whereas linearization did not change any of the fitting parameters of Temkin model. Both linear and nonlinear Temkin equations also gave the same model parameters for biosorption of methylene blue on an algae.[27] It is not surprising that due to the structure of Temkin model equation, same model parameters were observed for both linear and nonlinear equations. Fitting quality of the models was assessed using R^2, RMSE, and q (%) values (Table 1.2). In this study, compared to linear model, R^2 values slightly decreased for Freundlich nonlinear model. On the contrary, linearization was found to change the fitting quality of Langmuir model as revealed by lower R^2 value obtained. Other studies indicated no clear-cut correlation between linearization and R^2 values as linearization changes error distribution. Depending on the model, and adsorbent–adsorbate pair, both higher and lower R^2 values were obtained for nonlinear fitting.[28–32]

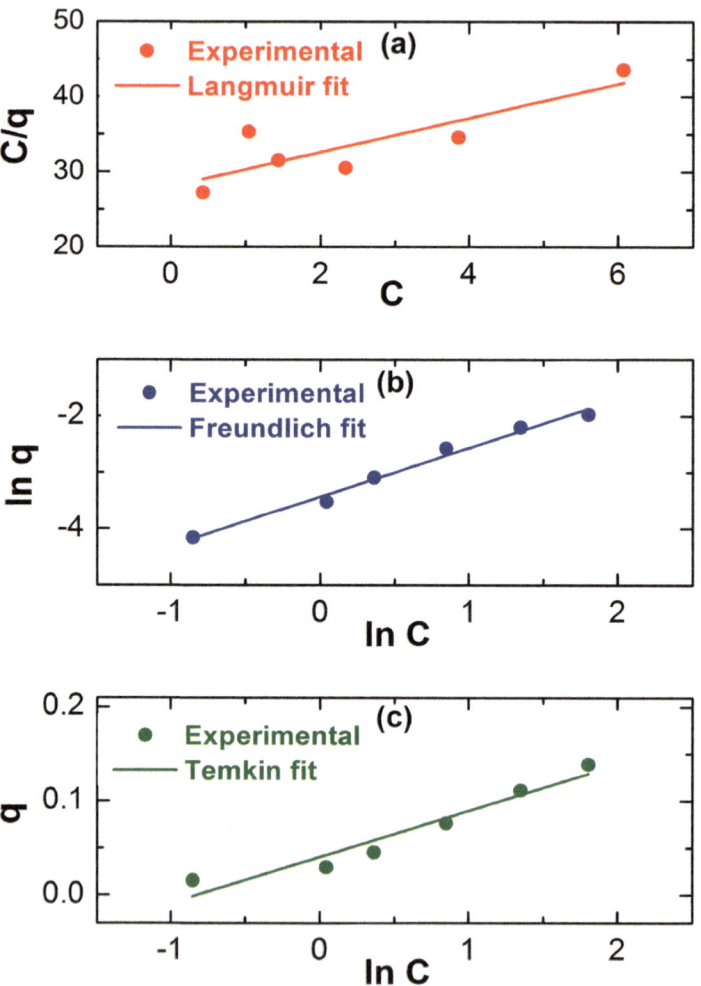

FIGURE 1.2 Linearized plots of (a) Langmuir, (b) Freundlich, and (c) Temkin adsorption equilibrium models.

TABLE 1.2 Fitting Parameters of Adsorption Equilibrium Models.

Isotherm/parameter	Linear model	Nonlinear model
Langmuir		
q_m (mg/g)	0.44	0.36
K_L (L/mg)	0.08	0.11
R^2	0.725	0.990
RMSE $\times 10^3$	4.93	4.37
Δq (%)	9.19	11.6
Freundlich		
N	0.87	0.77
K_F (mg/g)(L/mg)$^{1/n}$	0.032	0.036
R^2	0.986	0.979
RMSE $\times 10^3$	8.04	6.37
Δq (%)	9.97	16.4
Temkin		
B_T	0.05	0.05
K_T (L/mg)	2.27	2.27
R^2	0.934	0.934
RMSE $\times 10^3$	11.4	11.4
Δq (%)	53.8	53.8

Our study also indicated that for Langmuir and Freundlich models, nonlinear fitting decreased RMSE values suggesting an increase in the fitting quality. As opposed to this result, Δq (%) values increased in the case of nonlinear fitting. As it is depicted from the definition of Δq (%), this value is quite sensitive to the errors in the initial portion of experimental data. Thus, lower Δq (%) values obtained in linear fits are due to minimization of errors in the initial region as clearly seen in Figure 1.3a. Nonlinear curve fitting, on the other hand, provided much more evenly distributed errors (Fig. 1.3b). Thus, nonlinear models should be used to describe equilibrium data to avoid changing the error distribution associated with linearization as well as to obtain fair agreement with the experimental data points, as suggested in the other studies.[31,33]

According to fitting quality parameters obtained using nonlinear models, Langmuir model gave slightly better fitting than Freundlich model. Langmuir model assumes homogeneous adsorption in which all sites possess equal affinity for the adsorbate (constant heat of adsorption), whereas according

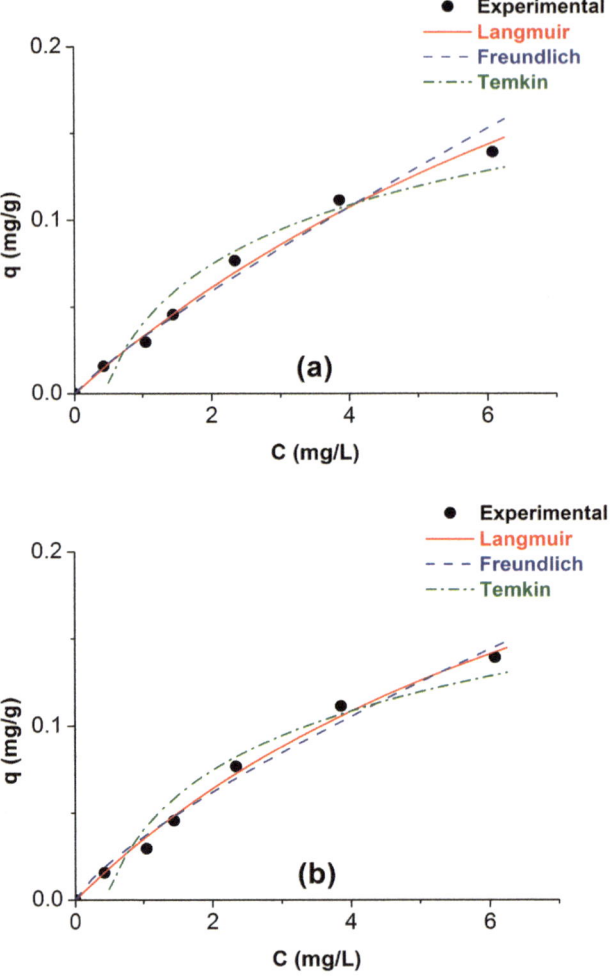

FIGURE 1.3   Predictions of (a) linear and (b) nonlinear adsorption isotherm models.

to Freundlich model, stronger binding sites are occupied first, and heat of adsorption decreases exponentially as a function of surface coverage.[33] Based on the curve fitting results, it is not possible to draw conclusions about the nature of adsorbent–adsorbate interactions, as none of the models could show significant superiority to the other in terms of fitting quality. Thus, another experimental technique such as adsorption microcalorimetry which relates surface coverage to heat of adsorption may confirm the nature of adsorbent–adsorbate interactions, and hence, the adsorption equilibrium model. Nevertheless, it can be fair to compare the model parameters with those in the previous studies as both Langmuir and Freundlich models have been frequently applied. In Freundlich model, the exponent, n, which takes a value between 0 and 1, gives a measure of surface heterogeneity. As the value of n gets closer to 0, surface heterogeneity increases.[23,33] In this study, value of n was obtained as 0.77, similar to the values, 0.78, 0.88, and 0.88, obtained for the adsorption of MG on the hydrogels prepared at pH values of 4.5, 7.0, and 9.5, respectively.[23]

In the Langmuir model, strength of attractive adsorbate–adsorbent interactions is correlated with K_L parameter, which is the ratio between the rates of adsorption and desorption or simply the equilibrium constant. From Langmuir isotherm, K_L value was obtained as 0.11 L/mg for MG adsorbed on the silica hydrogel. This value is comparable to K_L values obtained for MG adsorption on silica gels[34] (0.02–0.07 L/mg) and silica nanoparticles[35] (0.12–0.27 L/mg). It is noteworthy to state that in these studies, K_L values increased with the modifications with oxalic acid and citric acid,[34] or glycine.[35] Another Langmuir parameter, q_m is a measure of adsorption capacity. q_m value of the silica hydrogel was obtained as ~0.4 mg/g hydrogel for MG. Adsorption isotherm of MG on the silica hydrogels prepared at pH 4.5, 7.0, and 9.0 in which silica is mainly negatively charged revealed quite higher adsorption capacity (~20–25 μmol MG/g silica hydrogel or 8 ± 1 mg MG/g hydrogel)[23] compared to the one prepared at pH 1.0. This significant difference in the adsorption capacity of these silica hydrogels is mainly dictated by pH inside the hydrogel. For MG sorption of the hydrogel prepared at pH 1.0, pH values of the solution phase was measured between ~5 and ~2 decreasing as time proceeded. At that preparation pH of the hydrogel, silica was mainly positively charged ($Si–OH_2^+$) and during washing steps protons at the surface diffused out. Thus, the pH gradient of the hydrogel suggests that surface composition of the hydrogel is $Si–O^-$ and $Si–OH$ and inside the hydrogel $Si–OH$ and mostly $Si–OH_2^+$ species exist.[36] MG is cationic at initial pH 5. Thus, the early stage of the adsorption of the dye is mainly due to the electrostatic attractions between negatively charged silica and positively charged dye. However, as pH goes down, MG keeps its positive charge and at sufficiently low pH values, it becomes doubly positive charged.[37] As a result, during the course of adsorption, electrostatic attractions are replaced by van der Waals forces coupled with electrostatic repulsions by promoting desorption rate of MG, and hence decreasing extent of MG adsorption. BET surface areas of silica aerogels from anionic hydrogels, ~250–350 m²/g,[23] were reported to be lower than the silica from cationic hydrogel (540 m²/g).[24] These results clearly indicate the critical role of electrostatic interactions, rather than surface area, in the MG adsorption on silica hydrogels.

1.3.2 ADSORPTION KINETICS

Initial adsorption kinetic data (covering initial ~2 h period) were fitted to pseudo-first-order (PFO), pseudo-second-order (PSO), Elovich, and intra-particle models. PFO and PSO are the most popular kinetic models derived by assuming rate is a linear function of and proportional to square of the number of available adsorption sites, respectively, as follows:[38–40]

$$\frac{dq_t}{dt} = k_1 \left(q_e - q_t \right) \tag{1.4}$$

$$\frac{dq_t}{dt} = k_2 \left(q_e - q_t \right)^2, \tag{1.5}$$

where q_t and q_e are the amount of adsorbate in adsorbent phase at time t, and at equilibrium respectively, k_1 and k_2 are the respective PFO and PSO rate constants. By applying initial and boundary conditions, at $t = 0$; $q_t = 0$; and at $t = t$, $q_t = q_t$, and integrating PFO and PSO models given in Table 1.3 are obtained. Azizian[38] analyzed PFO and PSO models, and expressed rate constants of PFO and PSO models in terms of adsorption (v_a) and desorption rate constants (v_d) and initial concentration of solute by starting from the following equations in terms of surface coverage fraction (θ):

$$\frac{\partial \theta}{\partial t} = v_a - v_d \tag{1.6}$$

$$\frac{d\theta}{dt} = k_a \left(C_i - \beta\theta \right)\left(1 - \theta \right) - k_d \theta \tag{1.7}$$

$$\beta = \frac{C_i - C}{\theta} = \frac{mq_e}{M_w V} \tag{1.8}$$

In eq 1.8, m is the mass of the adsorbent, V is the volume of solution and M_w is the molecular weight of the adsorbate. Assuming $C_i \gg \beta\theta$ which corresponds to the experimental conditions with quite high initial concentration of solute compared to $\beta\theta$ leads to the derivation of PFO model. If $\beta\theta$ term is not neglected, PSO model can be derived. Accordingly, PFO rate constant, k_1, is a linear function of initial concentration (eq 1.9), whereas PSO rate constant, k_2, is a complex function of the initial concentration of solute.[38]

$$k_1 = k_a C_i + k_d \tag{1.9}$$

Elovich model, on the other hand, assumes rate of adsorption decreases exponentially as follows:

$$\frac{dq_t}{dt} = \alpha e^{-\beta q_t} \tag{1.10}$$

where α and β are Elovich constants. The constant, α is simply initial rate of adsorption since $dq/dt \rightarrow \alpha$ a as $q_t \rightarrow 0$. Linear form of the model was obtained by assuming $\alpha\beta t \gg 1$. Elovich model has been successfully applied to the systems involved in heterogeneous surfaces.[39,41]

Intraparticle or Weber–Morris model assumes that adsorption kinetic is controlled solely by intraparticle diffusion. In this model, uptake is proportional to the square root of time with a proportionality constant of intraparticle diffusion rate constant, k_p.[40]

Linear and nonlinear forms of these kinetic models are presented in Table 1.3. Model parameters along with R^2, RMSE, and Δq (%) values obtained for the initial kinetic data of MG on the silica hydrogel are given in Table 1.4. Linearized plots of these models and comparison of the linear and nonlinear models with the experimental data are shown in Figures 1.4 and 1.5, respectively.

TABLE 1.3 Linear and Nonlinear Forms of Adsorption Kinetic Models.

Kinetic model	Linear model	Nonlinear model
Pseudo-first order	$ln\left[\dfrac{q_e - q_t}{q_e}\right] = -k_1 t$	$q_t = q_e\left(1 - e^{-k_1 t}\right)$
Pseudo-second order	$\dfrac{t}{q_t} = \dfrac{1}{k_2 q_e^2} + \dfrac{1}{q_e}t$	$q_t = \dfrac{k_2 q_e^2 t}{1 + k_2 q_e t}$
Elovich	$q_t = \dfrac{1}{\beta}ln(\alpha\beta) + \dfrac{1}{\beta}ln(t)$ (if $\alpha\beta t \gg 1$)	$q_t = \dfrac{1}{\beta}ln(1 + \alpha\beta t)$
Intraparticle	$q_t = k_p \tau$ where $\tau = t^{0.5}$	$q_t = k_p t^{0.5}$

As revealed from Figure 1.4, R^2, RMSE, and Δq (%) values in Table 1.4, none of the linearized plots showed considerable agreement with the experimental data. In the linearized forms of PFO model, experimentally measured values of q_e were used. Especially, first portion of experimental data taken for $C_i = 1.8$ mg/L deviated from the model predictions significantly (Fig. 1.4a). According to the theoretical basis of PFO model developed by Azizian, this deviation is probably because the assumption $C_i \gg \beta\theta$ has not been validated. At high initial concentration, $C_i = 6.1$ mg/l, on the other hand, linearized PFO fit the experimental data better corroborating with this theory (Fig. 1.4a). On the contrary, PFO rate constants were found to decrease with initial concentration, suggesting linearized PFO model is not adequate to describe the experimental data. Likewise linearized PFO model, other linearized models failed to fit the experimental data as revealed by Figures 1.4b–d and 1.5a, b.

TABLE 1.4   Kinetic Models Fitting Parameters.

Model parameters	$C_i = 1.8$ mg/L		$C_i = 6.1$ mg/L	
	Linear fit	Nonlinear fit	Linear fit	Nonlinear fit
First-order model				
k_1, min^{-1}	0.0226	0.0394	0.0063	0.0134
$q_e \times 10^3$, mg g^{-1}	4.95*	4.34	17.6*	10.9
R^2	0.966	0.961	0.971	0.988
RMSE	4.44×10^{-4}	2.56×10^{-4}	5.05×10^{-4}	3.04×10^{-4}
Δq (%)	26.6	14.1	30.5	24.3
Second-order model				
k_2, L mg^{-1} min^{-1}	6.90	7.61	1.00	0.55
$q_e \times 10^3$, mg g^{-1}	5.49	5.35	10.4	16.8
R^2	0.984	0.982	0.795	0.988
RMSE	1.78×10^{-4}	1.74×10^{-4}	3.83×10^{-4}	3.01×10^{-4}
Δq (%)	13.4	13.8	23.9	23.8
Elovich model				
$\alpha \times 10^4$, mg g^{-1} min^{-1}	5.77	3.20	6.09	1.64
β, g mg^{-1}	962	736	476	160
R^2	0.974	0.992	0.823	0.988
RMSE	2.45×10^{-4}	1.18×10^{-4}	11.1×10^{-4}	3.02×10^{-4}
Δq (%)	47.6	22.4	52.2	23.2
Intraparticle model				
$k_p \times 10^4$, mg g^{-1} min$^{-0.5}$	4.61	4.61	7.45	7.45
R^2	0.955	0.959	0.939	0.942
RMSE	2.60×10^{-4}	2.60×10^{-4}	6.55×10^{-4}	6.55×10^{-4}
Δq (%)	42.8	42.8	31.3	31.3

*Experimentally determined values.

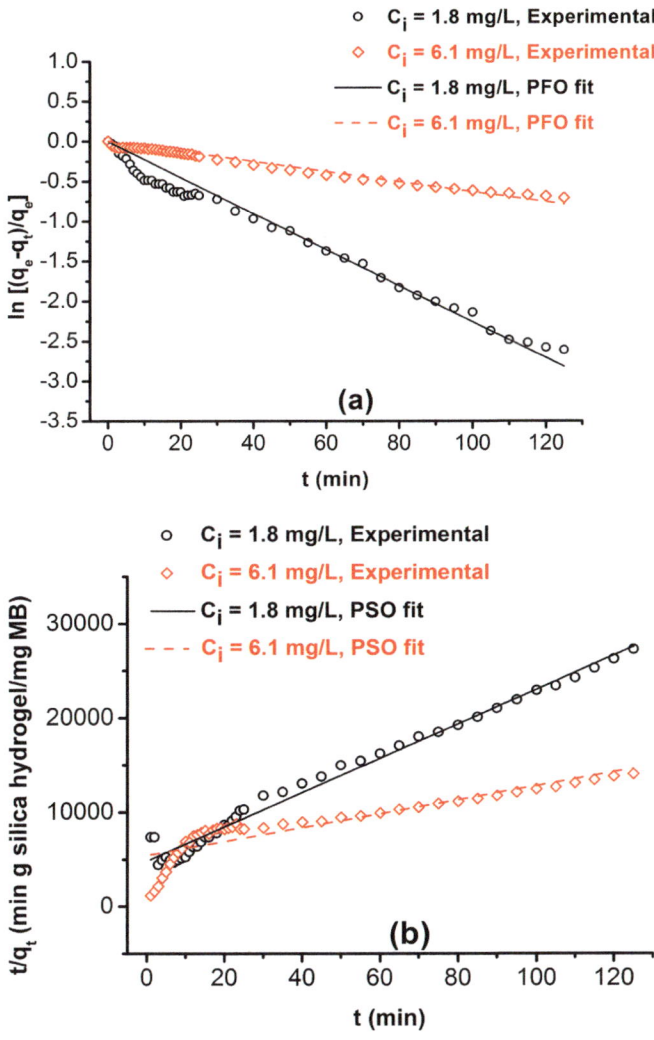

FIGURE 1.4 Linearized plots of (a) PFO, (b) PSO, (c) Temkin, and (d) intraparticle adsorption kinetic models.

For linear Elovich model, according to the model parameters obtained, $\alpha\beta t$ was not significantly greater than 1, not conforming to linearization assumption. Linear intraparticle model plots did not pass through the origin and showed multiple straight lines suggesting that intraparticle is not the only rate controlling step but film diffusion is also involved.[40] Due to the structure or simplicity of the intraparticle model equation, similar model parameters were obtained for both linearized and nonlinear models. For the

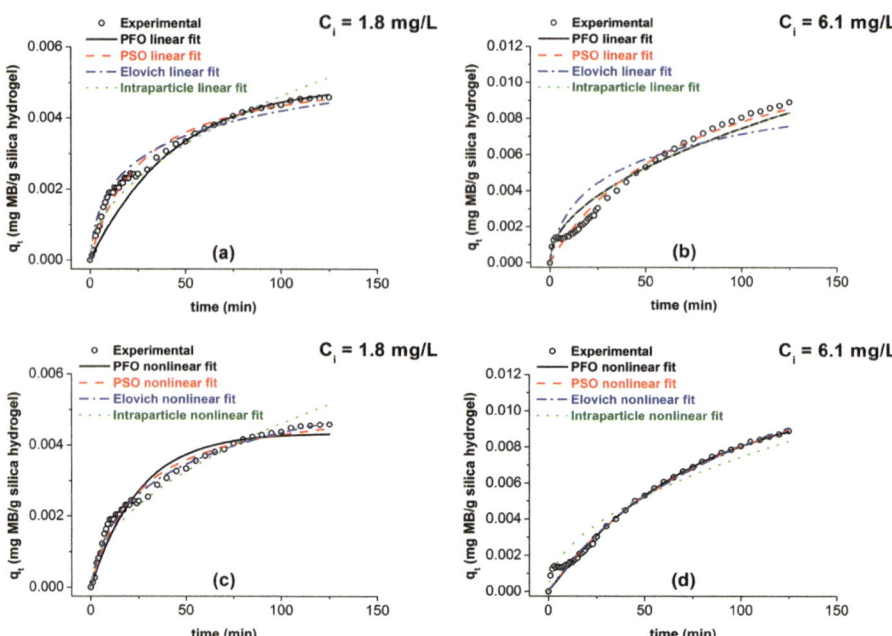

FIGURE 1.5 Predictions of linear adsorption kinetic models for (a) C_i = 8 mg/L and (b) C_i = 6.1 mg/L, nonlinear adsorption kinetic models for (c) C_i = 1.8 mg/L, and (d) C_i = 6.1 mg/L.

other models, model parameters changed upon linearization which can be attributed to the changes in the error distribution during transformation of actual data to linearized data.[42] Compared to its linearized form, nonlinear Elovich model exhibited better agreement with the adsorption kinetics as revealed by the decrease in RMSE and Δq (%) values and increase in R^2 values for both initial concentrations. No significant differences in fitting quality were obtained between nonlinear PFO, PSO, and Elovich models for C_i = 6.1 mg/L as indicated by similar R^2, RMSE, and Δq (%) values (Fig. 1.5d, Table 1.4). However, at C_i = 1.8 mg/L nonlinear Elovich model was observed to be superior to nonlinear PFO and PSO in the prediction of the experimental kinetic data (Fig. 1.5c). Thus, considering the analysis of the kinetics measurements carried out both initial concentrations, Elovich model was shown to corroborate with the initial kinetic data of MG–silica hydrogel system suggesting heterogeneous adsorption. On the other hand, it was reported that PFO and PSO models provided the best fitting for MG adsorption kinetics on silica nanoparticles,[35] and silica gels,[34] respectively, indicating that form of silica effects adsorption kinetics.

1.3.3 DIFFUSION COEFFICIENT MEASUREMENT

Apparent diffusion coefficient of MG through silica hydrogel was measured using digital pictures showing the coloration of the hydrogel. As depicted from Figure 1.6, diffusion of MG inside the hydrogel is quite slow. Hence, by measuring the representative time course movement of dye inside the hydrogel (diffusion length), it can be possible to measure the apparent diffusion coefficient using the following equation:[43]

$$X = (2Dt)^{0.5} \tag{1.11}$$

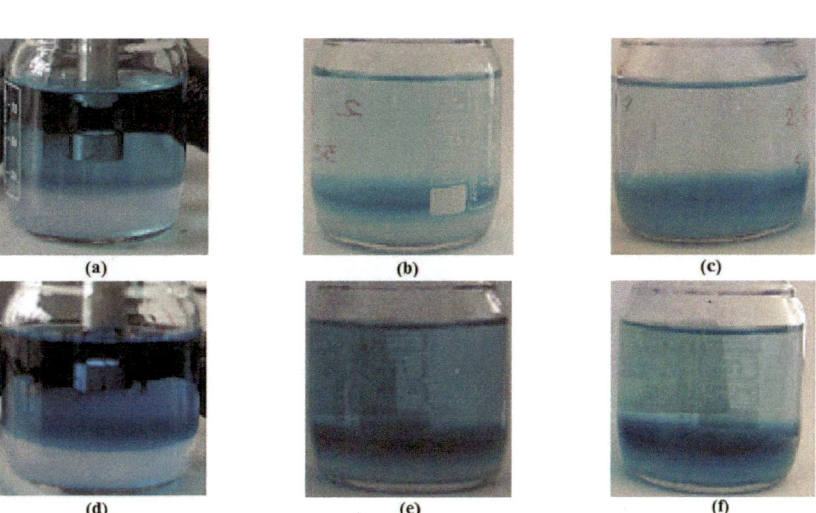

FIGURE 1.6 Pictures revealing the extent of the sorption of malachite green on silica hydrogel for $C_i = 1.8$ mg/L (a) initially, (b) after 1 week, (c) after 2 weeks and for $C_i = 6.1$ mg/L (d) initially (e) after 1 week, and (f) after 2 weeks.

For $C_i = 1.8$ mg/mL, considering much longer time period (days) than initial kinetic measurement interval (minutes), plot of measured diffusion length, X, of dye versus square root of time was observed to be linear giving an apparent diffusion coefficient (D) of 1.29×10^{-11} m²/s (Fig. 1.7).

Equation 1.11 has been derived from microscopic theory of diffusion.[44] Diffusion coefficients have been frequently estimated using the macroscopic theory via the solution of Fick's second law with appropriate initial and boundary conditions. Perullini et al.[23] measured the apparent diffusion coefficient for MG–silica hydrogel systems as ~0.5–3.5 × 10⁻¹¹ m²/s using the macroscopic theory of diffusion by neglecting adsorption. By considering the adsorption, on the other hand, effective diffusion coefficient of those systems was measured as $1.5 \pm 0.1 \times 10^{-10}$ m²/s. Low adsorption capacity of the cationic hydrogel suggests the apparent diffusion coefficient of the system would likely be in the order of the effective diffusion coefficient of MG–anionic hydrogel pair. Considering the different protocols and precursors used in the preparation of the hydrogels, slower diffusion of the cationic hydrogel can be due to its lower porosity and/or higher tortuosity and hence its different microstructure. Indeed, though both hydrogels have similar amount of silica (~10 wt%), microstructure of these anionic hydrogels contains both preformed silica nanoparticles and polymerized silica,[23] whereas network of the cationic hydrogel formed by the polymerization of silica only.

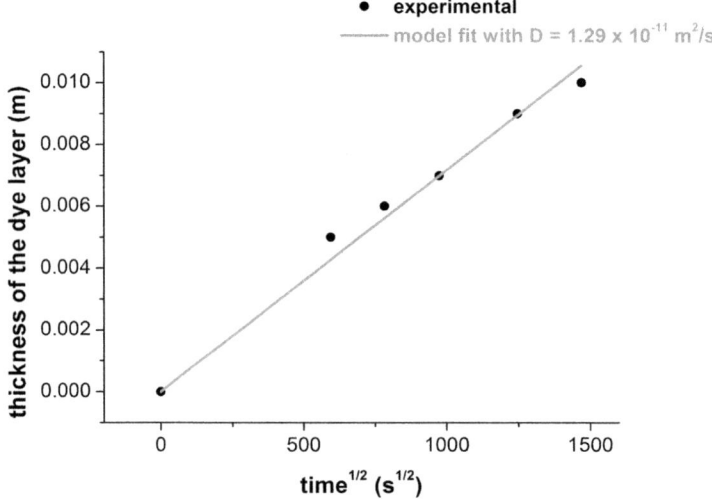

FIGURE 1.7   Thickness of the diffused dye layer as a function of square root of time for $C_i = 1.8$ mg/L.

1.4 CONCLUSIONS

High surface area silica hydrogel was prepared at pH 1 and its MG sorption behavior was tested. From Langmuir model, MG adsorption capacity of the hydrogel was estimated to be ~0.4 mg/g hydrogel. This value is almost 20 times lower than that of the anionic hydrogels prepared at higher pH, indicating the crucial role of electrostatic interactions in the adsorption behavior of silica hydrogels. One of the advantages of silica hydrogels is that their precursor hydrosol can be casted into various shapes including pellets with any geometry and films with desired thickness. Besides, the hydrogel is quite stable even in water maintaining its integrity for a long period of time and unlike aerogels and xerogels silica hydrogels do not cause any turbidity in aqueous phase. Therefore, silica hydrogels can be used as a potential adsorbent to remove contaminants from waste streams. However, as confirmed in this study, preparation pH should be optimized before using in the adsorption of charged species to get optimal performance.

Adsorption equilibrium and kinetic models in both linear and nonlinear forms were applied. It was shown that nonlinear forms of these models fitted to the experimental data more thoroughly than their linear forms by giving evenly distributed errors. Considering similar observations in the previous studies, nonlinear forms of the models should be used as they did not disrupt error distribution of the actual data. However, in some cases, even the use of nonlinear forms of different models gives the similar quality of fitting and hence requiring some other techniques to verify the most appropriate model describing the equilibrium or kinetics of adsorption.

Although adsorption capacity of the cationic hydrogel is low, apparent diffusion coefficient of the MG-cationic system was measured to be lower than effective diffusion coefficient of the dye–anionic hydrogel system with similar silica content measured by considering adsorption. Slower diffusion of the dye inside the cationic hydrogel can be attributed to the lower porosity and/or higher tortuosity of the hydrogel. Considering the ease of tuning microstructure and adsorption characteristics of silica hydrogels by simple changes in preparation protocols allows to control diffusion of the dye of interest, silica-based hydrogels may also have potential applications as dye diffusion-based time-monitoring systems in food and pharmaceutical industry.

KEYWORDS

- **malachite green**
- **adsorption**
- **silica hydrogels**
- **linear**
- **nonlinear forms**
- **equilibrium**
- **kinetics**

REFERENCES

1. Culp, S. J.; Beland, F. A. Malachite Green: A Toxicological Review. *Int. J. Toxicol.* **1996,** *15,* 219–238.
2. Mall, I. D.; Srivastava, V. C.; Agarwal, N. K.; Mishra, I. M. Adsorptive Removal of Malachite Green Dye from Aqueous Solution by Bagasse Fly Ash and Activated Carbon-kinetic Study and Equilibrium Isotherm Analyses. *Colloids Surf. A.* **2005,** *264,* 17–28.
3. Srivastava, S.; Sinha, R.; Roy, D. Toxicological Effects of Malachite Green. *Aquat. Toxicol.* **2004,** *66,* 319–329.
4. Stammati, A.; Nebbia, C.; De Angelis, I.; Albo, A. G.; Carletti, M.; Rebecchi, C.; Zampaglioni, F.; Dacasto, M. Effects of Malachite Green (MG) and Its Major Metabolite, Leucomalachite Green (LMG), in Two Human Cell Lines. *Toxicol. In Vitro.* **2005,** *19,* 853–858.
5. Tahir, S.; Rauf, N. Removal of a Cationic Dye from Aqueous Solutions by Adsorption onto Bentonite Clay. *Chemosphere.* **2006,** *63,*1842–1848.
6. Gupta, V.; Mittal, A.; Krishnan, L.; Gajbe, V. Adsorption Kinetics and Column Operations for the Removal and Recovery of Malachite Green from Wastewater Using Bottom Ash. *Sep. Purif. Technol.* **2004,** *40,* 87–96.
7. Bulut, E.; Ozacar, M.; Sengil, I. A. Adsorption of Malachite Green onto Bentonite: Equilibrium and Kinetic Studies and Process Design. *Micropor. Mesopor. Mat.* **2008,** *115,* 234–246.
8. Chowdhury, S.; Mishra, R.; Saha, P.; Kushwaha, P. Adsorption Thermodynamics, Kinetics and Isosteric Heat of Adsorption of Malachite Green onto Chemically Modified Rice Husk. *Desalination.* **2011,** *265,* 159–168.
9. Garg, V. K.; Kumar, R.; Gupta, R. Removal of Malachite Green Dye from Aqueous Solution by Adsorption Using Agro-industry Waste: A Case Study of Prosopis Cineraria. *Dyes Pigm.* **2004,** *62,* 1–10.
10. Han, R.; Wang, Y.; Sun, Q.; Wang, L.; Song, J.; He, X.; Dou, C. Malachite Green Adsorption onto Natural Zeolite and Reuse by Microwave Irradiation. *J. Hazard. Mater.* **2010,** *175,* 1056–1061.
11. Wang, S.; Ariyanto, E. Competitive Adsorption of Malachite Green and Pb Ions on Natural Zeolite. *J. Colloid Interface. Sci.* **2007,** *314,* 25–31.
12. Arellano, Cárdenas, S.; López, Cortez, S.; Cornejo, Mazón, M.; Mares, Gutiérrez, J. C. Study of Malachite Green Adsorption by Organically Modified Clay Using a Batch Method. *Appl. Surf. Sci.* **2013,** *280,* 74–78.
13. Suwandi, A. C.; Indraswati, N.; Ismadji, S. Adsorption of N-methylated Diaminotriphenilmethane Dye (Malachite Green) on Natural Rarasaponin Modified Kaolin. *Desalination Water Treat.* **2012,** *41,* 342–355.
14. Önal, Y.; Akmil, Başar, C.; Sarıcı, Özdemir, Ç. Investigation Kinetics Mechanisms of Adsorption Malachite Green onto Activated Carbon. *J. Hazard. Mater.* **2007,** *146,* 194–203.
15. Rahman, I.; Saad, B.; Shaidan, S.; Rizal, E. S. Adsorption Characteristics of Malachite Green on Activated Carbon Derived from Rice Husks Produced by Chemical–thermal Process. *Bioresour. Technol.* **2005,** *96,* 1578–1583.
16. Dubey, S.; Uma; Sujarittanonta, L.; Sharma, Y. C. Application of Fly Ash for Adsorptive Removal of Malachite Green from Aqueous Solutions. *Desalination Water Treat.* **2015,** *53,* 91–98.

17. Altınışik, A.; Gür, E.; Seki, Y. A Natural Sorbent, Luffa Cylindrica for the Removal of a Model Basic Dye. *J. Hazard. Mater.* **2010**, *179,* 658–664.

18. Hameed, B.; El-Khaiary, M. Batch Removal of Malachite Green from Aqueous Solutions by Adsorption on Oil Palm Trunk Fibre: Equilibrium Isotherms and Kinetic Studies. *J. Hazard. Mater.* **2008**, *154,* 237–244.

19. Khattri, S.; Singh, M. Removal of Malachite Green from Dye Wastewater Using Neem Sawdust by Adsorption. *J. Hazard. Mater.* **2009**, *167,* 1089–1094.

20. Samiey, B.; Toosi, A. R. Adsorption of Malachite Green on Silica Gel: Effects of NaCl, pH and 2-propanol. *J. Hazard. Mater.* **2010**, *184,* 739–745.

21. Cupane, A.; Levantino, M.; Santangelo, M. G. Near-infrared Spectra of Water Confined in Silica Hydrogels in the Temperature Interval 365–5 K. *J. Phys. Chem. B.* **2002**, *106,* 11323–11328.

22. Iler, R. K. *The Chemistry of Silica: Solubility, Polymerization, Colloid and Surface Properties, and Biochemistry;* 1st ed.; John Wiley & Sons, Inc.: New York, 1979.

23. Perullini, M.; Jobbágy, M.; Japas, M. L.; Bilmes, S. A. New Method for the Simultaneous Determination of Diffusion and Adsorption of Dyes in Silica Hydrogels. *J. Colloid Interface Sci.* **2014**, *425,* 91–95.

24. Ülkü, S.; Balköse, D.; Baltacioglu, H. Effect of Preparation pH on Pore Structure of Silica Gels. *Colloid Polym. Sci.* **1993**, *271,* 709–713.

25. Kothawala, D.; Moore, T.; Hendershot, W. Adsorption of Dissolved Organic Carbon to Mineral Soils: A Comparison of Four Isotherm Approaches. *Geoderma.* **2008**, *148,* 43–50.

26. Lin, J.; Wang, L. Comparison between Linear and Non-linear Forms of Pseudo-first-order and Pseudo-second-order Adsorption Kinetic Models for the Removal of Methylene Blue by Activated Carbon. *Front. Environ. Sci. Eng. Chin.* **2009**, *3,* 320–324.

27. Hammud, H. H.; Fayoumi, L.; Holail, H.; Mostafa, E. M. E. Biosorption Studies of Methylene Blue by Mediterranean Algae Carolina and its Chemically Modified Forms. Linear and Nonlinear Models' Prediction Based on Statistical Error Calculation. *Int. J. Chem.* **2011**, *3,* 147.

28. Armagan, B.; Toprak, F. Optimum Isotherm Parameters for Reactive Azo Dye onto Pistachio Nut Shells: Comparison of Linear and Non-linear Methods. *Pol. J. Environ. Stud.* **2013**, *22,*1007–1011.

29. Kumar, K. V. Comparative Analysis of Linear and Non-linear Method of Etimating the Sorption Isotherm Parameters for Malachite Green onto Activated Carbon. *J. Hazard. Mater.* **2006**, *136,* 197–202.

30. Kumar, K. V. Optimum Sorption Isotherm by Linear and Non-linear Methods for Malachite Green onto Lemon Peel. *Dyes Pigm.* **2007**, *74,* 595–597.

31. Kumar, K. V.; Porkodi, K.; Rocha, F. Isotherms and Thermodynamics by Linear and Non-linear Regression Analysis for the Sorption of Methylene Blue onto Activated Carbon: Comparison of Various Error Functions. *J. Hazard. Mater.* **2008**, *151,* 794–804.

32. Kumar, K. V.; Sivanesan, S. Prediction of Optimum Sorption Isotherm: Comparison of Linear and Non-linear Method. *J. Hazard. Mater.* **2005**, *126,* 198–201.

33. Foo, K.; Hameed, B. Insights into the Modeling of Adsorption Isotherm Systems. *Chem. Eng. J.* **2010**, *156,* 2–10.

34. Kushwaha, A. K.; Gupta, N.; Chattopadhyaya, M. Enhanced Adsorption of Malachite Green Dye on Chemically Modified Silica Gel. *J. Chem. Pharm. Res.* **2010**, *2,* 34–45.

35. Mansa, R. F.; Sipaut, C. S.; Rahman, I. A.; Yusof, N. S. M.; Jafarzadeh, M. Preparation of Glycine–modified Silica Nanoparticles for the Adsorption of Malachite Green Dye. *J. Porous Mater.* **2016**, *23,* 35–46.

36. Wu, S. H.; Mou, C. Y.; Lin, H. P. Synthesis of Mesoporous Silica Nanoparticles. *Chem. Soc. Rev.* **2013,** *42,* 3862–3875.
37. Cooksey, C. Quirks of Dye Nomenclature. 6. Malachite Green. *Biotech. Histochem.* **2016,** *91,* 438–444.
38. Azizian, S. Kinetic Models of Sorption: A Theoretical Analysis. *J. Colloid Interface Sci.* **2004,** *276,* 47–52.
39. Ho, Y. S. Review of Second-order Models for Adsorption Systems. *J. Hazard. Mater.* **2006,** *136,* 681–689.
40. Qiu, H.; Lv, L.; Pan, B. C.; Zhang, Q. J.; Zhang, W. M.; Zhang, Q. X. Critical Review in Adsorption Kinetic Models. *J. Zhejiang Univ. Sci.* A. **2009,** *10,* 716–724.
41. Chien, S.; Clayton, W. Application of Elovich Equation to the Kinetics of Phosphate Release and Sorption in Soils. *Soil Sci. Soc. Am. J.* **1980,** *44,* 265–268.
42. Kumar, K. V. Linear and Non-linear Regression Analysis for the Sorption Kinetics of Methylene Blue onto Activated Carbon. *J. Hazard. Mater.* **2006,** *137,* 1538–1544.
43. Galagan, Y.; Hsu, S. H.; Su, W. F. Monitoring Time and Temperature by Methylene Blue Containing Polyacrylate Film. *Sensor. Actuat. B.Chem.* **2010,** *144,* 49–55.
44. Berg, H. C. *Random Walks in Biology;* 1st ed.; Princeton University Press: Princeton, NJ, 1993.

CHAPTER 2

FULLERENES IN THE AIR OXIDATION ENVIRONMENT

ELDAR ZEYNALOV[1,*], MATANAT MAGERRAMOVA[1],
NAZILYA SALMANOVA[2], and AYTEN BAGHIYEVA[1]

[1]*Institute of Catalysis and Inorganic Chemistry, Azerbaijan National Academy of Sciences, H. Javid Ave. 113, AZ 1143, Baku, Republic of Azerbaijan*

[2]*Azerbaijan State University of Oil and Industry, Ministry of Education, 20, Azadlig Ave., AZ 1010, Baku, Republic of Azerbaijan*

*Corresponding author. E-mail: zeynalov_2000@yahoo.com

CONTENTS

ABSTRACT

The investigation was undertaken to determine the antioxidative activity of a range of fullerenes C_{60} and C_{70} generally manufactured in practice in order to rank them according to their comparative efficiency. The model reaction of cumene-initiated (2,2'-azo-bisisobutyronitrile [AIBN]) oxidation was employed herein to determine rate constants for addition of radicals to fullerenes. Kinetic measurements of oxidation rate in the presence of different fullerenes showed that the antioxidative activity as well as the mechanism and mode of inhibition were different for fullerenes C_{60} and C_{70} and fullerene soot. All fullerenes—C_{60} of gold grade, C_{60}/C_{70} (93/7, mix 1), C_{60}/C_{70} (80 ± 5/20 ± 5, mix 2), and C_{70} operated in the mode of an alkyl radical acceptor, whereas fullerene soot surprisingly retarded the model reaction by a dual mode similar to that for the fullerenes and with an induction period like many of the sterically hindered phenolic and amine antioxidants. For the C_{60} and C_{70}, the oxidation rates were found to depend linearly on the reciprocal square root of the concentration over a sufficiently wide range, thereby fitting the mechanism for the addition of cumyl alkyl R· radicals to the fullerene core. This is consistent with the gathered literature data of the more readily and rapid addition of alkyl and alkoxy radicals to the fullerenes compared with peroxy radicals. Rate constants for the addition of cumyl R· radicals to the fullerenes were determined to be $k_{(333K)}$ = (1.9 ± 0.2) × 10^8 (C_{60}); (2.3 ± 0.2) × 10^8 (C_{60}/C_{70}, mix 1); (2.7 ± 0.2) × 10^8 (C_{60}/C_{70}, mix 2); and (3.0 ± 0.3) × 10^8 (C_{70}) M^{-1} s^{-1}. The incremental C_{70} constituent in the fullerenes leads to a respective increase in the rate constant. The fullerene soot inhibits the model reaction according to the mechanism of trapping of peroxy radicals: The oxidation proceeds with a pronounced induction period and kinetic curves are linear at the semilogarithmic coordinates. For the first time, the effective concentration of inhibiting centers and inhibition rate constants for the fullerene soot have been determined to be fn[C_{60}-soot] = (2.0 ± 0.1) × 10^{-4} mol g^{-1} and k_{inh} = (6.5 ± 1.5) × 10^3 M^{-1} s^{-1}, respectively. The kinetic data obtained specify the level of antioxidative activity for the commercial fullerenes and scope of their rational use in different composites. The results may be helpful for designing an optimal profile of composites containing fullerenes.

2.1 RADICAL-QUENCHING EFFICIENCY OF FULLERENES C_{60}–C_{70}

The remarkable property of the fullerenes for trapping free radicals resulting from their high electron affinity (ca. 2.6–3.3 eV) is widely highlighted in the literature.[1–16] There are many evidences and observations of effective inhibition of radical polymerization, prevention of oxidative damage of polymers, and deterioration of cells and tissues in biological objects. For instance, fullerene C_{60} effectively inhibits the free radical polymerization of styrene,[17–21] methyl methacrylate,[17,22,23] vinyl acetate,[24] and acrylonitrile.[25]

The radical-scavenging nature of fullerenes is also utilized for improving their solubility in conventional polar solvents by grafting the macro-radicals—hydrophilic poly(ethylene oxide) moieties,[26] obtaining fullerene containing branched polymers,[11,27,28] including star-like polymers with the fullerene core and synthesis of controlled polymeric arms.[29–31]

The apparent stabilizing role of fullerenes during thermo-oxidative degradation of polymers was confirmed by a relatively recent set of investigations.[32,33] The stabilizing activity of fullerenes C_{60}, C_{60}/C_{70}, and C_{70} has been shown to be totally comparable to the activity of the well-known hindered phenolic stabilizers such as Irganox 1010, Irganox 1076, and Agerite white, by means of model reactions of cumene and styrene initiated oxidation as well as in accelerated tests of polystyrene (PS) and polydimethylsiloxane (PDMS) rubber with fullerene moieties.[34,35] The stabilizing effect of fullerenes C_{60} and C_{70} was also shown in the thermal and thermo-oxidative degradation of (1) poly(2,6-dimethyl-1,4-phenylene oxide) and its blends,[36] (2) natural rubber (cis-1,4-polyisoprene) and synthetic cis-1,4-polyisoprene,[37] (3) isotactic polypropylene,[38] (4) polymethylmethacrylates (PMMAs),[39–44] (5) poly-n-alkyl acrylates,[45] and (6) polyamide 6.[46]

Fullerenes have also found broad application in biomedicine as a quencher of free radicals for precluding cellular breakdown and protecting a liver against free radical damage.[47–51] Addition of the fullerene soot significantly hampered the peroxide formation and thus increased the oxidation stability of rapeseed vegetable oils.[52]

The facile addition of free radicals to fullerenes is also confirmed by high values of addition rate constants determined in a number of works. These data are accumulated in Table 2.1.

It is readily seen from the table and the literature material that among the scavenged radicals are preferably nucleophilic carbon centered and alkoxy species whereas peroxy radicals in turn have only a weak affinity to the addition mechanism. The analysis of literature material clearly shows that fullerenes preferably act as scavengers of carbon-centered free radicals, and therefore they appear to be inhibitors for the radical polymerization and thermal stabilizers for polymer materials. No distinct indication of any proved active or specific role for oxygen-centered radicals has been specified.[32] The low inherent rate constant for trapping of peroxy radicals by C_{60}

TABLE 2.1   Addition Rate Constants of Different Radicals to Fullerene C_{60}.

R	k, M^{-1} s^{-1}	Temperature, K	Techniques	References
$\cdot CH_3$	$(4.6 \pm 2.0) \times 10^8$	295	Pulse radiolysis	4
$\cdot C(CH_3)_3$	$(2.2 \pm 0.5) \times 10^7$	293	ESR (spin-trapping)	53
$\cdot C(CH_3)_3$	3.0×10^9	293	ESR (laser flash photolysis and pulse radiolysis)	5
$\cdot CCN(CH_3)_2$	3.0×10^6	295	ESR (spin-trapping)	54
$\cdot CCN(CH_3)_2$	1.8×10^9	–	ESR (pulse radiolysis)	55
$\cdot CH_2Ph$	4.5×10^6	241	ESR (photolysis)	56
$\cdot CH_2Ph$	6.7×10^6	263	ESR (photolysis)	56
$\cdot CH_2Ph$	9.3×10^8	293	ESR (laser flash photolysis and pulse radiolysis)	5
$\cdot CH_2Ph$	2.7×10^5	293	ESR (spin-trapping)	53
$\cdot CH_2Ph$	$(1.4 \pm 0.2) \times 10^7$	298	ESR (photolysis)	56
$\cdot C(CCl_3)_3$	$(2.7 \pm 0.5) \times 10^8$	293	ESR (laser flash photolysis and pulse radiolysis)	57
$\cdot C(CCl_3)_3$	1.3×10^9	–	Laser flash photolysis and pulse radiolysis)	58
$\cdot CH_2 CH_2Cl$	2.6×10^9	–	Laser flash photolysis and pulse radiolysis)	58
$\cdot C(CH_3)_2Ph$	$(2.0 \pm 0.8) \times 10^8$	333–353	Liquid model oxidation (cumene)	34
$\cdot OOC(CH_3)_2Ph$	$(3.1 \pm 1.1) \times 10^2$	303	Liquid model oxidation (cumene)	58
$\cdot C(CH_3)_2Ph$	8.1×10^5	295	ESR (spin-trapping)	54
$CCl_3CH_2\cdot CHPh$	8.5×10^5	295	ESR (spin-trapping)	54
$\cdot C_2H_5$	$(5.3 \pm 0.4) \times 10^6$	293	ESR (spin-trapping)	53
$\cdot CH_2 (CH_2)_3 CH_3$	$(4.5 \pm 0.4) \times 10^6$	293	ESR (spin-trapping)	53
$\cdot CHCH_3 C_2H_5$	$(4.8 \pm 0.2) \times 10^6$	293	ESR (spin-trapping)	53
$\cdot CH_2 CH{=}CH_2$	$(6.5 \pm 0.5) \times 10^5$	293	ESR (spin-trapping)	53
$\cdot C{=}CH_2Ph$ (addition to C_{60}/C_{70}:3/1)	$(9.0 \pm 1.5) \times 10^7$	333	Liquid model oxidation (styrene)	35

per se is $k_{(inh.)} = (3.1 \pm 1.1) \times 10^2\ M^{-1}\ s^{-1}$ obtained in the work[59] and the fact of inertness of peroxy radicals toward fullerene C_{60} established by the chemi-luminescence[60] indicates that fullerene C_{60} is an extremely weak acceptor of peroxy radicals. This is why the antioxidative function of fullerenes has become especially discernible in hypoxic condensed matrixes like solid polymers, that is, in the media having a certain restriction for the diffusion of oxygen.

It should be noted that the relative radical-scavenging efficiencies of different fullerenes depend on and may be altered due to a different number of graphitic bonds, difference in energy strain correlated to the degree of flatness, the electron affinities, HOMO–LUMO gaps, etc. Each of these factors could contribute to relative differences in the efficiency of different fullerenes to scavenge radicals, and affect the resulting utility of different fullerenes in different applications. It has been shown that different C_{60} fullerene derivatives which have different strains and/or electron affinities show significantly different radical-scavenging effi-ciencies.[61] Similarly, differences relative to the C_{60} cage resulting from changes in the number of carbons may give significant differences in radical-scavenging efficiency. Although the data of buckminsterfullerene and C_{70} are available to some extent it would be nevertheless helpful in practice to determine the inhibition rate constants for other different commercial fullerenes to rank them according to their comparative anti-oxidative efficiency.

To determine the kinetic parameters of antioxidative activity of the commercial fullerenes, a model oxidation reaction of cumene initiated (2,2′-azobisisobutyronitrile [AIBN]) in liquid phase has been employed in this study. This model reaction is designed to simulate the thermo-oxidative processes in carbon-chain polymers and allows one to assess the antioxi-dative capacity of tested compounds for possible extrapolation in polymer materials.

2.2   EXPERIMENTAL PROCEDURES

The model cumene oxidation was carried out at initiation rates: $W_i = 6.8 \times 10^{-8}$ mol/L·s, temperature $-60\ (\pm 0.02)°C$ and oxygen pressure was $Po_2 = 20$ kPa (air). This condition is the most suitable for the preliminary correct deter-mination of antioxidative activity of compounds.[62,63] Employed cumene has 98% purity ("Aldrich"). AIBN was employed as the initiator. The volume of the reaction mixture was 10 cm³ (25°C). In order to have the assigned initiation rate, 10 mg of AIBN had to be added.[62–64]

Rate constants for the cumene oxidation at 60°C are as follows: chain propagation, $k_3 = 1.75$; termination, $k_6 = 1.84 \times 10^5$ M⁻¹s⁻¹; and concentration of cumene, [RH] = 6.9 mol L⁻¹.[62,63,65,66] The rate of oxidation was evaluated from the amount of oxygen consumed, which was measured volumetrically with the simple equipment as described elsewhere.[63,67,68] Oxidation rates were assessed both from the slopes of the kinetic curves of oxygen consumption in the case of steady rate values and also by means of differentiating the curve in the case of an observed induction period.

Experiments were carried out at least in triplicate and the correctness of the oxidation rate values determined was within the range 1–5%. The induction period (τ) was evaluated from the graphical processing of kinetic curves of oxygen uptake, as described elsewhere.[62,63] Correctness in determining of the induction period was 5–10%. The fullerenes used were commercial samples of fullerene: (1) C_{60} "gold grade," (2) C_{60}/C_{70} in proportion 93/7 [C_{60}/C_{70} (mix 1)], (3) $C_{60}/C_{70} = 80 \pm 5/20 \pm 5$ [C_{60}/C_{70} (mix 2)], (4) fullerene C_{70} (5) fullerene C_{76}, and (6) fullerene soot. All samples were provided by Xzillion GmbH & Co. KG (it has been recently renamed to Proteome Sciences R&D GmbH & Co. KG).

2.3   RESULTS AND DISCUSSION

The results of kinetic experiments are exemplified in the Figures 2.1 and 2.2, where profiles of kinetic curves of the oxygen uptake in the presence of different fullerene samples are shown for cumene oxidation. It is apparent from the experimental data that all the fullerene samples retard the model cumene oxidation. However, in the case of the fullerene soot a mechanism of retardation is different from that of fullerenes C_{60}, C_{70}, and C_{76}. For the fullerenes, the oxidation proceeds with steady rate values whereas their soot induction periods are clearly observed. These results on the fullerene C_{76} and soot were quite surprising and reported for the first time.

In the earlier studies, we have already shown that the fullerenes C_{60} and C_{70} intervene in the initiation stage of the model oxidation and compete with very rapid formation of peroxy radicals from alkyl radicals according to the Scheme 2.1.[34,35]

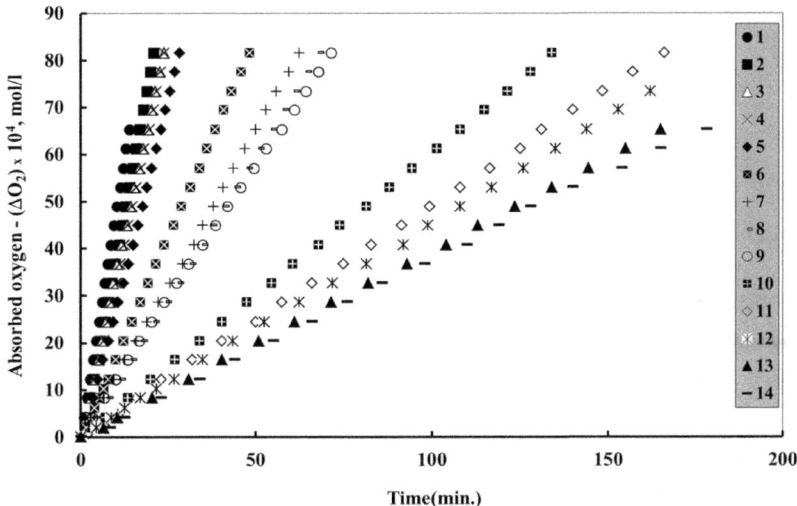

FIGURE 2.1 Kinetic lines of oxygen-uptake during aerobic-initiated oxidation of cumene in the absence (1) and presence of fullerenes: C_{60} "gold grade" (2, 6, 10); C_{60}/C_{70} mix1 (3, 7, 11); C_{60}/C_{70} mix 2 (4, 8, 12); C_{70} (5, 9, 13) and C_{76} (14). The initiator is AIBN, initiation rate: $W_i = 6.8 \times 10^{-8}$ mol/L s, reaction mixture volume 10 mL, oxygen pressure: $Po_2 = 20$ kPa (air), temperature 60°C. Concentration of fullerenes—$[C_{60}(C_{70})\,(C_{76})]$: (1) = 0, (2–5) = 1 × 10⁻⁵, (6–9) = 1 × 10⁻⁴, and (10–14) = 5 × 10⁻⁴ mol/L.

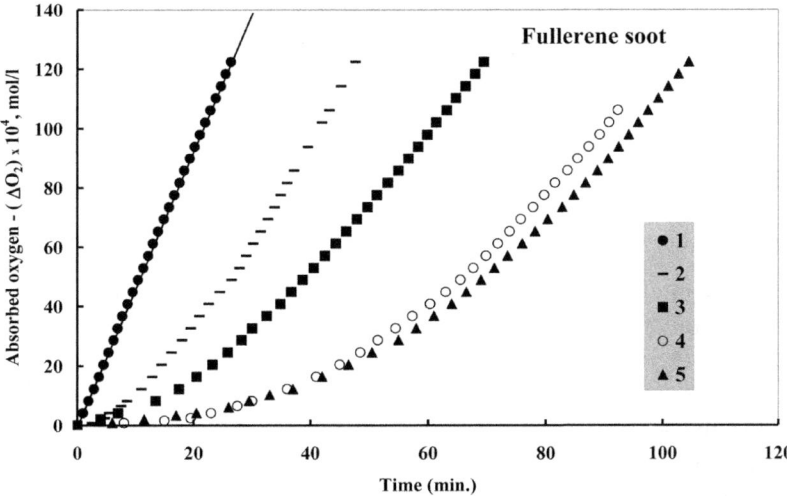

FIGURE 2.2 Kinetic dependencies of oxygen consumption for initiated oxidation of cumene in the absence (1) and presence of fullerene soot (2–5). The initiator is AIBN, initiation rate: $W_i = 6.8 \times 10^{-8}$ mol/L s, reaction mixture volume 10 mL, oxygen pressure: $Po_2 = 20$ kPa (air), temperature 60°C. Concentration of the fullerene soot added—$[C_{60}$ soot]: (1) = 0; (2) = 0.4; (3) = 0.7; and (4) = 1.3 g/L.

Scheme 2.1*

Chain initiation: AIBN ® r· (rO$_2$·) + RH ® R· (initiation rate is W$_i$)

$$R· + C_{60} ®·C_{60} R \qquad\qquad (1)$$

Chain propagation: R· + O$_2$ RO$_2$·+RH®ROOH + R·   (2)/(3)

Chain termination: 2 RO$_2$·®inactive products   (6)

(RH: cumene, R·: cumyl alkyl radical, RO$_2$·: cumyl peroxy radical, and ROOH: cumyl hydroperoxide).

*Here the generally accepted oxidation stage numbering is used.

By intercepting the alkyl radicals the fullerenes decrease the rate of initiation and accordingly the rate of oxidation. Consequently, no induction period is observed and the rate of oxidation has a steady value which is proportional to the decreased rate of initiation.

At the steady-state condition an oxidation reaction rate fitting Scheme 2.1 is described by the following equation[34,67,69]:

$$Wo_{2 (C60/C70)} = k_3 [RO_2·] [RH] = W_{i1}^{1/2} k_3 k_6^{-1/2}[RH], \qquad (2.1)$$

where Wo$_{2(C60/C70)}$ and W$_{i1}$ are oxidation and initiation rate in the presence of a fullerene, respectively.

$$W_{i1} = W_i - W_{(C60/C70)}, \qquad\qquad (2.2)$$

where

$$W_{(C60/C70)} = k_1[R·] [C_{60}/C_{70}] \qquad\qquad (2.3)$$

is the rate of interaction between fullerene and an alkyl radical.

It is obvious that our present experimental results obtained for all kinds of fullerenes C$_{60}$/C$_{70}$ are in full consistency with Scheme 2.1 and the related eq 2.1. In all cases over sufficient wide range of concentrations of the added fullerenes the kinetic curves of oxygen consumption do not exhibit any induction period and the model oxidation proceeds with a steady oxidation rate. The dependence of the Figure 2.3 shows quite a good linearity between the experimentally observed oxidation rates Wo$_{2 (C60/C70)}$ and the square root of the concentration of fullerenes over the range 5 × 10^{-5} to 5 × 10^{-4} mol/L. That is, Wo$_{2 (C60/C70)}$~[C$_{60}$/C$_{70}$]$^{1/2}$.

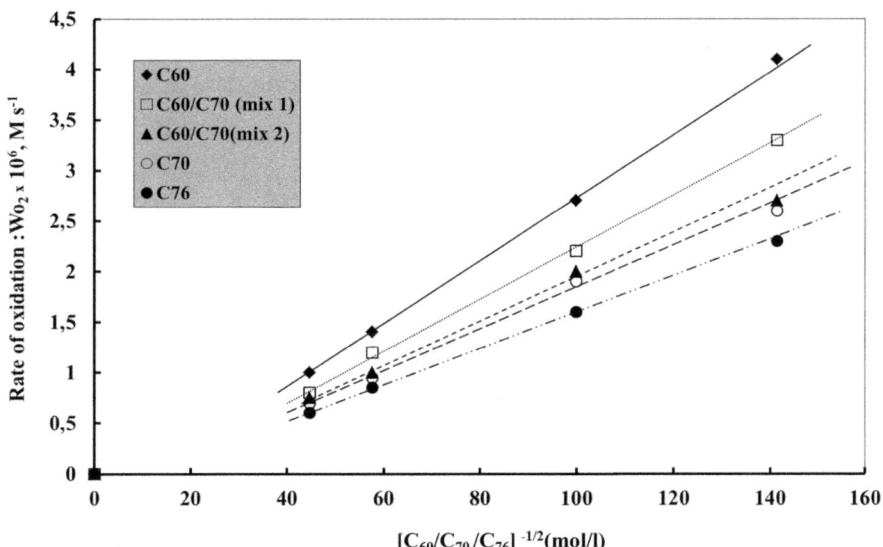

FIGURE 2.3 Rate of cumene-initiated model oxidation versus concentration of the fullerenes. $W_{i\,(AIBN)} = 6.8 \times 10^{-8}$ Ms^{-1}, $Po_2 = 20$ kPa (air), 60°C.

This behavior pattern is also in accordance with eqs 2.2 and 2.3 and conclusion that we have made previously upon the analysis of the literature data, that is, on the predominant interaction the fullerene molecules with alkyl radicals. As the model cumene oxidation is conducted in such conditions (the temperatures lower 100°C) where the contribution of hydroperoxide to the initiation rate is negligible, the alkoxy radicals which are known to be readily formed from the hydroperoxide decay are practically absent in the system.

From the experimentally observed values of oxidation rates, using the known rate constants for cumene oxidation, we may determine W_{i1} (eq 2.1) and afterwards from eqs 2.2 and 2.3 an inhibition rate constant (k_1) for the trapping cumyl alkyl radicals by the fullerenes:

$$k_1[R\cdot]\,[C_{60}/C_{70}] = W_i - [Wo_{2\,(C60/C70)}]^2\,k_6\,(k_3)^{-2}\,[RH]^{-2} \qquad (2.4)$$

At steady-state conditions of the oxidation for fairly long chains the following equations are valid[65,70]:

$$k_2\,[R\cdot][O_2] = k_3\,[RO_2\cdot]\,[RH] = Wo_{2(C60/C70)}. \qquad (2.5)$$

Using the known magnitudes of the constants for cumene oxidation (see the experimental section) and assuming $k_2 = 10^9$ L/mol·s and $[O_2] = 10^{-3}$ mol/L,[65,66,70,71] the following expressions can be obtained for calculations of k_1:

$$[R\cdot] = Wo_{2(C60/C70)} \approx k_2 [O_2] = Wo_{2(C60/C70)}/10^6,$$

$$k_1 = 10^6 \{W_i - [Wo_{2\,(C60/C70)}]^2 k_6 (k_3)^{-2} [RH]^{-2}\} [C_{60}/C_{70}]^{-1} [Wo_{2\,(C60/C70)}]^{-1}. \quad (2.6)$$

Rate constant values for the fullerenes C_{60} and C_{70} are calculated from eq 2.6 over the concentration range 1×10^{-4} to 3×10^{-4} mol/L and also those for known light antioxidants acting as alkyl radical acceptors are tabulated in the Table 2.2.

TABLE 2.2 Rate Constants for the Addition of Cumyl Alkyl Radicals to the Fullerenes and Some Light Stabilizers.

Antioxidant/stabilizer	The inhibition rate constant, $k_{(333K)}$ M^{-1} s^{-1}	References
C_{60}	$(2.0 \pm 0.8) \times 10^8$	34
C_{60}	$(1.9 \pm 0.2) \times 10^8$	Found
C_{60}/C_{70} (mix 1)	$(2.3 \pm 0.2) \times 10^8$	Found
C_{60}/C_{70} (mix 2)	$(2.7 \pm 0.2) \times 10^8$	Found
C_{60}/C_{70} (mix 2) (for styrylalkyl radicals)	$(9.0 \pm 1.5) \times 10^7$	35
C_{70}	$(3.0 \pm 0.3) \times 10^8$	Found
C_{76}	$(3.6 \pm 0.3) \times 10^8$	Found
Cyasorb 3529 (1,6-hexanediamine, N,N'-bis (2,2,6,6-tetramethyl-4-piperidinyl)-, Polymers with morpholine-2,4,6,-trichloro-1,3,5,-triazine)	$(2.0 \pm 0.8) \times 10^8$	69
Chimassorb 119 (1,3,5-Triazine-2,4,6,-triamine,N,N'''-[1,2-ethane-diyl-bis [[[4,6-bis-[butyl (1,2,2,6,6-pentamethyl-4-piperidinyl)amino]-1,3,5-triazine-2yl-]imino]-3,1-propanediyl]]bis N',N''-dibutyl-N', N''-bis (1,2,2,6,6-pentamethyl-4-piperidinyl)	$(1.2 \pm 0.2) \times 10^8$	72
Chimassorb 119FL	$(1.4 \pm 0.2) \times 10^8$	72
Chimassorb 2020 (1,6-Hexanediamine, N, N'-bis-(2,2,6,6-tetramethyl-4-piperidinyl)-polymer with 2,4,6-trichloro-1,3,5-triazine, reaction products with N-butyl-1-butanamine an N-butyl-2,2,6,6-tetramethyl-4-piperidinamine)	$(1.5 \pm 0.2) \times 10^7$	67

As may be seen from these data, the antioxidative activity of the fullerenes is high and comparable with the values for the commercial antioxidants.

The data show that the inhibition rate constants for fullerenes increase with increasing C_{70} moiety in the fullerenes. Finally, fullerenes C_{70} and C_{76} reveal the inhibition rate constant is 1.5–2.0 times higher than that of C_{60}. This is in accordance with the data of Shibaev et al.[36] indicating that the inhibiting effect of fullerene C_{70} is stronger compared to C_{60}. The datum for the fullerene C_{76} is obtained for the first time.

As to the profile the kinetic curves of the fullerene soot (Fig. 2.4), they resemble typical kinetic curves of oxidation in the presence of basic antioxidants—phenols or amines.[62,63,67,69,70,73,74] In this case, the following simple scheme and related kinetic expressions are valid[62,63,65,67,69]:

Scheme 2.2*

Chain initiation: Initiation with concomitant formation of cumyl alkyl radicals R• (Initiation rate is W_i)

Chain propagation: $R^· + O_2 \; RO_2^·$

Chain termination: $RO_2^· + InH$ (In) ®inactive products (rate constant $k_{inh.}$), where InH and/or In are inhibitors.

*$W_{inh.} = k_{inh.} \; [RO_2^·]\{InH \; (In)\} \gg W_{term.} = k_{term.} \; [RO_2^·]^2$ ($W_{term.}$ is the rate of recombination of peroxy radicals).

The fitting equations are:

$$W_{inh} O_2 = W_i \, k_3 \, [RH] / fn \, k_{inh.} \, [InH] \, ([In]) \tag{2.7}$$

$$\tau = fn[InH] \, ([In]) / W_i \tag{2.8}$$

$$\Delta(\, O_2)/[RH] = -k_3 \ln (1 - t/\tau)/k_{inh.}, \tag{2.9}$$

where $W_{inh} O_2$ is rate of inhibited oxidation; τ is induction period; n is number of functional groups in one molecule of an antioxidant ; f is inhibition coefficient, representing the number of $RO_2^·$ peroxy radicals deactivated per one antioxidizing functional group of one molecule of antioxidant or how many oxidation chains are terminated by one antioxidizing group of one molecule of antioxidant; [InH] ([In]) is concentration of an antioxidant ; $\Delta(O_2)$ is volume of absorbed oxygen; t is reaction time; and k_{inh} is rate constant of the inhibition.

The data in Figure 2.4 represent the dependency between graphically evaluated induction periods and the amounts of the fullerene soot added in the model reaction. The observed linear profile of the dependency implies that Scheme 2.2 and the kinetic expressions 2.7–2.9 are good and fulfilled in the presence of the fullerene soot and hence that may be applicable for further calculations and operations.

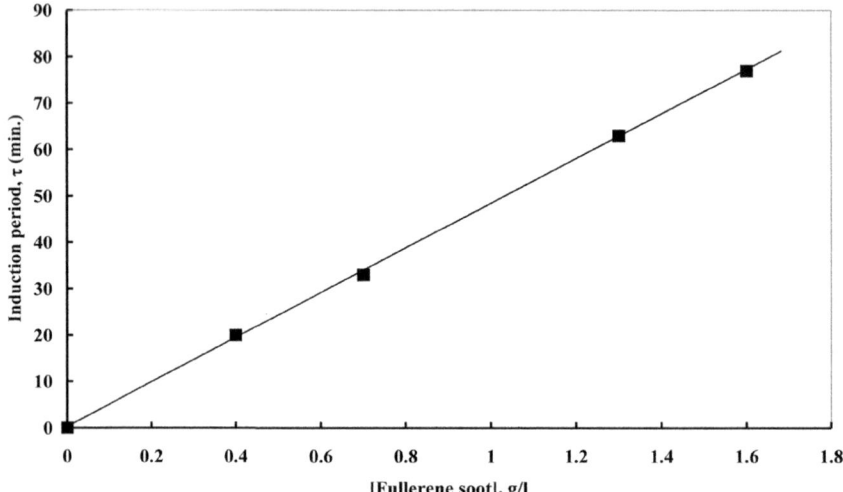

FIGURE 2.4   Induction times versus concentration of fullerene soot for the cumene-initiated oxidation. $W_{i\,(AIBN)} = 6.8 \times 10^{-8}$ Ms^{-1}, $Po_2 = 20$ kPa (air), 60°C.

The rate constant of inhibition k_{inh} for the fullerene soot may be determined from the diagram $\Delta(O_2)$–Lg $(1 - t/\tau)$ plotted in Figure 2.5. The experimental data of oxygen absorption of curves 2–5 (Fig. 2.3) were used therewith. The observed good linearity of the kinetic curves is also in favor that the eqs 2.1–2.3 are valid and the above-mentioned Scheme 2.2 is fulfilled in the presence of the fullerene soot.

The effective concentration of inhibiting centers and inhibition rate constant $k_{inh.}$ for the fullerene soot calculated from the obtained induction periods (Fig. 2.2) and the semilogarithmic plots of Figure 2.5 are equal— fn[C$_{60}$-soot] = $(2.0 \pm 0.1) \times 10^{-4}$mol g^{-1} and $k_{inh} = (6.5 \pm 1.5) \times 10^3$ M^{-1} s^{-1}. The obtained rate constant for the fullerene soot is not so high in comparison with the well-known sterically hindered phenolic and amine antioxidants: for Irganox 1010 the rate constant is $(1.6 \pm 0.1) \times 10^4$, BHT (2,6-di-tert-butyl-4-methylphenol)is $(2.1 \pm 0.1) \times 10^4$, and Neozon-D (N-phenyl-2-naphthyl-amine) is $(6.8 \pm 0.4) \times 10^4$ M^{-1} s^{-1}.[62,75]

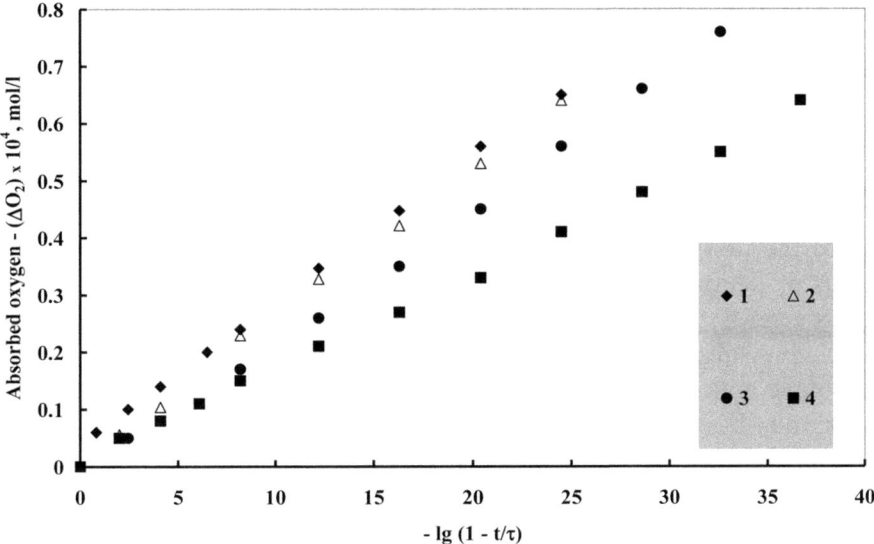

FIGURE 2.5 Semilogarithmic transformation of the kinetic curves for cumene-initiated oxidation in the presence of fullerene soot. $W_{i\,(AIBN)} = 6.8 \times 10^{-8}$ Ms^{-1}, Po$_2$ = 20 kPa (air), 60°C. The integers 1–4 are anamorphosis of the curves 2–5 (Fig. 2.2).

However, it should be noted that the fullerene soot inhibits the oxidation according to two different mechanisms: the actual post-induction rates of oxidation (Fig. 2.2) reveals lower values than those for the uninhibited oxidation. It is evident that a double antioxidative action of the fullerene soot as both a peroxy and alkyl radical acceptor is confirmed. The fullerene moiety of the fullerene soot is apparently acting as alkyl radical scavenger while the non-fullerene part inhibits the oxidation by the other mode.

There is related data to the fact that fullerene soot except the C$_{60}$ part may contain carbon monoxide, hydrogenated species (C$_{60}$ H$_2$, C$_{60}$ H$_4$, C$_{70}$ H$_2$, and C$_{60}$ · CH$_4$),[76] rounded amorphous particles of 20–300 nm, crumpled agglomerates of tangled graphene and multi-layered graphite structures derived from the graphene agglomerates,[77] high amount of bent and spheroidal carbon fragments and polyynic carbon chains.[78] The presence of heteroatoms in fullerene soot is also not excluded and strongly depends on the limited purity of graphite rods and helium gas used in the procedure of its synthesis. Each of these factors might contribute to the observed ability of the fullerene soot for quenching peroxy radicals. However, this item still requires further thorough explorations and indeed could be quite complex to evaluate in detail.

2.4 CONCLUSIONS

Thus the kinetic inhibition rate constants for a series of commercial fullerenes C_{60}–C_{70} have been obtained. These magnitudes allow one to set the compounds according to the following sequence of antioxidative efficiency $C_{70} > C_{60}/C_{70}$ (mix 2) $> C_{60}/C_{70}$ (mix 1) $> C_{60}$. The high susceptibility of C_{60}–C_{70} to the exclusive addition of nucleophilic alkyl radicals has been proved.

For the first time, it has been shown that the fullerene soot not only retards oxidation in the mode of an alkyl radical quencher but also operates as a peroxy radical scavenger. Despite the fact that the rate constant of interaction with peroxy radicals obtained for the fullerene soot is not so high, the double antioxidative function puts it in the front position for effective nanocarbon-based antioxidants.[79]

KEYWORDS

- **thermo-oxidative**
- **radical polymerization**
- **vegetable oils**
- **peroxy radicals**
- **cumene oxidation**

REFERENCES

1. Krusic, P. J.; Wasserman, E.; Keizer, P. N.; Morton, J. R.; Preston, K. F. Radical Reactions of C60. *Science.* **1991,** *254* (5035), 1183–1185.
2. Morton, J. R.; Preston, K. F.; Krusic, P. J.; Hill, S. A.; Wasserman, E. ESR Studies of the Reaction of Alkyl Radicals with C-60. *J. Phys. Chem.* **1992,** *96* (6), 3576–3578.
3. Taylor, R. The Pattern of Additions to Fullerenes. *Philos. Trans. Phys. Sci. Eng.* **1993,** *343* (1667), 87–101.
4. Guldi, D. M.; Hungerbuhler, H.; Janata, E.; Asmus, K. D. Radical-Induced Redox and Addition-Reactions with C-60 Studied by Pulse-Radiolysis. *J. Chem. Soc. Chem. Commun.* **1993,** *1,* 84–86.
5. Dimitrijevic, N. M.; Kamat, P. V.; Fessenden, R. W. Radical Adducts of Fullerenes C-60 and C-70 Studied by Laser Flash-Photolysis and Pulse-Radiolysis. *J. Phys. Chem.* **1993,** *97* (3), 615–618.
6. Cremonini, M. A.; Lunazzi, L.; Placucci, G., Krusic, P. J. Addition of Alkylthio and Alkoxy Radicals to C-60 Studied by ESR. *J. Org. Chem.* **1993,** *58* (17), 4735–4738.

7. Borghi, R.; Lunazzi, L.; Placucci. G.; Krusic, P. J.; Dixon, D. A.; Knight, L. B. Regiochemistry of Radical-Addition to C-70. *J. Phys. Chem.* **1994,** *98* (21), 5395–5398.

8. Morton, J. R.; Preston, K. F.; Krusic, P. J. EPR Spectroscopy of Fullerene Adducts. *Hyperfine Interact.* **1994,** *86* (1–4), 763–777.

9. Morton, J. R.; Negri, F.; Preston, K. F. Addition of Alkyl Radicals to C-60.3. The EPR-Spectra of R3C60 Radicals and a Theoretical Study of HC60 and H3C60 Radicals. *Can. J. Chem.* **1994,** *72* (3), 776–782.

10. Hirsch, A. Addition-Reactions of Buckminsterfullerene (C-60). *Synthesis Stuttgart.* **1995,** *8,* 895–913.

11. Pace, M. D. EPR of C-60 Thermal/Photochemical Reactions with Polystyrene and Polymethyl Methacrylate. *Appl. Magn. Reson.* **1996,** *11* (2), 253–261.

12. Borghi, R.; Lunazzi, L.; Placucci, G.; Cerioni, G.; Plumitallo, A. Photolysis of Dialkoxy Disulfides: A Convenient Source of Alkoxy Radicals for Addition to the Sphere of Fullerene C-60. *J. Org. Chem.* **1996,** *61* (10), 3327–3331.

13. Camp, A. G.; Ford, W. T.; Lary, A.; Sensharma, D. K.; Chang, Y. H.; Hercules, D. M.; Williams, J. B. Reaction of C-60 with Oxygen Initiated by Radicals from Azo(Bisisobutyronitrile). *Fullerene Sci. Technol.* **1997,** *5* (3), 527–545.

14. Morton, J. R.; Negri, F.; Preston, K. F. Addition of Free Radicals to C-60. *Acc. Chem. Res.* **1998,** *31* (2), 63–69.

15. Gross, R.; Dinse, K. P. Light-Induced Formation of C-60 Radical Adducts as Studied with Fourier Transform EPR. In *Recent Advances in the Chemistry and Physics of Fullerenes and Related Material;* Kamat, P. V., Guldi, D. M., Kadish, K. M. Eds.; Book Series; Electrochemical Society Series: Pennington, NJ, 1999; Vol. 7, pp 18–27.

16. Geckeler, K. E.; Samal, S. Rapid Assessment of the Free Radical Scavenging Property of Fullerenes. *Fullerene Sci. Technol.* **2001,** *9* (1), 17–23.

17. Camp, A. G.; Lary, A.; Ford, W. T. Free-Radical Polymerization of Methyl-Methacrylate and Styrene with C(60). *Macromolecules.* **1995,** *28* (23), 7959–7961.

18. Stewart, D.; Imrie, C. T. Role of C-60 in the Free Radical Polymerisation of Styrene. *Chem. Commun.* **1996,** *11,* 1383–1384.

19. Arsalani, N.; Geckeler, K. E. Radical Bulk Polymerization of Styrene in the Presence of Fullerene [60]. *Fullerene Sci. Technol. 1996. 4* (5), 897–912.

20. Cao T.; Webber, S. E. Free Radical Copolymerization of Styrene and $C_{60.}$ *Macromolecules* **1996,** *29* (11), 3826–3930.

21. Chen, Y.; Lin, K. C. Radical Polymerization of Styrene in the Presence of C-60. *J. Polym. Sci. A. Polym. Chem.* **1999,** *37* (15), 2969–2975.

22. Kirkwood, K.; Stewart, D.; Imrie, C. T. Role of C_{60} in the Free Radical Polymerization of Methyl Methacrylate. *J. Polym. Sci. A. Poly. Chem.* **1997,** *35* (15), 3323–3325.

23. Seno, M.; Fukunaga, H.; Sato, T. Kinetic and ESR Studies on Radical Polymerization of Methyl Methacrylate in the Presence of Fullerene. *J. Polym. Sci. A. Polym. Chem.* **1998,** *36* (16), 2905–2912.

24. Seno, M.; Maeda, M.; Sato, T. Effect of Fullerene on Radical Polymerization of Vinyl Acetate. *J. Polym. Sci. A Polym. Chem.* **2000,** *38* (14), 2572–2578.

25. Pabin-Szafko, B.; Wisniewska, E.; Szafko, J. Carbon Nanotubes and Fullerene in the Solution Polymerisation of Acrylonitrile. *Eur. Polym. J.* **2006,** *42* (7), 1516–1520.

26. Wakai, H.; Shinno, T.; Yamauchi, T.; Tsubokawa, N. Grafting of Poly(Ethylene Oxide) onto C-60 Fullerene Using Macroazo Initiators. *Polymer* **2007,** *48* (7), 1972–1980.

27. Ford, W. T.; Graham, T. D.; Mourey, T. H. Incorporation of C-60 into Poly(Methyl Methacrylate) and Polystyrene by Radical Chain Polymerization Produces Branched Structures. *Macromolecules* **1997,** *30* (21), 6422–6429.
28. Ford, W. T.; Nishioka, T.; McCleskey, S. C., Mourey, T. H.; Kahol, P. Structure and Radical Mechanism of Formation of Copolymers of C-60 with Styrene and with Methyl Methacrylate. *Macromolecules* **2000,** *33* (7), 2413–2423.
29. Ford, W. T.; Lary, A. L.; Mourey, T. H. Addition of Polystyryl Radicals from TEMPO-Terminated Polystyrene to C-60. *Macromolecules* **2001,** *34* (17), 5819–5826.
30. Audouin, F.; Renouard, T.; Schmaltz, B.; Nuffer, R.; Mathis, C. Asymmetric and Mikto-Arm Stars with a C-60 Core by Grafting of Macro-Radicals or Anionic Polymer Chains. *Polymer* **2005,** *46* (19), 8519–8527.
31. Mathis, C.; Schmaltz, B.; Brinkmann, M. Controlled Grafting of Polymer Chains onto C-60 and Thermal Stability of the Obtained Materials. *C. R. Chim.* **2006,** *9* (7–8), 1075–1084.
32. Zeynalov, E. B.; Friedrich, J. F. Anti-Radical Activity of Fullerenes and Carbon Nanotubes in Reactions of Radical Polymerization and Polymer Thermal/Thermooxidative Degradation: A Review. *Mater. Test.* **2007,** *49* (5), 265–270.
33. Ginzburg, B. M.; Shibaev, L. A.; Kireenko, O. F.; Shepelevskii, A. A.; Melenevskaya, E. Y.; Ugolkov, V. L. Thermal Degradation of Fullerene-Containing Polymer Systems and Formation of Tribopolymer Films. *Polym. Sci. A.* **2005,** *47* (2), 160–174.
34. Zeinalov, E. B.; Koβmehl, G. Fullerene C_{60} as an Antioxidant for Polymers. *Polym. Degrad. Stab.* **2001,** *71* (2), 197–202.
35. Zeynalov, E. B.; Magerramova, M. Y.; Ischenko, N. Y. Fullerenes C-60/C-70 and C-70 as Antioxidants for Polystyrene. *Iran. Polym. J.* **2004,** *13* (2), 143–148.
36. Shibaev, L. A.; Egorov, V. M.; Zgonnik, V. N.; Antonova, T. A.; Vinogradova, L. V.; Melenevskaya, E. Y.; Bershtein, V. A. An Enhanced Thermal Stability of Poly(2,6-Dimethyl-1,4-Phenylene Oxide) in the Presence of Small Additives of C_{60} and C_{70}. *Polym. Sci. A.* **2001,** *43* (2), 101–105.
37. Cataldo, F. On the Reactivity of C-60 Fullerene with Diene Rubber Macroradicals. I. The Case of Natural and Synthetic cis-1,4-polyisoprene under Anaerobic and Thermo-Oxidative Degradation Conditions. *Fullerene Sci. Technol.* **2001,** *9* (4), 407–513.
38. Jipa, S.; Zaharescu, T.; Santos, C.; Gigante, B.; Setnescu, R.; Setnescu, T.; Dumitru, M.; Kappel, W.; Gorghiu, L. M.; Mihalcea, I.; Olteanu, R. L. The Antioxidant Effect of Some Carbon Materials in Polypropylene. *Mater. Plast.* **2002,** *39* (1), 67–72.
39. Troitskii. B. B.; Troitskaya, L. S.; Dmitriev, A. A.; Yakhnov, A. S. Inhibition of Thermo-Oxidative Degradation of Poly(Methyl Methacrylate) and Polystyrene by C-60. *Eur. Polym. J.* **2000,** *36* (5), 1073–1084.
40. Troitskii, B. B.; Domrachev, G. A.; Khokhlova, L. V.; Anikina, L. I. Thermooxidative Degradation of Poly(methyl methacrylate) in the Presence of C-60 Fullerene. *Polym. Sci. A.* **2001,** *43* (9), 964–969.
41. Troitskii, B. B.; Domrachev, G. A.; Semchikov, Y. D.; Khokhlova, L. V.; Anikina, L. L.; Denisova, V. N.; Novikova, M. A.; Marsenova, Y. A.; Yashchuk, L. M. Fullerene-C-60, a New Effective Inhibitor of High-Temperature Thermooxidative Degradation of Methyl Methacrylate Copolymers. *Russ. J. Gen. Chem.* **2002,** *72* (8), 1276–1281.
42. Ginzburg, B. M.; Shibaev, L. A.; Ugolkov, V. L.; Bulatov, V. P. The Effect of Fullerene C-60 on the Thermooxidative Degradation of a Free-Radical PMMA Studied by Thermogravimetry and Calorimetry. *Techn. Phys. Lett.* **2001,** *27* (10), 806–809.

43. Ginzburg, B. M.; Shibaev, L. A.; Ugolkov, V. L. Effect of Fullerene C-60 on Thermal Oxidative Degradation of Polymethyl Methacrylate Prepared by Radical Polymerization. *Russ. J. Appl. Chem.* **2001,** *74* (8), 1329–1337.

44. Ginzburg, B. M.; Shibaev, L. A.; Ugolkov, V. L.; Bulatov, V. P. Influence of C-60 Fullerene on the Oxidative Degradation of a Free Radical Poly(Methyl Methacrylate). *J. Macromol. Sci. Phys. B.* **2003,** *42* (1), 139–166.

45. Zuev, V. V.; Bertini, F.; Audisio, G. Fullerene C$_{60}$ as Stabiliser for Acrylic Polymers. *Polym. Degrad. Stab.* **2005,** *90* (1), 28–33.

46. Kelar, K. Polyamide 6 Modified with Fullerenes, Prepared via Anionic Polymerization of Epsilon-Caprolactam. *Polimery* **2006,** *51* (6), 415–424.

47. Markovic, Z.; Trajkovic, V. Biomedical Potential of the Reactive Oxygen Species Generation and Quenching by Fullerenes (C-60). *Biomaterials.* **2008,** *29* (26), 3561–3573.

48. Lens, M.; Medenica, L.; Citernesi, U. Antioxidative Capacity of C-60 (Buckminster fullerene) and Newly Synthesized Fulleropyrrolidine Derivatives Encapsulated in Liposomes. *Biotechnol. Appl. Biochem.* **2008,** *51,* 135–140.

49. Jensen, A. W.; Wilson, S. R.; Schuster, D. I. Biological Applications of Fullerenes (Review Article). *Bioorg. Med. Chem.* **1996,** *4* (6), 767–779.

50. Bakry, R.; Vallant. R. M.; Najam-Ul-Haq, M.; Rainer, M.; Szabo, Z.; Huck, C. W.; Bonn, G. K. Medicinal Applications of Fullerenes. *Int. J. Nanomed.* **2007,** *2* (4), 639–649.

51. Gharbi, N.; Pressac, M.; Hadchouel, M.; Szwarc, H.; Wilson, S. R.; Moussa, F. [60] Fullerene is a Powerful Antioxidant In Vivo with No Acute or Subacute Toxicity. *Nano Lett.* **2005,** *5* (12), 2578–2785.

52. Bystrzejewski, M.; Huczko, A.; Lange, H.; Drabik, J.; Pawelec, E. Influence of C-60 and Fullerene Soot on the Oxidation Resistance of Vegetable Oils. *Fullerenes Nanotub. Car. Nanostruct.* **2007,** *15* (6), 427–438.

53. Gasanov, R. G.; Kalina, O. G.; Bashilov, V. V.; Tumanskii, B. L. Addition of Carbon-Centered Radicals to C-60. Determination of the Rate Constants by the Spin Trap Method. *Russ. Chem. Bull.* **1999,** *48* (12), 2344–2346.

54. Gasanov, R. G.; Tumanskii, B. L. Addition of (Me2CCN)-C-Center Dot, (Me2CPh)-C-Center Dot, and (CCl3CH2CHPh)-C-Center Dot Radicals to Fullerene C-60. *Russ. Chem. Bull.* **2002,** *51* (2), 240–242.

55. Guldi, D. M.; Ford, W. T.; Nishioka, T. Rate Constants of Reactions with 2-cyano-2-propyl Radical and Triplet State Lifetimes of Low Molar Mass and Polymeric Substituted [60]Fullerenes. In *Recent Advances in the Chemistry and Physics of Fullerenes and Related Material;* Kamat, P. V., Guldi, D. M., Kadish K. M., Eds.; Book Series; Electrochemical Society Series: Pennington, NJ, 1999; Vol. 7, pp 315–318.

56. Walbiner, M.; Fischer, H. Rate Constants for the Addition of Benzyl Radical to C-60 in Solution. *J. Phys. Chem.* **1993,** *97* (19), 4880–4881.

57. Dimitrijevic, N. M. Reaction of Trichloromethyl and Trichloromethylperoxyl Radicals with C-60—A Pulse-Radiolysis Study. *Chem. Phys. Lett.* **1992,** *194* (4–6), 457–460.

58. Ghosh, H. N.; Pal, H.; Sapre, A. V.; Mukherjee, T.; Mittal, J. P. Formation of Radical Adducts of C-60 with Alkyl and Halo-Alkyl Radicals—Transient Absorption and Emission Characteristics of the Adducts. *J. Chem. Soc. Faraday Trans.* **1996,** *92* (6), 941–944.

59. Enes, R. F.; Tome, A. C.; Cavaleiro, J. A. S.; Amorati, R.; Fumo, M. G.; Pedulli, G. F.; Valgimigli, L. Synthesis and Antioxidant Activity of [60]Fullerene-BHT Conjugates. *Chem. A Eur. J.* **2006,** *12* (17), 4646–4653.

60. Bulgakov, R. G.; Ponomareva, Y. G.; Maslennikov, S. I.; Nevyadovsky, E. Y.; Antipina, S. V. Inertness of C60 Fullerene toward $RO_2^{\cdot}$ Peroxy Radicals. *Russ. Chem. Bull.* **2005,** *54* (8), 1862–1865.

61. Chi, Y.; Bhonsle, J. B.; Canteenwala, T.; Huang, J. P.; Shiea, J.; Chen, B. J.; Chiang, L. Y. Novel Water-Soluble Hexa(sulfobutyl)Fullerenes as Potent Free Radical Scavengers. *Chem. Lett.* **1998,** *5,* 465–466.

62. Tsepalov, V. F.; Kharitonova, A. A.; Gladyshev, G. P.; Emanuel, N. M. Determination of the Rate Constants and Inhibition Coefficients of Phenol Antioxidants with the Aid of Model Chain Reactions/Determination of Rate Constants and Inhibition Coefficients of Inhibitors Using a Model Chain Reaction. *Kinet. Catal.* **1977,** *18* (5), 1034–1041/*18* (6), 1142–1151.

63. Zeynalov, E. B.; Vasnetsova, O. A. *Kinetic Screening of Inhibitors of Radical Reactions;* Elm: Baku, 1993.

64. Van Hook, J. P.; Tobolsky, A. V. The Thermal Decomposition of 2,2'-Azo-bisisobutyro Nitrile. *J. Am. Chem. Soc.* **1958,** *80,* 779–782.

65. Emanuel, N. M.; Denisov, E. T.; Maizus, Z. K. *Liquidphase Oxidation of Hydrocarbons;* Plenum Press: New York, 1967.

66. Gaponova, I. S.; Fedotova, T. V.; Tsepalov, V. F.; Shuvalov, V. F.; Lebedev, Y. S. Study of the Recombination of Cumylperoxy Radicals in Liquid and Supercooled Solutions. *Kinet. Katal.* **1971,** *12,* 1012–1018.

67. Zeynalov, E. B.; Allen, N. S. Simultaneous Determination of the Content and Activity of Sterically Hindered Phenolic and Amine Stabilizers by Means of an Oxidative Model Reaction. *Polym. Degrad. Stabil.* **2004,** *85* (2), 847–853.

68. Zeinalov, E. B.; Schroeder, H. F.; Bahr, H. In *Determination of Phenolic Antioxidant Stabilizers in PP and HDPE by Means of an Oxidative Model Reaction,* Proceedings of 6th International Plastics Additives and Modifiers Conference-Addcon World 2000; Paper 3.

69. Zeynalov, E. B.; Allen, N. S. Modelling Light Stabilizers as Thermal Antioxidants. *Polym. Degrad. Stabil.* **2006,** *91* (12), 3390–3396.

70. Scott, G., Ed. *Atmospheric Oxidation and Antioxidants;* 2nd ed.; Elsevier: New York, Amsterdam, 1993.

71. Maillard, B.; Ingold, K. U.; Scaiano, J. C. Rate Constants for the Reactions of Free Radicals with Oxygen in Solution. *J. Am. Chem. Soc.* **1983,** *105* (15), 5095–5099.

72. Zeynalov, E. B.; Allen, N. S. Effect of Micron and Nano-Grade Titanium Dioxides on the Efficiency of Hindered Piperidine Stabilizers in a Model Oxidative Reaction. *Polym. Degrad. Stabil.* **2006,** *91* (4), 931–939.

73. Roginsky, V. A. Phenolic Antioxidants: Efficiency and Reactivity. Nauka: Moscow, 1988.

74. Zeinalov, E. B.; Kossmehl, G.; Kimwomi, R. R. K. Synthesis and Reactivity of Antioxidants Based on Vernolic Acid and 3-(3,5-di-tert-butyl-4-hydroxyphenyl) Propionic Acid. *Macromol. Mater. Eng.* **1998,** *260,* 77–81.

75. Kimwomi, R. R. K.; Koβmehl, G.; Zeynalov, E. B.; Gitu, P. M.; Bhatt, B. P. Polymeric Antioxidants from Vernonia Oil. *Macromol. Chem. Phys.* **2001,** *202* (13), 2790–2796.

76. Anacleto, J. F.; Boyd, R. K.; Pleasance, S.; Quilliam, M. A.; Howard, J. B.; Lafleur, A. L.; Makarovsky, Y. Y. Analysis of Minor Constituents in Fullerene Soots by LC-MS Using a Heated Pneumatic Nebulizer Interface with Atmospheric-Pressure Chemical. *Can. J. Chem.* **1992,** *70* (10), 2558–2658.

77. Belz, T.; Schlogl, R. Characterization of Fullerene Soots and Carbon Arc Electrode Deposits. *Synth. Met.* **1996,** *77* (1–3), 223–226.
78. Kanowski, M.; Vieth, H. M.; Luders, K.; Buntkowsky, G.; Belz, T.; Werner, H.; Wohlers, M.; Schlogl, R. The Structure of Fullerene Black and the Incorporation of C-60 Investigated by C-13 NMR. *Carbon.* **1997,** *35* (5), 685–695.
79. Zeynalov, E. B.; Allen, N. S.; Salmanova, N. I. Radical Scavenging Efficiency of Different Fullerenes C60–C70 and Fullerene Soot. *Polym. Degrad. Stabil.* **2009,** *94* (8), 1183–1189.

CHAPTER 3

WATER VAPOR ADSORPTION BY ZEOLITES

ŞEFIKA ÇAĞLA SAYILGAN and SEMRA ÜLKÜ*

Chemical Engineering Department, Izmir Institute of Technology, Urla 35430, Izmir, Turkey

Corresponding author. E-mail: semraulku35@gmail.com

CONTENTS

ABSTRACT

Zeolites which are used as adsorbent, ion exchanger and catalyst in several industrial applications have high affinity to various gases and vapors. However, although pre-adsorbed water in the structure of zeolites affects all applications of zeolites, the researches on adsorption of zeolite–water pair are limited and heat and mass transfer properties are not sufficient.

In this chapter, the effect of pre-adsorbed water and regeneration conditions on zeolite-water pair adsorption and characterization studies for different zeolites were given. Furthermore, common adsorption equilibrium and kinetics models used for zeolite-water pairs were summarized.

3.1   INTRODUCTION

Zeolites, which are crystalline aluminosilicates, are used as adsorbent, catalysis, and ion exchanger in many industrial applications such as drying processes, water treatment and softening, agriculture and animal husbandry, mining and metallurgy, construction; and started to be used at energy recovery and storage systems.[39,82] In adsorption processes, the selection of working pair is the main task; and high affinity of the adsorbate and adsorbent is desired. Although zeolites have high affinity to various gases such as CO_2 and water vapor, the affinity to water has a special value for zeolites since the pre-adsorbed water affects the applications of zeolite as catalysis and in separation processes.

Besides affinity of pairs to each other, heat of adsorption, shape of isotherm, adsorption capacity, and mass diffusivity of adsorbate through the adsorbent are the parameters that affect the selection of working pair for the adsorption processes. There are several studies on adsorption equilibria of zeolite–water pairs. In most of these studies, Type I isotherm was given as characteristic curve for zeolite–water pair. However, due to the regeneration conditions, cation size, and location, impurity type and content of zeolites, the isotherm shape may change. In addition, various adsorption equilibrium models such as Langmuir, Dubinin–Radushkevich, experimental correlations were defined for zeolite–water pair.

Although several studies on adsorption equilibrium for zeolite–water pair appeared in literature, the studies on adsorption kinetics is limited and the discussions on the controlling mechanism for mass diffusivity still continue. Most of the researchers assume intraparticle diffusional resistance as controlling mechanism, and recently some researchers began to re-evaluate their studies by taking surface resistance into account.[78]

The aim of this study is to investigate the adsorption of water vapor on zeolites by pointing out the importance of pre-adsorbed water in the structure of zeolites and the effect of regeneration of zeolites. The studies on adsorption equilibrium and adsorption kinetics are also discussed.

3.2   ZEOLITES

Zeolites are porous crystalline aluminosilicates. Structurally, the primary building block of zeolite framework is tetrahedron, the center of which is occupied by a silicon or aluminum atom, with four atoms of oxygen at the corners. The oxygen atoms in the corners are shared between two tetrahedra and an infinitely extending three-dimensional network of AlO_4 and SiO_4 arises. The structural formula of zeolites can be represented as[14,97]:

$$M_{x/n}\left[\left(AlO_2\right)_x\left(SiO_2\right)_y\right].wH_2O$$

where n is the valence of cation M, w is the number of water molecules, the ratio y/x has the value of 1–5 depending on the structure and the sum of x and y defines the total number of tetrahedra in the unit cell.

According to the secondary building units (SBU), which relates the subunit of structure with a specific array of AlO_4 and SiO_4 tetrahedra, zeolites are classified in seven groups. While zeolite A is in the third group with an SBU of double 4-ring tetrahedra, zeolite X is a member of group 4 with double 6-ring tetrahedra. On the other hand, group 7 which includes clinoptilolite has a more complex structure.[14]

The crystal lattice of zeolites, which determines the micropore structure of zeolites, is precisely uniform, and it distinguishes zeolites from other adsorbents. The adsorption ability of zeolites is based on the Si/Al ratio and the adsorption ability of zeolites increases as the ratio decrease. Özkan mentioned that zeolites with Si/Al ratio smaller than 10 have high affinity to polar and polarized molecules.[72] Halasz et al. stated that although most of the zeolites are hydrophilic, aluminum-deficient zeolites with low hydroxyl content becomes hydrophobic.[42] Furthermore, since less energy is required to break the Al—O bonds instead of Si—O bonds, thermal stability also increases as Si/Al ratio increases.[70,97]

In zeolites, due to the lack of the electrical charge in the region of AlO_4 tetrahedra additional positive charges are required to balance the electrical charge and to obtain a stable crystal structure. This additional charge is provided by

cations such as Na, Ca, K, Mg, and Ti which are replaceable and easily remov-
able from the structure. The location, size, and number of cations, which are
the adsorption sites of zeolites, are effective on adsorption properties.

When zeolite is heated, the dehydration takes place and the water molecules in the structure of zeolite are removed in three steps without deterioration of the structure. The water which can be classified as external, loosely
bound and tightly bound are removed at different temperatures due to the
energy of cation–water bond distance and the exchangeable cation sites.[8]
In addition, the desorbed water can be adsorbed again when the zeolite is
cooled. Water, which is reversibly adsorbed and desorbed by zeolite, is
named as zeolitic water. In most zeolites, structural changes can be accom-
plished by removing zeolitic water and cations. Zeolitic water completes
the coordination of the cations in the cavities and minimize the electrostatic
attraction forces between the oxygen molecules.[72,97] Cavities, which occur
as a consequence of dehydration, cause zeolites act as sieves on a molecular
scale and named as molecular sieves.

The effect of cation size and location on enhancement of water
adsorption capacity on zeolites is also drawing attention of many
researchers.[6,31,33,47,49,66,72] For instance, Moise et al. investigated the water
adsorption capacities of BaX and BaY exchanged zeolites. They paid attention to the effect of the cation locations and sizes on adsorption capacity in
zeolite X and Y. They mentioned that while the adsorption capacity mainly
depended on the cation radius in zeolite X, it became strongly dependent on
cation location in zeolite Y.[66]

3.2.1 CHARACTERIZATION OF ZEOLITES

Prior to applications, zeolites should be characterized to detect the suitability
of the properties for the processes. Thermal and structural properties of zeolites
can be determined by different methods such as thermal gravimetric analysis
(TGA), scanning electron microscopy (SEM), X-ray diffractometer (XRD),
and infrared spectroscopy (IR). These characterization techniques were also
used by the authors in the characterization of different types of zeolites such
as 4A and 13X from Sigma-Aldrich Co. (4–8 mesh), 5A from Alfa Aeser Co.
(3–5 mm) and Clinoptilolite from Gördes in Turkey. The experimental conditions of the characterization study are summarized in Table 3.1.

Figure 3.1 shows the thermogravimetric analysis results for zeolites 4A,
5A, 13X, and clinoptilolite. The dehydration steps can be observed from
Figure 3.1. For instance, for zeolite 13X externally bound water, the first

inflection point was observed at ≈100°C. Other inflection point was detected at about ≈180°C where the loosely bound water was removed. Above ≈180°C, slow dehydration of tightly bound water took place. Furthermore, it was seen that there were deteriorations in the framework structure above 600°C which was consistent with the results given in literature.[14,97]

TABLE 3.1 Characterization Methods.

Methods	Apparatus	Conditions
TGA	Shimadzu TGA-51	Heating rate: 5°C/min
		N2 flow: 40 mL/min
		Temperature: 25–1000°C
SEM	SEM, FEI QUANTA 250 FEG	Magnification: 10,000×
XRD	X'pert Pro, Philips	2θ range of 5−50°
		Scanning speed: 0.139°/sn
FTIR	Shimadzu FTIR-8201 model	KBr pellet technique

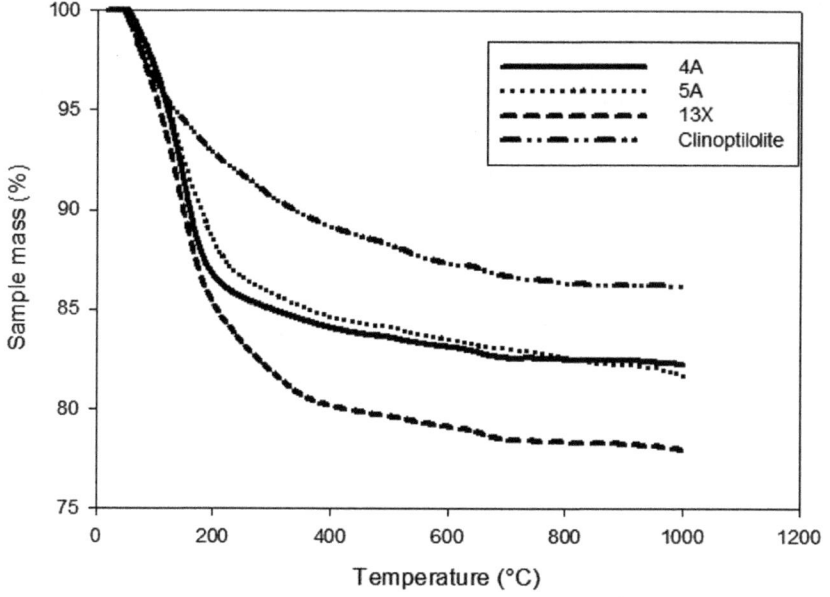

FIGURE 3.1 TGA curves of different zeolites.

Particle and surface morphologies, crystal size, and structures and elemental compositions were determined with SEM. The crystal structures of zeolites can be observed in Figure 3.2 with a magnification of 10,000×.

As it is seen from Figure 3.2, while zeolite 4A and 5A crystals have cubic, zeolite 13X crystals have spherical and clinoptilolite crystals have tubular/platy morphology.

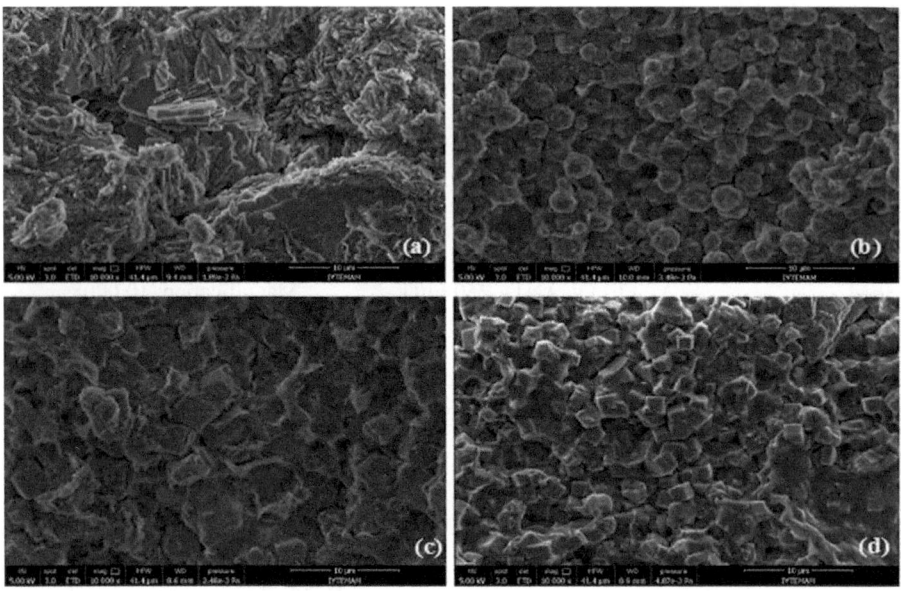

FIGURE 3.2 Representative SEM images of zeolites: (a) clinoptilolite, (b) 13X, (c) 5A, and (d) 4A.

The elemental compositions obtained from SEM–EDX are given in Table 3.2. The Si/Al ratio was given for zeolite A, zeolite X and clinoptilolite are in the range of 0.7–1.2, 1–1.5, and 4.25–5.25, respectively, as given by Breck.[14]

TABLE 3.2 Elemental Compositions of Zeolites.

Element	4A	5A	13X	Clinoptilolite
O	61.32	61.27	60.79	75.83
Na	5.81	3.69	10.06	0.31
Mg	1.77	1.79	1.23	0.44
Al	10.85	11.00	11.65	3.72
Si	14.89	16.06	16.28	18.03
K	4.82	–	–	0.81
Ca	0.53	6.19	–	0.85
Total	100.00	100	100	100

The structural characterization of zeolites was also provided by IR as a complementary to X-ray structural analysis as also suggested by Breck.[14] In the IR spectra obtained by KBr pellet technique, a typical IR pattern was observed for each zeolite and the spectra is given in Figure 3.3 for different zeolites. The characteristic spectra of zeolites can be seen in Figure 3.3 and are summarized as[14]:

- Asymmetric stretching due to internal vibrations (internal tetrahedra)→1250–950 cm^{-1},
- Internal tetrahedra symmetric stretching → 720–650 cm^{-1},
- Internal T—O bending → 500–420 cm^{-1},
- Double ring due to external linkages → 650–500 cm^{-1},
- Pore opening → 300–420 cm^{-1},
- External linkages symmetric stretching → 750–820 cm^{-1}, and
- External linkages asymmetric stretching → 1150–1050.

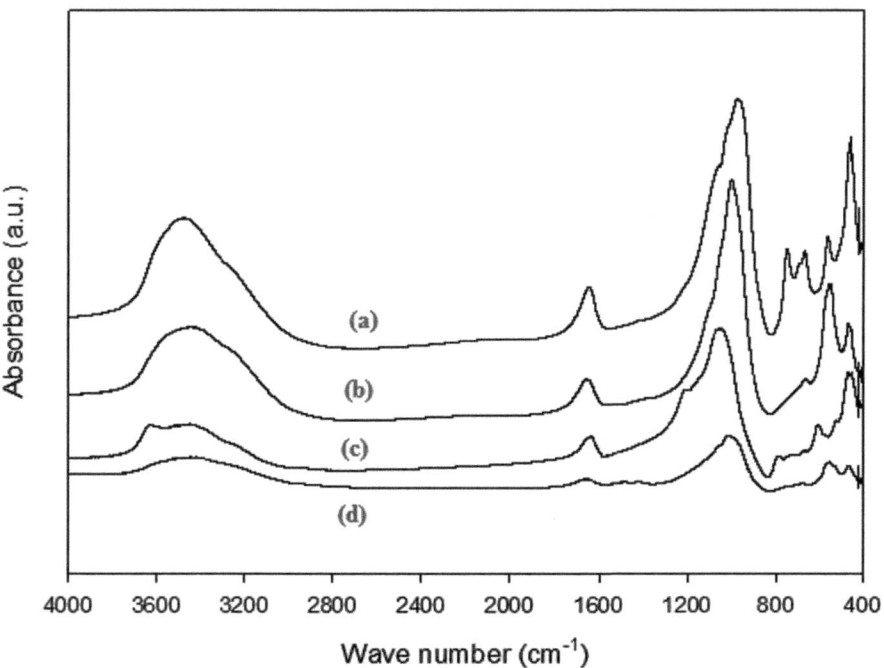

FIGURE 3.3 FTIR spectra of zeolites: (a) 13X, (b) 4A, (c) clinoptilolite, and (d) 5A.

The position of internal tetrahedra asymmetric stretching band depends on the Si/Al ratio and is considered as indicative for the aluminum content in the framework. Furthermore, the peaks at approximately 1645 and 3400 cm⁻¹ which are characteristics of bending vibration of water molecule and hydrogen bonded OH, respectively, were also observed.

There exists several researches on the characterization of crystallinity and mineral composition of zeolite samples with X-ray powder diffractometer.[8,16,35,68,69,83,87,89,111] In general, while the sharp peaks indicate the crystallinity, the narrow peaks represent the amorphous nature of the sample in XRDs. The X-ray diffractograms of different zeolite samples are represented in Figure 3.4. Furthermore, the impurity content can also be identified by this method. For instance, Narin et al. made characterization of clinoptilolite-rich sample and obtained characteristic peaks at 2θ= 9.93°, 22.48°, and 30.18°. Additionally, they indicated that sample contained 80% clinoptilolite, 5–8% opal-CT, 4–5% feldspar, 3% quartz, and 1–2% biotite.[69]

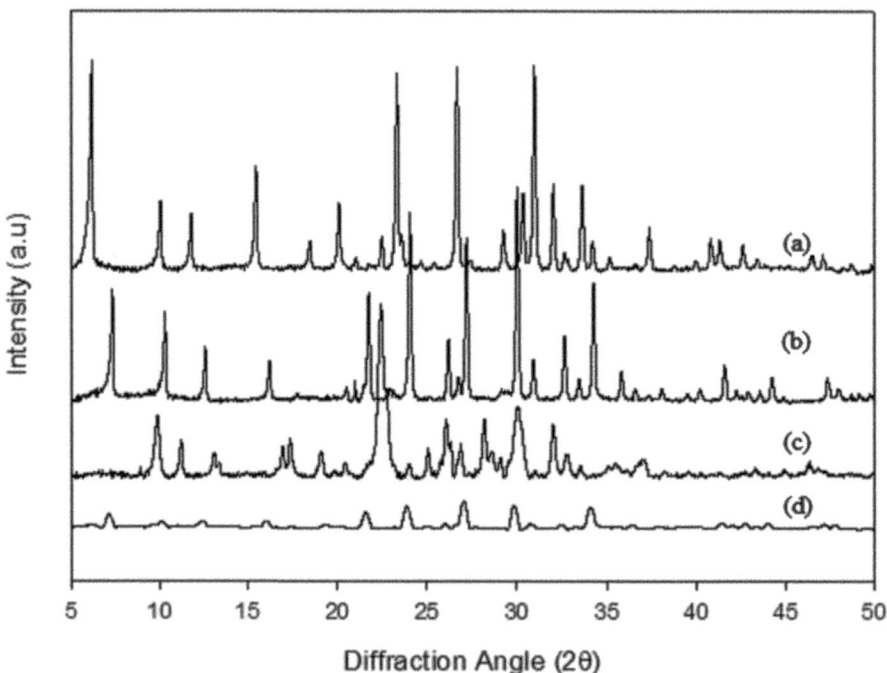

FIGURE 3.4 X-ray diffractograms for different zeolites: (a) 13X, (b) 4A, (c) clinoptilolite, and (d) 5A.

3.2.2 PRE-ADSORBED WATER AND REGENERATION OF ZEOLITES

Zeolites are used as adsorbent, catalyst, and ion exchanger and they have high affinity for various gases and vapor. Besides, water vapor has a special importance since pre-adsorbed water affects the properties of zeolites such as adsorptive, catalytic, and molecular sieve effect. The effect of pre-adsorbed water on adsorption of various gases such as CO_2 and hydrocarbons on zeolite samples was studied by several researchers.[13,33,63,92] They all agreed that the pre-adsorbed water decreases the adsorption capacity especially at low adsorbate concentration. Malka-Edery et al. also mentioned that the heat of adsorption and kinetics of alkenes are reduced by water loading on zeolite NaX.[63] On the other hand, Brandani and Ruthven highlighted that water is strongly polar and reducing the strength and heterogeneity of electric field, so CO_2 which has a high quadrupole moment; and is significantly affected from the changes in electric field while these changes have smaller impact on C_3H_8 which is a nonpolar molecule.[13]

Therefore, the dehydration behavior of zeolites becomes important and it requires information about thermal behavior of zeolites which depends on the type, amount, and position of cations within the structure, the coordination of cations with water molecules, Si/Al ratio, dehydration temperature, water vapor pressure and heating rate.[9,10]

Zeolites are classified into two groups according to the structural changes and continuity of dehydration curves. While zeolite A, X, Y, and chabazite, clinoptilolite which remain stable up to temperature range of 700–1000°C take place in the first group and natrolite, scolecite, and mesolite which transform into a metastable phase after dehydration are categorized in the second group. Although there are not any topographic changes in the framework structure of zeolites, the cation locations may change during reversible and continuous dehydration process.[9,14,69,75,97] For instance, Na$^+$ ions, which are located in the 8-ring and displaced about 1.2 Å from the center of dehydrated zeolite A, causes blocking of pores and affects the adsorption ability of zeolite A.

On the other hand, although zeolite X is categorized in reversible and continuous dehydration, in some cases irreversible adsorption may be detected.[14] Sticher mentioned that for the samples with Si/Si+Al ratio in the range of 0.5–0.55 and with high potassium content, the dehydration is reversible up to temperature of 180°C. Above 180°C, the crystal structure is destroyed and the dehydration becomes irreversible.[86]

The complete regeneration of zeolites is generally achieved at temperatures above 350°C under vacuum, but it needs attention since the aluminum-rich zeolites have poor hydrothermal stability and the destruction in crystallinity occurs even at low temperatures in the presence of water. Thus, the dehydration should be relatively slow at moderate temperature under a good vacuum.[76] Yucel and Ruthven indicated that the temperature should be increased 2–3°C/min during regeneration under vacuum.[110] Furthermore, the adsorption capacity and mass diffusivity was affected from regeneration conditions.[78,82]

3.3 ADSORPTION

Adsorption is a surface phenomenon which occurs at fluid/solid interface due to the molecular or atomic interactions. The schematic view and terminology of adsorption phenomena is given in Figure 3.5. The reverse process of adsorption which is called desorption is performed by increasing the temperature or decreasing the pressure.

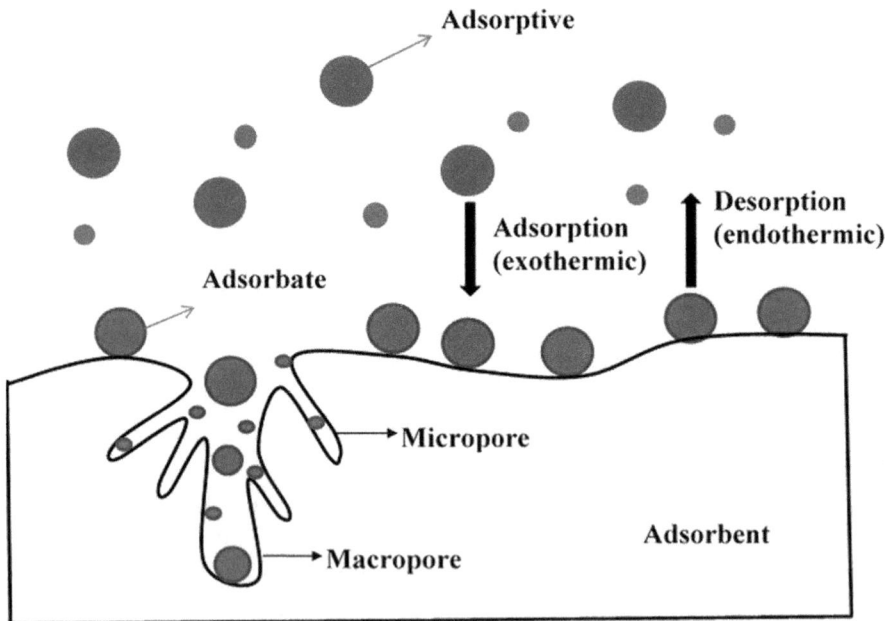

FIGURE 3.5   Schematic view of adsorption phenomena.

Two different types of adsorption may take place according to the interactions of the adsorbent–adsorbate pairs. The first one is the physical adsorption (physisorption or Van der Waals) which relates the nonspecific interactions and the other is chemical adsorption (chemisorption) where the electrons are shared or transferred between two phases. In chemisorption monolayer is observed, interactions are very strong compared to physisorption since a new chemical compound is formed. Additionally, the amount of heat released, in chemical adsorption processes is higher than physical adsorption processes due to the specific interactions between adsorbate and adsorbent pair. Nevertheless, the heat of adsorption values of physical and chemical adsorption processes varies between 5–50 and 40–800 kJ/mol, respectively.[46,54,96]

The overall free energy change for adsorption process is negative since the adsorption on solid surface occurs spontaneously. In addition, since the adsorbing molecules lose degree of freedom during the adsorption process, the entropy decreases. Considering the negativity of entropy and free energy changes, the enthalpy should also be negative which gives exothermic characteristic to adsorption processes. Nevertheless, in some cases such as protein adsorption processes, endothermic behavior can also be observed due to lateral protein–protein interactions and conformational changes in the adsorbed protein.[53]

From daily life to industrial applications, adsorption has broad range of usage: such as, wastewater treatment, drying processes, catalytic reactions and medical applications. In recent years, the use of adsorption in heat recovery and thermal storage systems has also gained attention of researchers. The selection of an appropriate adsorbent for a particular adsorbate is the most important task in industrial application of adsorption. The affinity of the pairs to each other, shape of isotherm, adsorption capacity, thermal conductivity, heat of adsorption and mass diffusivity of the adsorbate through the adsorbent are the important parameters that affect the selection of the working pair.[82] Among the pairs, water–zeolite pair has a special importance not only as working pair for a specific application, but also due to the effect of adsorbed water on the success of other applications.

3.3.1 ADSORPTION EQUILIBRIUM

Adsorption equilibrium can be described as the amount of adsorbate sorbed by unit mass of adsorbent at specific temperature and pressure. Surface characteristics of the adsorbent, working temperature and pressure, thermophysical properties of the adsorbent and adsorbate concentration can define the adsorption equilibrium.[34,81]

According to the relationship between temperature, adsorptive pressure (adsorptive concentration) and adsorbate concentration the adsorption equilibrium can be illustrated with plots as isobar, isotherm, and isoster. According to the IUPAC classification, the isotherms are divided into six groups considering the adsorbate–adsorbent interactions and they are represented in Table 3.3 for gaseous adsorptive and vapor. Additionally, the fundamental and empirical models representing adsorption equilibrium are given in Table 3.4.

TABLE 3.3   Types of Adsorption Isotherms.

Isotherm type	Assumptions
Type I	Adsorption on microporous adsorbents having relatively small external surface area
	Multilayer coverage
Type II	Adsorption on nonporous or macroporous adsorbent
	Point B indicates the end of monolayer coverage and start of multilayer coverage
Type III	Adsorption on nonporous or macroporous adsorbent
	Multilayer adsorption
	Weak interactions between adsorbate and adsorbent
Type IV	Hysteresis loop due to the capillary condensation in mesopores
	Multilayer adsorption

TABLE 3.3 *(Continued)*

Isotherm type	Assumptions
Type V Amount of adsorbed gas — P/P^sat	Hysteresis loop due to the capillary condensation in mesopores
	Multilayer adsorption
	Weak interactions between adsorbate and adsorbent
Type VI Amount of adsorbed gas — P/P^sat	Adsorption on nonporous adsorbent
	Stepwise multilayer adsorption
	Height of steps indicates the capacity of monolayer adsorption

TABLE 3.4 Adsorption Equilibrium Models Used for Zeolites in Literature.

Model	Equation	Assumption	References
Henry's relationship	$q = \left(K_0 e^{\Delta H/RT} \right) P$	All adsorbate molecules are isolated from their neighbors at low pressure of adsorptive	76, 91, 105, 111
Langmuir relationship	$q = q_m \dfrac{bP}{1+bP}$	The surface is homogenous. Heat of adsorption is independent of surface coverage. Monolayer coverage is observed. Adsorption rate is equal to desorption rate	19, 29, 38, 40, 42, 55, 74, 76
Dubinin–Radushkevich relationship	$q = q_0 \exp\left[-\gamma \left(T \ln \dfrac{P^{sat}}{P} \right)^2 \right]$	Based on Polanyi potential theory. Adsorption on microporous adsorbents	17, 38, 50, 58, 88, 95, 102
Dubinin–Astakhov relationship	$q = q_0 \exp\left[-D \left(T \ln \dfrac{P^{sat}}{P} \right)^n \right]$	Same with D–R relationship. Surface heterogeneity is high ($n > 2$)	23, 28, 88, 93, 106
Modified Dubinin–Astakhov relationship	$q = q_0 \exp\left[-K \left(\dfrac{T_z}{T^{sat}} - 1 \right)^n \right]$	Same with D–R relationship. Surface heterogeneity is high ($n > 2$)	93

TABLE 3.4 *(Continued)*

Model	Equation	Assumption	References
Experimental correlations	$\ln P = a(q) + \dfrac{b(q)}{T}$ $a(q) = a_0 + a_1 q + a_2 q^2 + a_3 q^3$ $b(q) = b_0 + b_1 q + b_2 q^2 + b_3 q^3$	The difference in the heat capacity of the adsorbate in the adsorbed and vapor phase is neglected	15, 65, 80, 99, 100
Three-term Langmuir relationship	$q = \dfrac{q_{s1} b_1 P}{1 + b_1 P} + \dfrac{q_{s2} b_2 P}{1 + b_2 P} + \dfrac{q_{s3} b_3 P}{1 + b_3 P}$	There are two or three sites for adsorption with energy of adsorption constant at each site	7, 23, 59, 90

Among them in the study by Cakıcıoglu-Ozkan and Ulku, adsorption of water vapor on acid treated clinoptilolite was investigated. They obtained Type I isotherm, but there was an increase in adsorption at high relative pressures ($P/P_0 > 0.6$) which they related it with the presence of the impurity or extra framework formation on the crystal surface as expected for natural zeolites. Cakıcıoglu-Ozkan and Ulku also stated that Dubinin–Radushkevich relationship defined the equilibrium data better than Langmuir relationship for acid treated zeolite–water pair.[17]

Additionally, Ülkü et al. used both Dubinin–Radushkevich and experimental correlations in determination of adsorption equilibrium of clinoptilolite water pair for air drying in packed bed adsorbers. They concluded that adsorption equilibria was better represented by experimental correlations which are tested in the range of $q = 0.04$–0.115 kg H_2O/kg dry zeolite.[100]

On the other hand, instead of traditional adsorption isotherm models, and correlations, there are alternative models used in literature to illustrate the adsorption equilibria of zeolites such as Hill's, Toth's, Aranovich–Donohue, and Frenkel–Halsey–Hill models.[43,56,108]

In the Hill's model, the adsorption is stated by the absolute activity of adsorbate and canonical partition function of sites with a variable number of molecules adsorbed (eq 3.2).[43]

$$\frac{C}{C_s} = \frac{1}{m}\left[\frac{\displaystyle\sum_{n=0}^{m} n q_n \lambda_a^n}{\displaystyle\sum_{n=0}^{m} q_n \lambda_a^n}\right] \tag{3.2}$$

where λ_a is the absolute activity of the adsorbate, q_n is the canonical partition function of a subsystem with n molecules adsorbed, C is the amount adsorbed at equilibrium by the unit mass of adsorbent and C_s is the specific uptake at the saturation limit.

Furthermore, Hill's model is modified by several researchers to determine adsorption isotherm.[11,60,76] Llano-Restrepo and Mosquera used generalized statistical thermodynamic adsorption model which is modified from Hill's model to define the adsorption equilibria of zeolite 3A. In their model, adsorption in sites was expressed in terms of pressure and a set temperature dependent parameters (eq 3.3).

$$\frac{C}{C_s} = \frac{1}{m}\left[\frac{\sum_{n=1}^{m} nK_n P^n}{1 + \sum_{n=1}^{m} K_n P^n}\right] \tag{3.3}$$

where the coefficients K_n are temperature-dependent adjustable parameters.

In 2009, Loughlin reanalyzed the data of Gorbach et al.[37] and Morris[67] by taking α and β cages in zeolite 4A. He claimed that due to the ratio of these cages, which is approximately 5–1, the adsorption is changing in each cavity significantly. Thus, he rearranged the traditional adsorption isotherm models according to the two site hypothesis. For instance, Henry's model is modified as[61]:

$$\frac{q}{q_s} = 0.162 K_\beta P + 0.0838 K_\beta P \tag{3.4}$$

Where the Henry's constant for both cages is defined as $H_i = f_i K_i q_s$ and K_i is equilibrium parameter, P is pressure, q is adsorbed 1phase concentration, and q_s is saturated adsorbed phase concentration.

3.3.1.1 *DETERMINATION OF ADSORPTION ISOTHERM EXPERIMENTALLY*

The heat of adsorption value and maximum adsorption capacity of the working pair is obtained from the adsorption equilibrium data. Thus, the adsorption equilibrium should be well defined to select the appropriate working pair. The experimental methods such gravimetric, volumetric, and calorimetric can be used to obtain equilibrium data as a function of pressure and temperature.

The most common technique is volumetric method in which the pressure changes before and after adsorption in a closed system are measured.[22,55,71,72,82,98,109] The main components of a volumetric system are an adsorption vessel, a gas storage vessel and a vacuum pump. The amount of adsorbate adsorbed is generally determined by applying the equation of ideal gas law.

In gravimetric method, the amount of adsorbate is directly calculated from the weight change of adsorbent as a function of time by using a micro-balance system.[2,18,24,79,98]

On the other hand, in calorimetric technique heat which might be generated, consumed or simply dissipated by a sample is measured. Tian–Calvet microcalorimetry which consists of a sample bed, heat sink and thermopile is the most suited system for isothermal gas adsorption processes.[26,27,30,32,85,104]

3.3.2   ADSORPTION KINETICS

The movement of the fluid through a porous adsorbent particle takes place in five steps:

1. Diffusion from bulk solution to the external surface of adsorbent particle,
2. Diffusion of adsorptive molecules through the macropores of adsorbent particle,
3. Diffusion of adsorptive molecules through the micropores of adsorbent particle,
4. Adsorption due to the physical and chemical interactions between adsorbate and adsorbent,
5. Diffusion of adsorbed molecules in the sorbed state.

Although, the resistances related to the given adsorption steps such as diffusional resistance in bulk fluid, diffusional resistance in laminar fluid film on the particle surface, skin resistance at the surface of the particle, diffusional resistance in the meso- and macropores, a possible barrier to mass transfer at the external surface of the microparticle, and diffusional resistance in the micropores within the microparticles are all effective, one of these steps may control the transport process.[1,81]

In the analysis of the mass and heat transfer controlling mechanisms, a dimensionless parameter, Biot number, plays an important role. Mass transfer Biot number (eq 3.5) gives information about the concentration distribution inside the particle. For the cases where the Biot number is smaller than 1 (Bi<<1), external mass transfer resistance is controlling mechanism according to the homogeneous concentration profile inside the particle. On the other hand, if Bi>>1, a concentration gradient arises inside the particle, then the intraparticle diffusion becomes controlling mechanism.[1,26,45,102]

$$Bi_m = \frac{k_f r_p}{3 \varepsilon D_m} = \frac{Sh}{6} \frac{D_p}{\varepsilon D_m} \tag{3.5}$$

3.3.2.1   EXTERNAL FILM AND SURFACE (SKIN) RESISTANCES

External resistance is dominated when the adsorbent is nonporous or there is more than one component in the fluid surrounding the particle. In such cases, the adsorption rate may be controlled by the diffusional resistance through the laminar fluid film.[52,81] The properties of fluid, the particle size and particle surface roughness are the parameters that affect the fluid film resistance.

Based on the assumptions that there is no concentration gradient through the adsorbent and the adsorptive concentration is in equilibrium with the adsorbed phase concentration throughout the particle, the fluid film resistance equation can be given as[52]:

$$\frac{d\overline{q}}{dt} = k_f a \left(C - C^* \right) \tag{3.6}$$

For the case of linear adsorption equilibrium relationship ($q^* = KC$), eq 3.6 becomes:

$$\frac{d\overline{q}}{dt} = \frac{3k_f}{KR_p} \left(q^* - \overline{q} \right) \tag{3.7}$$

$$\frac{\overline{q}}{q_\infty} = 1 - \exp \left[-\frac{3k_f t}{KR_p} \right] \tag{3.8}$$

In general, the external film diffusion is faster than intraparticle diffusion except nonporous adsorbents. Thus, intraparticle control mechanism is generally assumed to be the controlling mechanism in adsorption rate. However, an additional resistance (skin resistance) due to the constriction of the pore mouth, blockage of the large pores near the surface of the particle or from the deposition of the extrinsic materials at the crystal surface may also arise.[77] The analytical solution of skin resistance has a similar form with fluid film resistance and is represented as:

$$\frac{\overline{q}}{q_\infty} = 1 - \exp\left[-\frac{3k_s t}{R_p}\right] \qquad (3.9)$$

where $k_s = D_s/\delta$ is the ratio of the effective diffusivity and the thickness of the solid surface film.[52]

3.3.2.2 INTRAPARTICLE DIFFUSIONAL RESISTANCE

3.2.2.2.1 Diffusional Resistance in Micropores

In micropores, the interactions between the pore wall and fluid are domi-nated since the size of the adsorptive and the microporous adsorbent is almost equal. In such systems, the adsorptive cannot escape from the force field of the pore walls and the diffusion arises from a random jump of the adsorbate molecules from one site to next site. Therefore, the fluid phase in micropores can be thought as the adsorbed phase or a single phase and the diffusion is named as "intracrystalline diffusion" or "micropore diffusion." Furthermore, the micropore diffusion becomes independent of particle size and the mass transfer equation can be given by Fick's law as:

$$\frac{\partial q}{\partial t} = \frac{1}{r^2}\frac{\partial}{\partial r}\left(r^2 D_c \frac{\partial q}{\partial r}\right) \qquad (3.10)$$

where r is the pore radius, D_c is the intracrystalline diffusivity, and $q\,(r,t)$ is the adsorbed phase concentration.

In the estimation of the micropore diffusion coefficient from the uptake curves, generally simplified models based on the Fick's law are used.[3,12,21,25,52] Assuming:

- Isothermal system,
- Infinite system volume,
- Constant initial concentration of adsorbate in the fluid, q_∞, and
- Spherical particle.

The fractional attainment of equilibrium for microporous adsorbent with constant diffusivity can be given as:

$$\frac{m_t}{m_\infty} = \frac{\overline{q} - q_0}{q_\infty - q_0} = 1 - \frac{6}{\pi^2}\sum_{n=1}^{\infty}\frac{1}{n^2}\exp\left(-\frac{n^2\pi^2 D_c t}{r_c^2}\right) \qquad (3.11)$$

where r_c is the crystal radius. On the other hand, the particle radius can be considered instead of crystal radius when the diffusion coefficient is taken as effective diffusivity (eq 3.12).[12,18,73]

$$\frac{m_t}{m_\infty} = \frac{\bar{q} - q_0}{q_\infty - q_0} = 1 - \frac{6}{\pi^2} \sum_{n=1}^{\infty} \frac{1}{n^2} \exp\left(-\frac{n^2 \pi^2 D_{eff} t}{R_p^2}\right) \tag{3.12}$$

In the long time period (generally $m_t/m_\infty > 0.7$), the higher terms of the summation become insignificant and the analytical solution of mass transfer equation becomes:

$$\frac{m_t}{m_\infty} = 1 - \frac{6}{\pi^2} \exp\left(-\frac{\pi^2 D_{eff} t}{R_p^2}\right) \tag{3.13}$$

For the intraparticle diffusion control, the plot of $\ln (1 - m_t/m_\infty)$ vs t graph should be linear for long time period with an intercept of $\ln (6/\pi^2)$. This is the main way to distinguish intraparticle diffusion and surface diffusion resistances from each other since the plot should pass through the origin when the surface diffusion resistance is the main controlling mechanism.

For the short time period (generally $m_t/m_\infty < 0.3$), the analytical solution of the mass transfer equation becomes:

$$\frac{m_t}{m_\infty} = \frac{6}{\sqrt{\pi}} \left(\frac{D_{eff} t}{R_p^2}\right)^{1/2} \tag{3.14}$$

Nevertheless, eq 3.11 is valid when the uptake within the particle is too small compared to the adsorption capacity of the system (infinite volume). Carman and Haul[21] derived an analytical solution for the limited volume with the assumption of linear equilibrium relationship ($q_\infty = KC$) as:

$$\frac{m_t}{m_\infty} = 1 - 6 \sum_{n=1}^{\infty} \frac{\lambda(1+\lambda)\exp\left(-D_c q_n^2 t / r_c^2\right)}{9(1+\lambda) + \lambda^2 q_n^2} \tag{3.15}$$

where q_n is given by the nonzero root of following equation:

$$\tan q_n = \frac{3q_n}{3 + \lambda q_n^2} \tag{3.16}$$

λ is the ratio of the adsorptive concentration to the adsorbate concentration on the solid surface between the initial to final steps of a pulse.[21,82]

Sayılgan et al. determined the effective diffusivity of zeolite 13X–water pair for different adsorption and regeneration temperatures with a volumetric system. It was observed that the effective diffusivity was in the range of 4×10^{-9} to 6×10^{-8} m²/s and decreased with increasing adsorption capacity in the long time period. Additionally, they indicated that while the regeneration temperature did not significantly influence the diffusivity in the long time period, the change in initial adsorbate concentration had a significant effect on effective diffusivity.[82]

As distinct from Sayılgan et al.,[82] Karger and Pfeifer gave the diffusion coefficient for zeolite NaX–water pair in the range of 4×10^{-10} to 2×10^{-9} m²/s and they mentioned that due to the decrease in the relative number of strong adsorption binding with increasing adsorption loading, the overall mobility would increase.[51]

The effect of thermal and mass diffusivities of zeolite 4A–water pair on the performance of adsorption heat pumps with zeolite coating was investigated by Tatlıer and Erdem-Şenatalar. They calculated the effective diffusivity by using intraparticle diffusion model in the range of 4×10^{-11} to 1×10^{-9} m²/s. They concluded that the mass diffusivity is the limiting parameter for the adsorption heat pump in determining the zeolite coating thickness on metal support.[94]

Although there are several studies on adsorption equilibria zeolite–water pair (Table 3.5), the kinetic studies are still not enough. Most researchers used linear driving force model (eq 3.17) to define the mass transfer coefficient due to its simplicity compared to detailed pore models.[48,64,79,84] In this model, it was assumed that the mean internal concentration rate is directly proportional to the difference between the surface concentration and the mean internal concentration.[36]

$$\frac{d\bar{q}}{dt} = \frac{15D}{r^2}(q - \bar{q}) \qquad (3.17)$$

Ryu et al. performed a study to determine the adsorption equilibrium and kinetics of water on zeolite 13X. Different kinetic models such as intraparticle diffusion model (solid diffusion model), LDF, Vermeulen and Nakao & Suzuki models were defined and compared for the working pair. It was stated that instead of solid diffusion model, Nakao & Suzuki model, which is a modified form of LDF model, is better fit to experimental data.

Simo et al. examined the effects of temperature, pressure, water concentration, bed velocity and pellet size on breakthrough curves to

TABLE 3.5 Previous Studies on Adsorption of Water on Zeolites.

Year	Researcher	Zeolite	Temperature (°C)	Adsorption equilibrium Model	Max. adsorption capacity (g/g)	Diffusion coefficient (m²/s)
1961a–1961b	Barrer and Fender[4,5]	Chabazite	30–80	Langmuir	0.28	4.8×10^{-10}
		Gmelinite			0.27	1.5×10^{-10}
		Heulandite			0.19	1.6×10^{-10}
1969	Dzhigit et al.	LiNaX	23–400	—	0.36	—
		NaX			0.31	
		KNaX			0.28	
		RbNaX			0.22	
		CsNaX			0.19	
1986	Ülkü[99]	Clinoptilolite-rich zeolitic tuff	25–250	Experimental correlation	0.14	—
1986a	Ülkü et al.[100]	Clinoptilolite	20–220	Dubinin–Radushkevich	0.08–0.115	—
				Experimental correlation	0.04–0.115	
1986b	Ülkü et al.[101]	Clinoptilolite	20–200	Experimental correlation	0.12	—
1987	Karger and Pfeifer[51]	NaX	20		—	4×10^{-10} to 2×10^{-9}
1992	Ülkü et al.[103]	Clinoptilolite	25–200		—	5×10^{-10}
1995	Cacciola and Restuccia[15]	13X	5–55	Experimental correlation	—	—
		4A				
1995	Sun et al.[90]	13X	20–200	Three-term Langmuir	—	—
1997	Hunger et al.[44]	NaA			0.26	—
		NaX			0.30	—

TABLE 3.5 *(Continued)*

Year	Researcher	Zeolite	Temperature (°C)	Adsorption equilibrium Model	Max. adsorption capacity (g/g)	Diffusion coefficient (m²/s)
		NaY			0.29	4×10^{-11} to 1×10^{-12}
		Na,K-erionite			0.16	
		Na-mordenite			0.13	
		NaZSM-5			0.08	
1999	Tatler and Erden-Şenatalar[94]	4A	20–1500		—	
2001	Ryu et al.[79]	13X	25–340	Langmuir–Freundlich model	0.30	1×10^{-11} to 1×10^{-10}
2001	Moise et al.[66]	BaX	25–400	Dubinin–Radushkevich	0.28	—
		BaY			0.28	
2003	Cakicioglu-Ozkan and Ulku[16]	Ba-rich clinoptilolite	25–400	Dubinin–Astakhov	0.13	—
2003	Lu et al.[62]	13X	10–250	Dubinin–Astakhov	0.26	—
2004	Kwapinski and Tsotsas[57]	4A	25–350	Experimental correlation	0.24	2.8×10^{-10} to 1×10^{-9}
2004	Janchen et al.[49]	NaX	20–350		0.192	—
		MgNaX			0.212	
		LiX			0.244	
		CaNaA			0.162	
2005	Liu and Leong[59]	13X	45–200	Three-term Langmuir	—	—

TABLE 3.5 (Continued)

Year	Researcher	Zeolite	Temperature (°C)	Adsorption equilibrium Model	Max. adsorption capacity (g/g)	Diffusion coefficient (m²/s)
2006	Ülkü et al.[104]	Zeolitic tuff 4A	20–300	Langmuir	—	—
2007	Caputo et al.[20]	Zeolitic tuff	20–250	Dubinin–Astakhov	0.022–0.117	—
2008	San and Lin[80]	13X	30–120	Experimental correlation	0.236	—
2008	Çakıcıoğlu-Ozkan and Ülkü[18]	Clinoptilolite-rich zeolitic tuff	18–400		0.15	1.2×10^{-8} to 3.154×10^{-6}
2009	Simo et al.[84]	3A	100–270	Langmuir	0.07–0.16	—
2009	Wang and LeVan[107]	5A 13X	25–300	Toth's	0.13–0.24 0.15–0.26	—
2009	Llano-Restrepo and Mosquera[60]	3A	0–100	Generalized statistical thermodynamic model	0.03–0.23	—
2010	Ivanova et al.[48]	Clinoptilolite	20–200		0.016	2.5×10^{-12} to 4.4×10^{-12}
2010	Cortes et al.[24]	13X	50–450	Dubinin–Radushkevich	0.20–0.25	—
2014	Mette et al.[64]	13X	20–350	Dubinin–Astakhov	0.02–0.32	—
2016	Sayılgan et al.[82]	13X	35–200		0.19–0.23	4×10^{-9} to 6×10^{-8}

define the mass transfer mechanisms of zeolite 3A–water and ethanol pairs. They stated that while diffusion in micropores is controlling mechanism for the cases of the pressure decreased and water concentration increased, the mechanism became macropore diffusion control at high temperatures.[84]

Çakıcıoğlu-Ozkan and Ulku reviewed the diffusion mechanism of water in zeolitic tuff rich in clinoptilolite and found that the intraparticle mass transfer was controlling mechanism at initial periods of adsorption and external heat transfer also became controlling mechanism above the Henry's law region (up to 0.1 mg/mg amount adsorbent).[18]

3.4 CONCLUSIONS

The characterization of different zeolites, the effect of pre-adsorbed water, and regeneration conditions of zeolites were discussed in this chapter. Furthermore, the previous studies on adsorption equilibrium and kinetics were presented.

It was observed that although most researchers defined the adsorption equilibrium isotherm as Type I, different types of isotherms may be seen for zeolite–water pair due to the cation size and location and regeneration conditions. Furthermore, different adsorption equilibrium models were used for this pair.

On the other hand, the information about the mass diffusivity of zeolite–water pair is limited and the controlling mechanism is still contentious. Thus, the adsorption kinetics of zeolite–water pair should be further investigated and well discussed in the later studies.

KEYWORDS

- **crystalline aluminosilicates**
- **clinoptilolite**
- **water adsorption**
- **adsorption kinetics**
- **surface morphologies**

REFERENCES

1. Altıok, E. Recovery of Phytochemicals (Having Antimicrobial and Antioxidant Characteristics) From Local Plants; Ph.D. Dissertation: Izmir Institute of Technology, İzmir, Turkey, 2010.
2. Aristov, Y. I.; Tokarev, M. M.; Freni, A.; Glaznev, I. S.; Restuccia, G. Kinetics of Water Adsorption on Silica Fuji Davison Rd. *Micropor. Mesopor. Mater.* **2006,** *96* (1–3), 65–71.
3. Barrer, R. M. *Diffusion in and Through Solids;* Cambridge University Press: Cambridge, 1941.
4. Barrer, R. M.; Fender, B. E. F. The Diffusion and Sorption of Water in Zeolites-I. Sorption. *J. Phys. Solids.* **1961a,** *21* (1–2), 1–11.
5. Barrer, R. M.; Fender, B. E. F. The Diffusion and Sorption of Water in Zeolites-II. Intrinsic and Self Diffusion. *J. Phys. Solids.* **1961b,** *21* (1–2), 12–24.
6. Benaliouche, F.; Hidous, N.; Guerza, M.; Zouad, Y.; Boucheffa, Y. Charazterization and Water Adsorption Properties of Ag- and Zn-exchanged A Zeolites. *Micropor. Mesopor. Mater.* **2015,** *209,* 184–188.
7. Ben Amar, N.; Sun, L. M.; Meunier, F. Numerical Study on Adsorptive Temperature Wave Regenerative Heat Pump. *Appl. Therm. Eng.* **1996,** *16* (5), 405–418.
8. Bish, D. L. Effects of exchangeable Cation Composition on the Thermal Expansion/Contraction of Clinoptilolite. *Clays Clay Miner.* **1984,** *32* (6), 444–452.
9. Bish, D. L. Thermal Behavior of Natural Zeolites. In *Natural Zeolites'93;* Ming, D. W., Mumpton F. A. Eds. Int. Comm.Natural Zeolites; Brockport, New York; 1995; pp 259–269.
10. Bish, D. L.; Wang, H. Phase Transition in Natural Zeolites and the Importance of P_{H2O}. *Philos. Mag.* **2010,** *90* (17–18), 2425–2441.
11. Boddenberg, B.; Rakhmatkariev, G. U.; Hufnagel, S.; Salimov, Z. A Calorimetric and Statistical Mechanics Study of Water Aadsorption in Zeolite NaY. *Phys. Chem. Chem. Phys.* **2002,** *4,* 4172–4180.
12. Boyd, G. E.; Adamson, A. W.; Myers, L. S. The Exchange Adsorption of Ions from Aqueous Solutions by Organic Zeolites. II. Kinetics. *J. Am. Chem. Soc.* **1947,** *69,* 2836–2848.
13. Brandani, F.; Ruthven, D. M. The Effect of Water on the Adsorption of CO_2 and C_3H_8 on Type X Zeolites. *Ind. Eng. Chem. Res.* **2004,** *43,* 8339–8344.
14. Breck, D. W. *Zeolite Molecular Sieves: Structure, Chemistry and Use;* John Wiley & Sons: New York, 1974.
15. Cacciola, G.; Restuccia, G. Reversible Adsorption Heat-Pump—A Thermodynamic Model. *Int. J. Refrig.* **1995,** *18* (2), 100–106.
16. Cakicioglu-Ozkan, F.; Ulku, S. Adsorption Characteristics of Lead-, Barium- and Hydrogen-rich Clinoptilolite Mineral. *Adsorp. Sci. Technol.* **2003,** *21* (4), 309–317.
17. Cakicioglu-Ozkan, F.; Ulku, S. The Effect of HCl Treatment on Water Vapor Adsorption Characteristics of Clinoptilolite Rich Natural Zeolite. *Micropor. Mesopor. Mat.* **2005,** *77* (1), 47–53.
18. Cakicioglu-Ozkan, F.; Ulku, S. Diffusion Mechanism of Water Vapour in a Zeolitic Tuff Rich in Clinoptilolite. *J. Therm. Anal. Calorim.* **2008,** *94* (3), 699–702.
19. Cansever-Erdoğan, B.; Ülkü, S. Ammonium Sorption by Gördes Clinoptilolite Rich Mineral Specimen. *Appl. Clay Sci.* **2011,** *54,* 217–225.

20. Caputo, D.; Iucolano, F.; Pepe, F.; Colella, C. Modelling of Water and Ethanol Adsorption Data on a Commercial Zeolite-rich Tuff and Prediction of the Relevant Binary Isotherm. *Micropor. Mesopor. Mat.* **2007**, *105*, 260–267.
21. Carman, P. C.; Haul R. A. W. Measurement of Diffusion Coefficients. *Proc. Royal Soc. London. Series A. Math. Phys. Sci.* **1954**, *222*, 109–118.
22. Chua, H. T.; Ng, K. C.; Chakraborty, A.; Oo, N. M.; Othman, M. A. Adsorption Characteristics of Silica Gel Plus Water Systems. *J. Chem. Eng. Data.* **2002**, *47*(5), 1177–1181.
23. Clausse, M.; Meunier, F.; Coulie, J.; Herail, E. Comparison of Adsorption Systems Using Natural Gas Fired Fuel Cell as Heat Source, for Residual Air Conditioning. *Int. J. Refrig.* **2009**, *32*, 712–719.
24. Cortes, F. B.; Chejne, F.; Carrasco-Marin, F.; Moreno-Castilla, C.; Perez-Cadenas, A. F. Water Adsorption on Zeolite 13x: Comparison of the Two Methods Based on Mass Spectrometry and Thermogravimetry. *Adsorpt. J. Int. Adsorpt. Soc.* **2010**, *16* (3), 141–146.
25. Crank, J. *The Mathematics of Diffusion;* 2nd ed.; Oxford University Press: London, 1975.
26. Demir, H. An Experimental and Theoretical Study on the Improvement of Adsorption Heat Pump Performance. Ph.D. Dissertation, Izmir Institute of Technology, İzmir, Turkey, 2008.
27. Demir, H.; Mobedi, M.; Ulku, S. Microcalorimetric Investigation of Water Vapor Adsorption on Silica Gel. *J. Therm. Anal. Calorim.* **2011**, *105* (1), 375–382.
28. Dieng, A. O.; Wang, R. Z. Literature Review on Solar Adsorption Technologies for Ice-Making and Air-Conditioning Purposes and Recent Developments in Solar Technology. *Renew. Sust. Energ. Rev.* **2001**, *5*, 313–342.
29. Do, D. D.; Jordi, R. G.; Ruthven, D. M. Sorption Kinetics in Zeolite Crystals with Finite Intracrystal Mass Exchange - Isothermal Systems. *J. Chem. Soc. Faraday Trans.* **1992**, *88* (1), 121–131.
30. Dunne, J. A.; Rao, M.; Sircar, S.; Gorte, R. J.; Myers, A. L. Calorimetric Heats of Adsorption and Adsorption Isotherms. 2. O2, N2, Ar, Co2, CH4, C2H6, and SF6 on Nax, H-Zsm-5, and Na-Zsm-5 Zeolites. *Langmuir.* **1996**, *12* (24), 5896–5904.
31. Dzhigit, O. M.; Kiselev, A. V., Mikos, K. N.; Muttik G. G.; Rahmanova, T. A. Heats of Adsorption of Water Vapour on X-Zeolites Containing Li+, Na+, K+, Rb+ and Cs+ Cations. *Trans. Faraday Soc.* **1971**, *67* (578), 458–467.
32. Ertan, A. CO$_2$, N$_2$ and Ar Adsorption on Zeolites. Ph.D. Dissertation, Izmır Institute of Technology, Izmir, Turkey, 2004.
33. Fan, M.; Panezai, H.; Sun, J.; Bai, S.; Wu, X. Thermal and Kinetic Performance of Water Desorption for N$_2$ Adsorption in Li-LSX Zeolite. *J. Phys. Chem. C.* **2014**, *118* (41), 23761–23767.
34. Gediz İliş, G. An Experimental and Numerical Study on Heat and Mass Transfer in Adsorbent Bed of an Adsorption Heat Pump. Ph.D. Dissertation, Izmir Institute of Technology, Izmir, Turkey, 2012.
35. Ghosh, A.; Ma, L.; Gao, C. Zeolite Molecular Sieve 5A Acts as a Reinforcing Filler, Altering the Morphological, Mechanical, and Thermal Properties of Chitosan. *J. Mater. Sci.* **2013**, *48*, 3926–3935.
36. Glueckauf, E. Theory of Chromatography Part.10- Formula for Diffusion into Spheres and Their Application to Chromatography. *Trans. Faraday Soc.* **1955**, *51*, 1540–1551.
37. Gorbach, A.; Stegmaier, M.; Eigenberger, G. Measurement and Modelling of Water Vapor Adsorption on Zeolite 4A-Equilibria and Kinetics. *Adsorption.* **2004**, *10* (1), 29–46.

38. Gregg, S. J.; Sing, K. S. W. *Adsorption, Surface Area and Porosity;* 2nd ed.; Academic Press Inc.: London, 1982.
39. Gülen, J.; Zorbay, F.; Arslan, S. Zeolitler Ve Kullanım Alanları. *Karaelmas Sci. Eng. J.* **2012,** *2* (1), 63–68.
40. Hamamoto, Y.; Alam, K. C. A.; Saha, B. B.; Koyama, S.; Akisawa, A.; Kashiwagi, T. Study on Adsorption Refrigeration Cycle Utilizing Activated Carbon Fibers. Part 1. Adsorption Characteristics. *Int. J. Refrig.* **2006,** *29* (2), 305–314.
41. Halasz, I.; Kim, S.; Marcus B. Hydrophilic and Hydrophobic Adsorption on Y Zeolites. *Mol. Phys.* **2002,** *100* (19), 3123–3132.
42. Hall, K. R.; Eagleton, L. C.; Acrivos, A.; Vermeulen, T. Pore and Solid Diffusion Kinetics in Fixed Bed Adsorption under Constant Pattern Conditions. *Ind. Eng. Chem. Fundamen.* **1966,** *5* (2), 212–223.
43. Hill, T. L. *Introduction to Statistical Thermodynamics;* Dover Publications: New York, 1960.
44. Hunger, B.; Matyski, S.; Heuchel, M.; Geidel, E.; Toufar, H. Adsorption of Water on Zeolites of Different Types. *J. Therm. Anal.* **1997,** *49,* 553–565.
45. Incropera, P. F.; DeWitt, P. D. *Fundamentals of Heat and Mass Transfer;* 6th ed.; John Wiley and Sons: New York, 2006.
46. Inglezakis, V. J.; Poulopoulos, S. G. *Adsorption, Ion Exchange and Catalysis: Design of Operations and Environmental Applications;* Elsevier Publisher: Amsterdam, Netherland, 2006.
47. Ivanova T. N.; Sarakhov, A. I.; Dubinin M. M. Change in the Parameters of the Unit Cells of Type A Zeolites in Various Ion Exchange Forms as a Result of the Adsorption of Water Vapors. *Russ. Chem. Bull.* **1975,** *24,* 1361.
48. Ivanova, E. P.; Kostova, M. A.; Koumanova, B. K. Kinetics of Water and Alcohol Vapors Adsorption on Natural Zeolite. *Asia-Pac. J. Chem. Eng.* **2010,** *5,* 869–881.
49. Janchen, J.; Ackermann, D.; Stach, H.; Brösicke, W. Studies on the Water Adsorption on Zeolites and Modified Mesoporous Materials for Seasonal Storage of Solar Heat. *Sol. Energ.* **2004,** *76,* 339–344.
50. Jaroniec, M. Fifty Years of the Theory of the Volume Filling of Micropores. *Adsorption-J. Int. Adsorption Soc.* **1997,** *3* (3), 187–188.
51. Karger, J.; Pfeifer, H. N.m.r self-Diffusion Studies in Zeolite Science and Technology. *Zeolites.* **1987,** *7,* 90–107.
52. Karger, J.; Ruthven, D. M. *Diffusion in Zeolites and Other Microporous Solids*; John Wiley & Sons: New York, 1992.
53. Katiyar, A.; Thiel, S. W.; Guliants, V. V.; Pinto, N. G. Investigation of the Mechanism of Protein Adsorption on Ordered Mesoporous Silica Using Flow Microcalorimetry. *J. Chromatogr. A.* **2010,** *1217* (10), 1583–1588.
54. Keller, J. U. *Gas Adsorption Equilibria Experimental Methods and Adsorptive Isotherms*; Springer: Boston, 2005.
55. Kim, J. H.; Lee, C. H.; Kim, W. S.; Lee, J. S.; Kim, J. T.; Suh, J. K.; Lee, J. M. Adsorption Equilibria of Water Vapor on Alumina, Zeolite 13x, and a Zeolite X/Activated Carbon Composite. *J. Chem. Eng. Data.* **2003,** *48* (1), 137–141.
56. Kim, K. M.; Oh, H. T.; Lim, S. J.; Ho, K.; Park, Y.; Lee, C. H. Adsorption Equilibria of Water Vapor on Zeolite 3A, Zeolite 13X and Dealuminated Y Zeolite. *J. Chem. Eng. Data.* **2016,** *61,* 1547–1554.
57. Kwapinski, W.; Tsotsas, E. Determination of Kinetics and Equilibria for Adsorption of Water Vapor on Single Zeolite Particle by a Magnetic Suspension Balance. *Chem. Eng. Technol.* **2004,** *27* (6), 681–686.

58. Lavanchy, A.; Stockli, M.; Wirz C.; Stoeckli F. Binary Adsorption of Vapours in Active Carbons Described by the Dubinin Equation. *Adsorption Sci. Technol.* **1996,** *13* (6), 537–545.

59. Liu, Y.; Leong, K. C. The Effect of Operating Conditions on the Performance of Zeolite/Water Adsorption Cooling Systems. *Appl. Therm. Eng.* **2005,** *25* (10), 1403–1418.

60. Llano-Restrepo, M.; Mosquera, M. A. Accurate Correlation, Thermochemistry, and Structural Interpretation of Equilibrium Adsorption Isotherms of Water Vapor in Zeolite 3A by Means of a Generalized Statistical Thermodynamic Adsorption Model. *Fluid Phase Equilib.* **2009,** *283,* 73–88.

61. Loughlin, K. F. Water Isotherm Models for 4A (NaA) Zeolite. *Adsorption.* **2009,** *15,* 337–353.

62. Lu, Y. Z.; Wang, R. Z.; Zhang, M.; Jiangzhou, S. Adsorption Cold Storage System with Zeolite-Water Working Pair Used for Locomotive Air Conditioning. *Energ. Convers. Manage.* **2003,** *44,* 1733–1743.

63. Malke-Edery, A.; Abdallah, K.; Grenier, P.H.; Meunier, F. Influence of Traces of Water on Adsorption of Hydrocarbons in NaX Zeolite. *Adsorption.* **2001,** *7,* 17–25.

64. Mette, B.; Kerskes, H.; Drück, H.; Müller-Steinhagen, H. Experimental and Numerical Investigations on the Water Vapor Adsorption Isotherms and Kinetics of Binderless Zeolite 13X. *Int. J. Heat Mass Transf.* **2014,** *71,* 555–561.

65. Mobedi, M. Adsorpsiyonlu Isı Pompaları Üzerinde Teorik Ve Deneysel Bir Çalışma. MSc Dissertation, Dokuz Eylul University, Izmir, Turkey, 1987.

66. Moise J. C.; Bellat, J. P.; Methivier, A. Adsorption of Water Vapor on X and Y Zeolites Exchanged with Barium. *Micropor. Mesopor. Mater.* **2001,** *43,* 91–101.

67. Morris, B. Heats of Sorption in the Crystalline Linde-A Zeolite-Water Vapor System. *J. Colloid Interface Sci.* **1968,** *28,* 149–155.

68. Murali, R. S.; Ismail, A. F.; Rahman, M. A.; Sridhar, S. Mixed Matrix Membranes of Pebax-1657 Loaded with 4A Zeolite for Gaseous Separations. *Sep. Purif. Technol.* **2014,** *129,* 1–8.

69. Narin, G.; Yılmaz, S.; Ülkü, S. A Chromatographic Study of Carbon Monoxide Adsorption on a Clinoptilolite-containing Natural Zeolitic Mineral. *Chem. Eng. Commun.* **2004,** *191* (11), 1525–1538.

70. Narin, G.; Balköse, D.; Ülkü, S. Characterization and Dehydration Behavior of a Natural, Ammonium Hydroxide, and Thermally Treated Zeolitic Tuff. *Drying Technol Int. J.* **2011,** *29* (5), 553–565.

71. Ng, K. C.; Chua, H. T.; Chung, C. Y.; Loke, C. H.; Kashiwagi, T.; Akisawa, A.; Saha B. B.; Experimental Investigation of the Silica Gel-Water Adsorption Isotherm Characteristics. *Appl. Therm. Eng.* **2001,** *21* (16), 1631–1642.

72. Özkan, S. F. Adsorbent Yatakların Dinamik Davranışının İncelenmesi Ve Doğal Kaynakların Adsorbent Olarak Değerlendirilmesi. Ph.D. Dissertation, Ege University, Izmir, Turkey, 1996.

73. Özkan, F.; Ülkü, S.; *Enerji Depolama Sisteminde Yerel Doğal Zeolit Mineralinin Klinoptilolit Kullanılması;* Ulusal Isı Bilimi ve Tekniği Kongresi: Edirne, Can, A., Ataer, E. Eds.; 1997; pp 43–52.

74. Parfitt, R. L. *Anion Adsorption by Soils and Soil Materials.* In *Advances in Agronomy;* Brady, N. C. Ed.: Academic Press Inc.: New Zealand, Vol. 30, 1978.

75. Pinchon, C.; Rebours, B.; Paoli, H.; Bataille, T.; Lynch, J. Influence of the Dehydration on Location of Both Sodium and Calcium Cations in 5A Zeolite—An In Situ X-ray Diffraction Study. *Mater. Sci. Forum.* **2004,** *443–444,* 315–318.

76. Ruthven, D. M. *Principles of Adsorption and Adsorption Processes;* John Wiley & Sons, Canada, 1984.

77. Ruthven, D. M.; Heinke, L.; Karger, J. Sorption Kinetics for Surface Resistance Controlled Systems. *Micropor. Mesopor. Mater.* **2010,** *132* (1–2), 94–102.

78. Ruthven, D. M. Diffusion in Type A Zeolites: New Insights From Old Data. *Micropor. Mesopor. Mater.* **2012,** *162,* 69–79.

79. Ryu, Y. K.; Lee, S. J.; Kim, J. W.; Lee, C. H. Adsorption Equilibrium and Kinetics of H_2O on Zeolite 13x. *Korean J. Chem. Eng.* **2001,** *18* (4), 525–530.

80. San, J. Y.; Lin, W. M. Comparison Among Three Adsorption Pairs for Using as the Working Substances in a Multi-Bed Adsorption Heat Pump. *Appl. Therm. Eng.* **2008,** *28* (8–9), 988–997.

81. Sayılgan, Ş. Ç. Determination of Characteristics of Adsorbent for Adsorption Heat Pumps. M.Sc. Dissertation, Izmir Institute of Technology, Izmir, Turkey, 2013.

82. Sayılgan, Ş. Ç.; Mobedi, M.; Ülkü, S. Effect of Regeneration Temperature on Adsorption Equilibrium and Mass Diffusivity of Zeolite 13X-water Pair. *Micropor. Mesopor. Mater.* **2016,** 224, 9–16.

83. Shams, K.; Mirmohammadi, S. J. Preparation of 5A Zeolite Monolith Granular Extrudates Using Kaolin: Investigation of the Effect of Binder on Sieving/adsorption Properties Using a Mixture of Linear and Branched Paraffin Hydrocarbons. *Micropor. Mesopor. Mater.* **2007,** *106,* 268–277.

84. Simo, M.; Sivashanmugam, S.; Brown, C. J.; Hlavacek, V. Adsorption/Desorption of Water and Ethanol on 3A Zeolite in Near-Adiabatic Fixed Bed. *Ind. Eng. Chem. Res.* **2009,** *48,* 9247–9260.

85. Spiewak, B. E.; Dumesic, J. A. Microcalorimetric Measurements of Differential Heats of Adsorption on Reactive Catalyst Surfaces. *Thermochim. Acta.* **1997,** *290* (1), 43–53.

86. Sticher, H. Thermal Analysis of Synthetic Near-Chabazite Zeolites with Different Si/Al Ratios. *Thermochim. Acta.* **1974,** *10,* 305–311.

87. Storch, G.; Reichenauer, G.; Scheffler, F.; Hauer, A. Hydrothermal Stability of Pelletized Zeolite 13X for Energy Storage Applications. *Adsorption.* **2008,** *14,* 275–281.

88. Sumathy, K.; Yeung, K. H.; Yong, L.; Technology Development in the Solar Adsorption Refrigeration Systems*. Prog. Energy Combust. Sci.* **2003,** *29,* 301–327.

89. Sun, H.; Shen, B. Experimental Study on Coking, Deactivation and Regeneration of Binderless 5A Zeolite During 1-hexane Adsorption. *Adsorption.* **2013,** *19,* 111–120.

90. Sun, L. M.; Benamar, N.; Meunier, F. Numerical Study on Coupled Heat and Mass Transfers in an Adsorber with External Fluid Heating. *Heat Recovery Syst. Chp.* **1995,** *15*(1), 19–29.

91. Suzuki, M. *Adsorption Engineering;* Elsevier Science: New York, Vol. 25, 1990.

92. Szanyi, J.; Kwak, J. H.; Peden, C. H. F. The Effect of Water on the Adsorption of NO_2 in Na- and Ba-Y, FAU Zeolites: A Combined FTIR and TPD Investigation. *J. Phys. Chem, B.* **2004,** *108,* 3746–3753.

93. TamainotTelto, Z.; Critoph, R. E. Adsorption Refrigerator Using Monolithic Carbon-Ammonia Pair. *Int. J. Refrig.* **1997,** *20* (2), 146–155.

94. Tatlıer, M.; Erdem-Şenatalar, A. The Effects of Thermal and Mass Diffusivity on the Performance of Adsorption Heat Pumps Employing Zeolite Synthesized on Metal Supports. *Micropor. Mesopor. Mater.* **1999,** *28,* 195–203.

95. Teng, Y.; Wang, R. Z.; Wu, J. Y. Study of the Fundamentals of Adsorption Systems. *Appl. Therm. Eng.* **1996,** *17,* 327–338.

96. Thomas, W. J.; Crittenden, B. *Adsorption Technology and Design;* Butterworth-Heinemann: Oxford and Boston, 1998.

97. Tsistsishvili, G. V.; Andronikashvili, T. G.; Kirov, G. N.; Filizova L. D. *Natural Zeolites;* Ellis Horwood Limited: Chichester, England, 1992.

98. Ulku, S.; Balkose, D.; Caga, T.; Ozkan, F.; Ulutan, S. A Study of Adsorption of Water Vapour on Wool under Static and Dynamic Conditions. *Adsorpt. J. Int. Adsorpt. Soc.* **1998,** *4* (1), 63–73.

99. Ülkü, S. Natural Zeolites in Energy Storage and Heat Pumps. *Stud. Surf. Sci. Catal.;New Developments in Science and Technology,* **1986,** *28,* 1047–1054.

100. Ülkü, S.; Kıvrak, Z.; Mobedi, M. *Air Drying in Packed Bed Adsorbers. Drying 86;* Hemisphere Publishing Corporation: Washington, 1986a; pp 807–812.

101. Ülkü, S.; Beba, S.; Kıvrak, Z.; Seyrek, B. Enerji Depolama ve Hava Kurutmada Doğal Zeolitlerden Yararlanma. *Isı Bilim ve Tek. Derg.* **1986b,** *8* (4), 23–28.

102. Ülkü, S.; Mobedi, M. Adsorption in Energy Storage. In *Energy Storage Systems;* Springer: New York, 1989; pp 487–507.

103. Ülkü, S.; Balköse, D.; Baltacıoğlu, H.; Özkan, F.; Yıldırım, A. Natural Zeolites in Air Drying. *Drying Technol.* **1992,** *10* (2), 475–490.

104. Ülkü, S.; Balköse, D.; Alp, B. Dynamic Heat of Adsorption of Water Vapour on Zeolitic Tuff and Zeolite 4A by Flow Microcalorimetry. *Oxid. Commun.* **2006,** *29* (1), 204–215.

105. Valsaraj, K. T.; Thibodeaux, L. J.; On the Linear Driving Force Model for Sorption Kinetics of Organic Compounds on Suspended Sediment Particles. *Environ. Toxicol. Chem.* **1999,** *18* (8), 1679–1685.

106. Wang, L.; Zhu, D.; Tan, Y.; Heat Transfer Enhancement of the Adsorber of an Adsorption Heat Pump. *Adsorption* **1999,** *5,* 279–286.

107. Wang, L. W.; Wang, R. Z.; Oliveira, R. G. A Review on Adsorption Working Pairs for Refrigeration. *Renew. Sust. Energ. Rev.* **2009,** *13* (3), 518–534.

108. Wang, Y.; LeVan, M. D. Adsorption Equilibrium of Carbon Dioxide and Water Vapor on Zeolite 5A and 13X and Silica Gel: Pure Components. *J. Chem. Eng. Data.* **2009,** *54,* 2839–2844.

109. Yıldırım, Z. E. A Study on Isotherm Characteristics of Adsorbent-Adsorbate Pairs Used in Adsorption Heat Pumps. MSc Dissertation, Izmir Institute of Technology, Izmir, Turkey, 2011.

110. Yucel, H.; Ruthven, D. M. Diffusion of CO_2 in 4A and 5A Zeolite Crystals. *J. Colloid Interface Sci.* **1980,** *74* (1), 186–195.

111. Zheng, H.; Han, L.; Ma, H.; Zheng, Y.; Zhang, H.; Liu, D.; Liang, S. Adsorption Characteristics of Ammonium Ion by Zeolite 13X. *J. Hazard. Mater.* **2008,** *158,* 577–584.

CHAPTER 4

DEGRADATION AND STABILIZATION ISSUES OF POLYETHYLENE IN OPEN AIR APPLICATIONS

GÜNEŞ BORU IZMIRLI[1], SEVGI ULUTAN[1,*], and PINAR TÜZÜM DEMIR[1,2]

[1]Department of Chemical Engineering, Faculty of Engineering, Ege University, Bornova TR35100, Izmir, Turkey

[2]Department of Chemical Engineering, Faculty of Engineering, Usak University, Usak TR64200, Turkey

*Corresponding author. E-mail: sevgi.ulutan@gmail.com

CONTENTS

ABSTRACT

In this study, degradation behavior of Low density polyethylene (LDPE) pipes which are unstabilized and stabilized against light has been taken under investigation in the course of ultraviolet (UV) light exposure. Commercial light stabilizers have been used for the production of real irrigation pipes. Thus the pipes with hindered amine light stabilizer (HALS) and carbon black (CB) as well as without UV light stabilization agents were produced and exposed to natural and artificial weathering processes. The influence of the stabilizers on UV degradation behavior of LDPE pipes has been investigated by monitoring the changes in chemical, structural, and mechanical properties of the pipes. The oxidative degradation of PE was determined in terms of oxidation index (OI) through fourier transform infrared (FTIR) spectroscopic analysis and oxidation induction time (OIT) assessment, utilizing differential scanning calorimetry (DSC). The assessment of crystallinity change was performed by means of DSC and X-ray diffraction (XRD) analyses. Mechanical properties of the pipes were analyzed through hydrostatic pressure resistance test (HPRT) as well as hardness measurement and tensile tests. Scanning electron microscopy (SEM) pictures were taken from the broken surfaces after the tensile tests to illuminate the structural changes due to the aging of pipes.

4.1 INTRODUCTION

Polyethylene (PE) outdoor applications, where they receive radiation from the sun, are growing in number, and require knowledge of the anticipated rate of deterioration during their service life. Besides this, the results may be directed toward decomposition of PE following exposure to the weather, which would facilitate its disposal since the accumulation of waste PE is a major component of environmental pollution.[1,2]

In the following section, the problem, the solutions, and related studies in recent literature have been compiled.

4.2 PIPE FORMULATIONS FOR OPEN AIR APPLICATIONS

The outdoor weathering of PE results in several undesirable things such as discoloration, embrittlement, chain scission, crosslinking, loss of mechanical strength, etc.[3,4] Significant degradation occurs in PE in the presence of oxygen when the materials are exposed to heat or UV light radiation.[3] The susceptible sites are the tertiary hydrogens that are on the branches of PE molecule. Light radiation contributes very actively to polymer aging, especially when oxygen is present, which is a normal situation for most polymers. UV light radiation energy is high enough to break chemical bonds in polymers and produce degradation reactions. As an example, carbon–carbon (C—C) bonds have 348 kJ/mol energy and therefore, 300 nm wavelength having 400 kJ energy per mol is enough to break them.[5] In the initial step of degradation, macro-radicals are formed which react with the oxygen of air. In the chain propagation sequence, the ROO radicals separate hydrogen from the polymer resulting in the formation of hydroperoxides.[6] It is generally accepted that hydroperoxides are the key compounds in the mechanism of photo-oxidation of LDPE.[7] Their production is generally followed by their photochemical decomposition. Under UV exposure, the quantum yield of hydroperoxide decomposition is recognized to be high. Their decomposition may lead to several photo-products, such as carboxylic acid, alcohol, ketone, ester, etc.[6] Prediction of the remaining lifetime of PE pipes after 30 years in use was investigated.[8]

4.2.1   COMMON PIPE FORMULATIONS FOR IRRIGATION ON SOIL

PE pipes represent one of the major agricultural applications. Many polymers are prone to weather degradation caused by photochemical reactions involving sunlight ultraviolet photons and atmospheric oxygen, or even thermal reactions. A good knowledge of these weather conditions can lead to the use of a proper stabilization system for the polymeric material and consequently a longer lifetime in service.[9] In order to assess the durability of materials, ageing factors of importance have to be identified under various natural service conditions.[10] The radiation amount differs from zone to zone on the earth. Izmir city, lying at Aegean coast, of Turkey takes about 1600 KWh/m^2 solar radiation energy per year.

4.2.2   STABILIZATION OF POLYETHYLENE AGAINST UV LIGHT

There are various alternatives available for dealing with the problem of UV stabilization of PE against photo-thermal oxidation. UV absorbers should transform the absorbed energy into harmless energy, that is, energy corresponding to high wavelengths or thermal energy that does not heat the polymer above its decomposition temperature.[11,12]

The UV stabilizers can be classified according to their chemical structure such as derivatives of 2-hydroxybenzophenone, esters of aromatic acids and aromatic alcohols and hydroxyphenyl benzotriazoles, substituted acrylonitriles, metallic complexes based on nickel and cobalt and inorganic pigments. Stabilizers are commonly defined with their protection mechanism as the primary antioxidants, which trap free radicals, and secondary antioxidants which decompose hydroperoxides into more stable molecules. The stabilizers mainly belong to four main groups: Hindered phenol, phosphites, thio-synergists, and HALS.[13] As the CB is considered, its UV stabilizing ability is associated with both the UV absorbing capacity of CB and photo-generated ketonic species present on particles. It can be tolerated in some formulations where color is not a criterion. The pigment not only absorbs light, it is reactive with those free-radical species that might be formed.[14]

4.2.2.1   UV STABILIZATION OF PE WITH CB

CB has been used as a light-stabilizing additive in polyolefins for many years. It functions as a simple physical screen, a UV absorber, a radical trap, and a terminator of the free radical chains through which the photo-oxidative reactions are propagated.[14,15] The active sites on the CB surface are stronger acceptors than the free radicals. Addition of CB to PE delays the onset of oxidation suppressing thermal oxidation by means of oxygen containing group on CB surface.[16] The resistance of a polyolefin against UV degradation is usually related to the type and particle size of the CB used, as well as the concentration of the CB in the matrix. For example, good results have been obtained for PE light radiation stability by the use of activated CBs with particle sizes below 25 nm.[12] Although the small particle-sized carbon blacks are known to have the greatest UV stabilizing efficiency, they tend to agglomerate into aggregates or clusters which are not easily dispersed.

Liu and Horrocks[14] found that CBs of a range of particle properties show significant improvement in UV stabilization of linear low-density polyethylene (LLDPE) compared with clear films. They detected the slowest rates of photo-oxidation in the smallest particle-sized CB having films under artificial weathering conditions. The improvement of UV stabilization as the CB amount increased from 1.5 to 3.5%w/w was obvious, as well.

4.2.2.2 UV STABILIZATION OF PE WITH HALS

HALS protects polymer coatings against photo-oxidative damage through the formation of free radicals.[17,18]

High molecular mass HALS compounds reveal a pronounced contribution to long-term thermal stability of polyolefins.[11] However, low molecular mass HALS have some disadvantages such as high migration rate and moderate resistance to extraction while the oligomeric and polymeric type HALS have the advantage of limited migration and consequent any loss of this additive, under processing and application.[17] The polymeric HALS were developed to overcome such problems protecting the polymer properties and extending service life time is unbroken. New developments in the traditional area of processing, long-term thermal stability of polyolefins take place where stabilizers with improved performance or at low use concentrations are needed to provide an effect.[19]

HALS can interfere with some primary initiation reactions on photo-oxidation of PEs. Once free radicals are formed, the HALS can still trap them. This trapping is based essentially on complex formation between a peroxy radical and an HALS molecule.[20] A couple of reactions in the course of oxidation are as below:

A scheme of polymer oxidation reactions

$$-[CH_2 - CH_2]_n^{\cdot} + \quad O_2 \qquad \longrightarrow -[CH_2 - CH_2]_n^{-} OO^{\cdot}$$

$$-[CH_2 - CH_2]_n^{-} OO^{\cdot} + -[CH_2 - CH_2]_n^{-} H \longrightarrow -[CH_2 - CH_2]_n^{-} OOH + -[CH_2 - CH_2]_n^{\cdot}$$

$$-[CH_2 - CH_2]_n^{-} OO^{\cdot} + -[CH_2 - CH_2]_n^{-} OO^{\cdot} \longrightarrow \text{Products}$$

The possible effect of some HALS on primary initiation by air oxygen involves the adduct formation in the main reaction of stabilizing process. The reaction of this complex with a second peroxy radical leads to termination between the two peroxy radicals and liberation of the HALS as one of the major stabilization reactions involving peroxy radicals.[20]

4.2.2.3 SYNERGISTIC MIXTURES OF STABILIZERS

Synergism of stabilizer couples is defined with respect to one or more aspects of their stabilizing effect. The theoretical stabilizing effect of the combination is calculated, considering the effects of pure components and their amounts. If Y and r designate the effect and the relative amount of components, respectively, and the subscripts 1 and 2 designate the individual components, the additive effect of such a combination is given by eq 4.1.

$$Y_{\text{additive}} = r_1 \times Y_1 + r_2 \times Y_2 \qquad (4.1)$$

If this value is larger than the experimental value of the same effect, it is designated as synergism while the smaller effect than the experimental one is defined as antagonism.[11]

Physical mixtures of stabilizers are used to obtain synergistic effect in which the observed effect of the combination is greater than that of the simple addition of the effects of the individual stabilizers. Pena et al.[15] reported such interactions between two types of CB and two commercial HALS compounds in LDPE thermal oxidation. They reported that the CB presence alone resulted in a very low stabilizing effect compared with the presence of HALS. Small differences in stabilizing performance were found between formulations containing two types of CB because of the feature similarities such as particle size, tinting (coloring) strength and so on. The HALS (Tinuvin 622) stabilizer does not possess reactive labile hydrogen, which results in a lower reactivity and therefore in a different type of adsorption or interaction with the CB, compared with HALS a polymeric HALS (Chimmasorb 944).

Liauw et al.[21] determined the synergism between two UV stabilizers, namely, a sterically hindered phenolic antioxidant (Irganox 1010) and a polymeric HALS (Chimmasorb 944). They proved that their stabilizing performance for UV and thermal stability is virtually doubled when silica is present in a polyolefin (LLDPE) film.

4.2.3 OXIDATIVE DEGRADATION OF POLYETHYLENE UNDER UV LIGHT EXPOSURE

Degradation constitutes the main limit to the active life of synthetic polymers. This undesirable degradation may be accelerated by exposure to UV radiation. It is hardly possible to predict the service life of polymers under natural conditions. There is a real need to perform natural weathering of polymers. There are well-known weathering sites in Southern France, like Sanary sur Mer and Miami in Florida. However, the weathering conditions as direct weathering and weathering in glass vessels made difference in the lifetime of the polymer samples.[22] Natural weathering is inherently a time-consuming process and has unsteady external factors such as the intensity of sunlight, temperature, and humidity.[7] Therefore, the accelerated weathering gains importance to provide comparable experimental conditions. There are some aging strategies making use of special instruments providing constant UV light in literature.[5,23]

4.2.4 STRUCTURAL CHANGES ON POLYETHYLENE UPON OXIDATION DUE TO WEATHERING

The structural changes upon UV exposure of PEs are obvious, with their infrared spectra.[24,25] It is generally accepted that the material characteristics of the polymer dictate its characteristics and behavior under various environmental conditions.[26] LDPE is a semicrystalline material, in which spherulites, spherical shape having crystalline regions are separated by amorphous regions of polymer. Since the crystalline areas are typically more densely packed than the amorphous areas, these regions impart improved stiffness and chemical resistance to the polymer. On the other hand, degradation processes mainly take place in the amorphous phase of the material.

4.2.4.1 ASSESSMENT OF OXIDATION THROUGH ABSORPTIONS IN THE INFRARED REGION

Oxygen absorption from air in a polymer is a diffusion controlled event which occurs faster at elevated temperatures as a result of solar energy in open-air applications. The mobility of polymer chains is greater at the surface than that inside the bulk material. The weathering of high density polyethylene (HDPE) with and without additives when subjected to the natural atmosphere of Rio de Janerio City[24] resulted in several absorption bands in the infrared region. The bands at 888, 909–991, 964 cm^{-1} are attributed to vinyledene, terminal vinyl, and trans-vinylene, respectively. In addition, a shoulder appeared at 991 cm^{-1} and the peaks that appeared at about 1010 and 1030–1040 cm^{-1} are attributed to the stretching of the C—O group in primary alcohol and/or O—H deformation in alcohol, peroxide, or hydroperoxide.

The bands due to the C—O stretching vibration of peroxides and hydroperoxides occur in the region between 1000 and 1300 cm^{-1}. The O—H stretching vibration of free hydroperoxides is located in the region between 3560 and 3530 cm^{-1}, whereas the hydrogen bonded OH vibration absorbs in the region between 3550 and 3230 cm^{-1}.[6]

Carbonyl groups are of importance for the investigation of photo-oxidation which indicate degradation through 1720 cm^{-1} band, and appear as a new band on infrared spectrum.[25] The ratio of the absorbance of carbonyl band to reference band (1370 cm^{-1}) is defined as the oxidation (or, sometimes called carbonyl) index (OI) (eq.4.2).

$$OI = \frac{A_{C=0\ (1720\ cm^{-1})}}{A_{ref\ (1370\ cm^{-1})}} \times 100 \qquad (4.2)$$

Liu and Horrocks[14] studied the effect of UVB exposure on carbonyl index. The increases in absorbances of associated ketonic groups at 1712 cm^{-1} with exposure time. As expected, the unfilled film shows dramatic increases in the absorbance indices of functional groups with the UV exposure time, compared with the filled films. The large particle sized CB having films show the fastest rates of increase in the formation of these groups among the filled films. The film containing small particle-sized CB shows the slowest formation of the associated carbonyl groups, which would exhibit the best UV resistance.

4.2.4.2   ASSESSMENT OF OXIDATION THROUGH CALORIMETRIC MEASUREMENTS

The degradation of PE conceptually involves three distinct stages as generally accepted. In the first stage, depletion of antioxidants is detected in OIT tests. Then an induction period starts after which the oxidation is sufficient to cause a measurable change in the engineering properties such as melt index and tensile properties. Thereafter, the polymer eventually undergoes failure in the third stage as it reaches the end of its service life when it no longer meets the original design requirements.[13]

The efficiency of the stabilizers or stabilizing systems used in a polyolefinic material can simply be determined through the OIT[16] or oxidation induction temperature (OIT*) tests of the molten material. The thermal stability of a polymer can be assessed by measuring the time or temperature needed to start degradation in an oxidative atmosphere through OIT and OIT* measurements. The methods are well established for quality-control purposes as a quick screening method to check the activity of the stabilization system used under high temperature conditions above melting point.[9,27,28] Besides the phase transition and oxidation temperatures and enthalpies, the oxidation periods are determined through calorimetric studies. It is not practical to perform field tests to study the service life of a long-term with-standing material. The standard OIT tests (ASTM D5885) are conducted using a differential scanning calorimeter. OIT tests are generally used for the qualitative assessment of the level (or degree) of stabilization of the material tested typically as a quality-control measure to monitor the stabilization level in formulated resin prior to extrusion. OIT measures the time period for the oxidation of molten polymer. Depletion of antioxidants[13] and synergism between UV stabilizers[15,21] are some other issues to be determined by using OIT tests, as well.

4.2.5 CRYSTALLINITY OF POLYETHYLENE IN THE COURSE OF DEGRADATIVE PROCESSES

TGA is a viable technique to obtain quantitative information on the decomposition rates of polyolefins.[29] It is performed under isothermal conditions and constant heating rate to obtain kinetic information from the decomposition of polyolefins. The DSC analyses are used to determine the phase transition, oxidation temperatures, oxidation periods, and enthalpy changes. The enthalpy changes can be used to assess the changes in crystallinity. On the other hand, DSC instrument can be used to evaluate the thermo-mechanical and UV degradation, as well.[30,31]

The degree of crystallinity (DC) may give a gist about the change in the structure of the polymer. The percentage of crystallinity is determined using the melting enthalpy Δn_{sample}, which is obtained by means of DSC and using eq. 4.3.

The sunlight exposition, which heats the polymer, leads to an increase in molecular mobility, which is mainly attributed to chain scission of the polymer and in turn, to the increase in crystallinity.[24] Although a measurable change in melting temperatures was not determined during natural weathering of HDPE with and without additive, a progressive increment was observed in the crystallinity degree with exposure time for the unstabilized HDPE, which is lower in stabilized HDPE.[24] The $\Delta H_{crystal}$ is found in literature as 289.74 J/g for 100% crystalline PE.[32]

$$DC\%[DSC\ method] = \left(\frac{\Delta_{sample}}{\Delta a_{crystal}}\right) \times 100 \tag{4.3}$$

XRD data are used for the determination of the degree of crystallinity, as well. The eq. 4.4 given below, utilizes the area under the peak belonging to the crystalline region ($A_{\text{crystalline region}}$) and the total area ($A_{\text{total}}$) on the XRD pattern, as intensity versus 2θ curve.

$$\text{DC}\%[\text{XRD method}] = \left(\frac{A_{\text{crystalline region}}}{A_{\text{total}}}\right) \times 100 \tag{4.4}$$

The monoclinic (001) reflection and the orthorhombic (110) and (200) reflections, as well as broad noncrystalline region diffraction peaks of a PE reactor powder were observed on a wide angle X-ray scattering (WAXS) diffraction pattern.[33] The orthorhombic form is characterized by the peaks at 21.5° and 23.8° while the reflection at 19.4° belongs to the monoclinic form. Joo et al.[33] observed that the crystallinity found from XRD analysis is lower than that found from the calorimetry, although its dependence on polymerization temperature was in qualitative agreement.

Russell et al.[34] studied the monoclinic modification of linear PE. They attempted to maximize the monoclinic content of a variety of PE samples by subjecting them to a massive trauma by striking them sharply in a diamond mortar. Trauma that produces the monoclinic phase also increases the amorphous content. In general, the greater the trauma, as indicated by the change in dimensions of the sample, the greater the resulting monoclinic content.

4.2.6 CHANGE OF MECHANICAL STRENGTH OF POLYETHYLENE ON THE COURSE OF AGING

The relationship between the stress and strain which is presented as the stress–strain curve reveals many of the properties of a material. Usually, the ultimate strength of a material is the stress at or near failure, which is catastrophic with a complete break. However, some materials, especially spherulitic crystalline polymers, reach a point where a large inelastic deformation stress (yielding) occurs, but continue to deform and absorb energy, long beyond that point.[35]

The failure theories of materials are based on stress, strain, and energy. The applicability of a certain failure theory mainly depends on the nature and behavior of the material and the environmental conditions.[36]

The mechanical properties, yield strength and the ductility of PE are very sensitive to the weather ageing.[5]

HPRT is a method specified for determining the lifetime of a plastic pipe.[37] The test conditions simulated 50 years lifetime, since the pipes have been designed for 50 years lifetime.

The minimum hydrostatic pressure (applied pressure, P_a) that should be applied to a pipe is calculated by using eq. 4.5.

$$P_a = \left(\frac{2xtx\sigma}{D-t} \right) \tag{4.5}$$

where t and D refer to the thickness and diameter of the pipe, respectively. The minimum required tension, σ, is given as 7 MPa for LDPE pipes tested 1 h at 20°C according to EN 12201 standard.

Hardness testing is convenient for quality control and characterization, since it is easily and rapidly carried out.[38] The Shore D hardness scale is used for hard plastics to measure the resistance of plastics toward indentation, and provide an empirical hardness value.[39,40] PE samples thermally aged at 100°C up to 7200 h exhibit a slightly higher shore hardness than that of unaged sample.[40]

4.2.7 INVESTIGATION OF DEGRADATION THROUGH SEM AND EDS ANALYSES

Some contaminations are thought to be present on the samples, either in open air for natural weathering (NW) or under salt spray for AW purposes. Hassini et al.[41] found that sand wind does not affect oxygen permeation in PE greenhouse cover, aged under simulated sub-Saharan climatic conditions. Therefore, oxygen thought to come from photo-oxidative degradation. Gulmine et al.[42] suggested that sodium chloride (NaCl), silicon oxide (SiO_2), and aluminum oxide (Al_2O_3) might come from the incomplete purification of the sprayed water by the ion exchange resin of the accelerated weathering equipment.

The energy dispersive spectroscopy (EDS) analyses were done to obtain elemental information and identify foreign materials on the surface of pipes during natural and accelerated weathering period. Contaminants from rain and wind may accumulate on the surface and combine with condensation to create an environment which could be harmful to the stabilizing system in the polymer film.[22]

4.3 MATERIALS AND METHODS

The influences of natural and artificial weathering on stabilized and nonstabilized LDPE pipes were studied herein by means of well-known instrumental analyzing techniques.

4.3.1 MATERIALS

A typical commercial LDPE resin (type G03-5, density 0.921 g/cm³, Melt Flow Index 0.30 g/10 min Petkim Co., Turkey), was used to produce pipes appropriate for irrigation on the soil. The UV stabilizers such as masterbatchs(1:1 mass basis) are HALS (CIBA Co., Switzerland), equal mass mixture of Chimassorb 944 and Tinuvin 622) in Linear LDPE and CB(60–70 nm particle size, Kritilen Black 350, Plastika Kritissa Co., Greece) are provided by their vendors.

4.3.2 EXPERIMENTAL METHODS

The resin and both of the stabilizer masterbatchs were characterized by means of infrared spectroscopic and DSC analyses prior to the experiments. Thereafter, the stabilized and nonstabilized LDPE pipes were produced on *ad hoc* basis and then the aging investigations were performed.

Sample production and coding: The pipes having about 25.0 ± 0.1 mm diameter were produced through the extrusion of the granules of P;lE resin with and without UV stabilizer using Battenfeld single screw extruder with L/D = 30 and having a screw speed as 75 rpm and a heat set of 150°C along the cylinder at three zones and at the die.

The pipe samples were coded as PE, PE/HALS, and PE/CB, referring to the LDPE having no UV stabilizer, 1 phr (part per hundred resin) HALS masterbatch, and 5 phr CB masterbatch, respectively.

Aging Studies: The pipe pieces of about 6 m length were subjected to natural weathering in open air, lying on a terrace of Izmir, Turkey. The external factors as temperature, barometric pressure, and wind speed were recorded during the study and the average temperatures were found to be 21°C and 10°C for day and night, respectively. The pipe pieces of 0.20 m length were subjected to the accelerated aging by fixing them on a sample holder in a QUV tester with LU-0819 model (Q-Panel Company). The fluorescent tube source was a UVA lamp (340 nm maximum intensity) with a radiation intensity of 0.68 W/m²-nm at 40°C. A spray of de-ionized water was applied at each 4 h lasting for 4 h during UV exposure.

4.3.3   ANALYSIS METHODS

All samples were stored in the refrigerator to minimize oxidation prior to analyses. The following experimental methods were applied in this study, in line with the organization scheme shown in Figure 4.1.

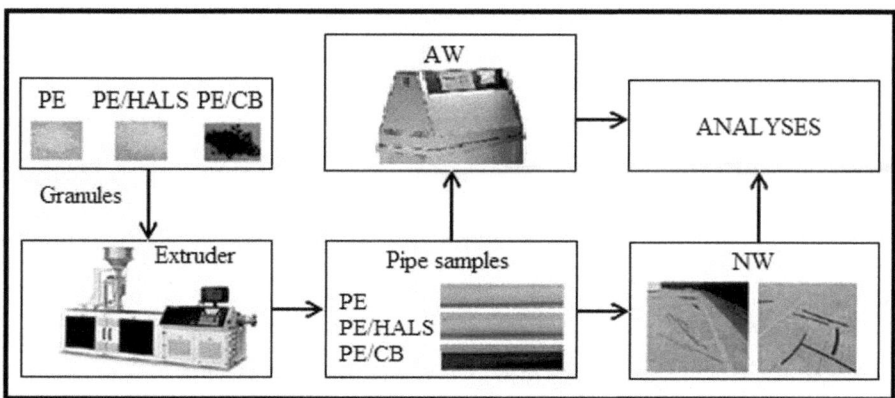

FIGURE 4.1   Preparation of pipe samples, aging experiments, and analyses.

The FTIR spectra of the samples were obtained by employing perkin elmer spectrum one having diamond crystal, by using its attenuated total reflection (ATR) attachment. The spectra were taken from the same point of the samples exposed to UV light.

OIT analyses have been carried out with a Setaram DSC 92 instrument, according to ASTM D5885 with 20°C/min heating rate. After reaching 200°C, the atmosphere was changed from N_2 to O_2.

TGA/DTA study was carried out on 14–24 mg samples by using a Shimadzu DTG-60H model TGA/DTA instrument, operated under nitrogen atmosphere having a heating rate of 10°C/min from room temperature to 500°C.

The conventional DSC analyses were carried out on ~4 mg pipe samples using a Perkin Elmer DSC instrument operated under nitrogen atmosphere. Heating from room temperature to 200°C at a rate of 10°C/min is followed by a 2 min isotherm at 200°C to erase the thermal history and then cooling is done at a rate of ~10°C/min, down to 25°C.

XRD measurements were done with Rigaku D/MAX-2200/PC X-Ray Diffractometer using CuKα radiation (λ = 1.5418 nm) operating at 40 kV and 36 mA.

The SEM and EDS analyses were performed by coating the samples with a layer of gold and using a Jeol JSM-6060 SEM instrument equipped with an EDS unit.

The pipe samples prior to, and after NA were subjected to HPRT in accordance with EN 921 standard under the applied pressures (P_a) as 29.0–34.9 atm. The pipe samples which have thicknesses of 4.6 ± 0.3 mm were kept in a water bath at 20°C for 1 h under 7 MPa tension. Figure 4.2 is a snapshot in the course of sample preparation in front of the test system.

FIGURE 4.2   A snapshot in the course of sample preparation in front of the HPRT system.

The tensile tests were applied at 100 mm/min extension rate on a Shimadzu AG-IS Autograph tensile tester equipped with an extensometer. The samples of 20 cm pieces were supported from the inside by inserting solid steel rods having the diameter equal to the inner diameter of the pipes, so as to prevent collapsing of hollow pipes.

The Shore D Hardness tests were performed on five samples, each with the pieces cleaned by using sand paper, prior to the experiments in accordance with ASTM D2240.

4.4   RESULTS AND DISCUSSION

In this study, degradation behavior of LDPE pipes produced herein on *ad hoc* basis and exposed to natural and artificial weathering were scrutinized. The influence of the presence and the type of light stabilizer as HALS and CB were evaluated mainly with the crystallinity change, correlating with the mechanical strength and interface structure of rupture surfaces. Since the driving force of this study is the illumination of UV stabilization effect of CB and HALS on PE, methods for illumination of degradation and stabilization and their assessment should be discussed.

The FTIR characterization of stabilizers as the initial components of the pipes was confirmed their structures. The MFI of the pipe sample, prior to weathering experiments is 0.33 g/10 min, which is in compliance with original LDPE given on its datasheet by the producing company. The addition of UV stabilizer slightly ascends MFI to 0.44 g/10 min for the samples coded as PE/HALS and PE/CB.

4.4.1   DEGRADATION FOLLOWED THROUGH OXIDATION OF POLYETHYLENE

Commencement and progression of the degradation are detected by means of spectroscopic measurements following the bands and peaks related to the oxidation of polymer. OIT test is as measure of oxidation in the polymer, although it is typically used as a quality-control tool for the formulation prior to extrusion.

4.4.1.1   ASSESSMENT OF OXIDATION BY MEANS OF INFRARED SPECTROSCOPY

The FTIR characterization of stabilizers as the initial components of the pipes was confirmed by their structures. FTIR spectra of the pipe samples on the course of NW and artificial Weathering (AW) were given in Figure 4.3. The spectra reveal the effect of exposure time since the carbonyl peak at 1720 cm^{-1} on each spectrum became more evident after 720 days NW. A progressive increase was observed at the peaks around 1000 cm^{-1} on each spectrum after UV exposure, in line with the study of Mendes et al.[24] Another aspect that deserves to be commented on, is related to the bands in the region of 1100–1300 cm^{-1}, that is more pronounced for all samples.

The band between 1250 and 1150 cm^{-1} likely belonging to long linear esters revealed increasing intensity with aging time on FTIR spectra of artificially weathered LDPE films.[6]

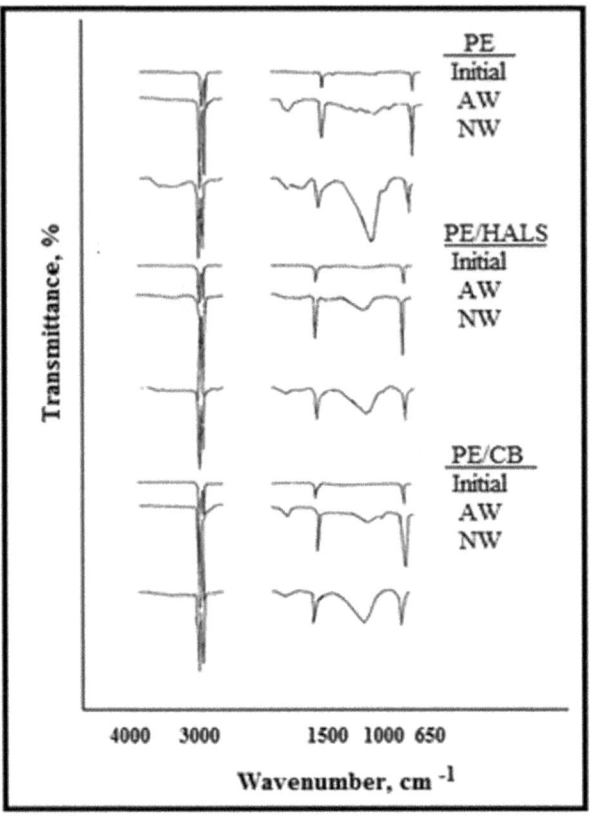

FIGURE 4.3 FTIR spectra of PE samples before and after aging (201 days AW and 720 days NW, respectively).

The carbonyl peak became huge on the spectrum of PE after 201 days of AW. The pipe sample of PE started to degrade after 120 days of weathering while the PE/HALS and PE/CB were not affected by the UV exposure. The absorptions between 1100 and 1000 cm^{-1} can be attributed to the C—O stretching vibration of hydroperoxides, confirmed by weak O—H bands above 3000 cm^{-1}.[6] The peaks at around 1000 cm^{-1} which were observed during NW were also observed on each spectrum during AW. Additional peaks were observed at around 1600 cm^{-1} on the spectrum of the PE/HALS before AW. These peaks, which were not observed on the spectra of PE pipe sample, can be assigned to HALS.[2] Its disappearance after weathering suggests a possible migration and consumption of stabilizer during UV exposure. The effect of UV exposure became more evident on 720 days NW.

The OI values of all the samples after aging calculated by using eq. 4.2 were shown in Figure 4.4. While there is greater change at carbonyl region of PE/CB samples after 720 days NW, the PE/HALS samples have the minimum OI value. The pipe without stabilizer (PE sample) has the maximum OI value as expected. Its OI value increased by 1476%, which indicates a really high degree of degradation at such a long exposure period. The minimum OI increment, maximum resistance against AW conditions, was determined with PE/HALS sample as 89% during 201 days of AW. A series of CB having a range of particle properties shows significant improvement in UV stabilization compared with clear LLDPE films under AW conditions. The large particle size of about 45–60 nm CB shows the fastest rates of increase in formation of functional groups in PE. An increase in CB mass percent from 1.5 to 3.5% improved UV stabilization as well.[14] The 60 nm particle size of CB used in this study might be a reason for the increased OI of pipe PE/CB in comparison with PE/HALS during AW. During outdoor applications, the rate of oxygen uptake and CO and CO_2 formation, and the increase in oxygen index depend on the climate factors to which the polymer is exposed.[22] From this point of view, one would expect faster degradation in the accelerated weathering conditions which has 1.55

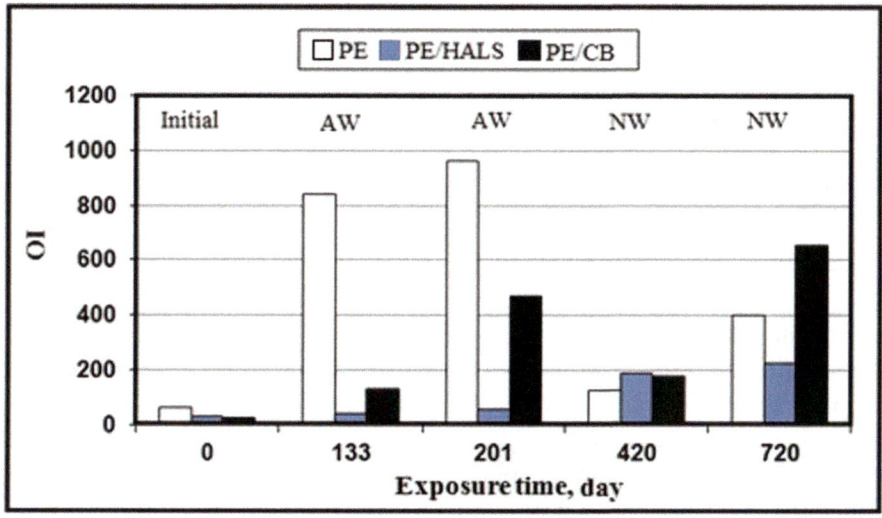

FIGURE 4.4 Oxidation index of PE samples before and after aging.

times greater irradiation energy than natural weathering conditions in this study. The OI value of accelerated weathered unstabilized pipe was found to be more than twice that of the natural weathered one, as expected.

4.4.1.2 ASSESSMENT OF OXIDATION THROUGH CALORIMETRIC MEASUREMENTS

OIT of the pipes soon after the extrusion and after 180 days natural weathering were searched and no severe degradation was found even after 180 day weathering. Thus it was concluded that, for thermo-mechanical degradation, the performances of HALS and CB showed no difference.

4.4.2 THERMAL BEHAVIOR OF POLYETHYLENE IN THE COURSE OF DEGRADATIVE PROCESSES

The thermal degradation evaluation was performed by means of TGA/ DTA curves. The degradation and melting processes were expressed by the temperatures at starting and ending points (Table 4.1).

TABLE 4.1   TGA/DTA Results of PE Samples before and after 420 Days of Natural Weathering.

Sample	Aging time, day	Start, °C	End, °C	Mass loss, %	Melting period		Decomposition period	
					Start, °C	End, °C	Start, °C	End, °C
PE	0	403.0	551.30	99.69	31.51	146.60	403.79	512.69
	420	402.0	551.73	98.83	37.79	143.34	408.84	509.80
PE/ HALS	0	417.0	551.27	97.47	41.09	142.54	403.06	510.03
	420	402.0	550.58	98.87	31.73	155.30	462.14	508.95
PE/CB	0	400.0	551.67	98.42	31.93	146.48	460.37	506.04
	420	410.0	551.41	98.00	63.10	155.37	380.65	507.51

The mass loss of weathered samples started at 36–38 min and over 400°C temperature. No considerable difference is detected in thermal behavior of samples before and after NW from TGA/DTA investigation. The change in melting temperature of sample without stabilizer is higher than that of samples with stabilizer and the melting points are lower than that of samples with stabilizer for the same UV exposure time period.

4.4.3 EVALUATION OF CRYSTALLINITY OF POLYETHYLENE IN THE COURSE OF DEGRADATIVE PROCESSES

LDPE is a semicrystalline material, in which spherulites, spherical shape having crystalline regions, are separated by an amorphous polymer. Since the crystalline areas are the denser areas than the amorphous ones, these regions impart improved stiffness and chemical resistance to the polymer. On the other hand, degradation processes mainly take place in the amorphous phase of the material.[26]

There is no considerable difference between XRD curves of samples, before and after 420 days weathering. However, some crystal peaks of LDPE vanished after 720 days weathering. The XRD results represent the crystallinity at the state of nonentangled conformation, while the calorimetric measurements may involve certain conformational changes during the heating process.

4.4.3.1 CALORIMETRIC INVESTIGATION OF DEGRADATION ON LDPE PIPES

In this study, DSC curves of the pipe samples were taken in the course of degradation under NW and by the end of 201 days AW. The melting peak (T_m) and the melting enthalpy (ΔH_m) were obtained from the DSC curves and then the percentages of crystallinity were determined by taking melting enthalpy of 100% crystalline PE $(DH_{crystal})$ as 289.74 J/g[6] in eq. 4.3. The values obtained and calculated from DSC curves were given in Table 4.2.

As the melting temperatures of the samples with and without stabilizer were compared during NW and AW (Table 4.2), they exhibited no considerable difference.

The determined degree of crystallinity of all samples exhibited a sharp increase with exposure time period. These behaviors of samples indicate that the molecular mobility and chain scission started, and the samples suffered from oxidative degradation. The obtained values from NW and AW are in agreement that the sample with HALS stabilizer showed better performance compared with the samples without stabilizer, and with CB stabilizer.

Yenigül and Tuna[43] studied pyrolysis of waste PE on waxes. In their study, PE films were subjected to thermal degradation to obtain waxy products under nitrogen atmosphere. As they took the DSC curves of scrap greenhouse PE wax after 15 and 30 min of degradation, they found that the melting temperature of the product decreases as the degradation period gets longer. Therefore, UV degradation and thermal degradation are thought to be in compliance with each other.

TABLE 4.2 DSC Analysis Results and Crystallinity of Weathered LDPE Samples.

Evaluation	Aging time, days	PE	PE/HALS	PE/CB
T_m °C	0	114.20	110.90	113.64
	105	109.71	110.86	110.33
	180	109.98	110.31	–
	420	110.29	110.30	114.25
	720	110.89	110.16	109.34
	201 (AW)	110.15	109.66	110.81
$-\Delta H_m$, J/g	0	89.01	61.70	55.10
	105	53.61	82.62	33.69
	180	55.02	36.23	–
	420	55.57	51.70	50.44
	720	106.68	97.73	92.51
	201 (AW)	139.53	104.53	98.42
DC, %	0	30.7	21.3	19.0
	105	18.5	28.5	11.6
	180	19.0	12.5	–
	420	19.2	17.8	17.4
	720	36.8	33.7	31.9
	201 (AW)	48.2	36.0	34.0

The crystallinity of samples was found to be decreasing until 420 days NW surprisingly. As the molecules undergo degradation, the crystallinity was expected to increase since the smaller molecules could take place in crystal lattice easier. However much, the crystallinity of 720 days NW samples showed an increasing tendency by 90% of obtained values after 420 days weathering, the XRD measurements were also done to determine the influence of aging on crystallinity. Similar results were found by Dintcheva et al.[30] for post-consumer LDPE films (PF). Indeed, the PF crystalline content was about 16% higher than that of virgin film (VF). This increased crystallinity can be ascribed to the lowering of the molecular weight of PF, if compared with VF, as a consequence of the photo-oxidative degradation.

4.4.3.2 INVESTIGATION OF LDPE PIPE DEGRADATION BY MEANS OF X-RAY DIFFRACTION (XRD) ANALYSIS

After searching for the changes on the differently formulated LDPE, pipe surfaces subjected to UV exposure in terms of oxygen uptake as a measure of chemical degradation, namely, the oxidation behavior and contamination with silica, XRD patterns of samples were studied to probe the crystalline and amorphous phase structures and compared that with those of fresh samples.

The XRD patterns of samples subjected to NW and AW were given in Table 4.3. There is no considerable difference between XRD curves of samples before and after 420 days NW. However, some crystal peaks of LDPE could not be observed after 720 days NW.

Russell et al.[34] observed that relaxation in battered PE begins immediately after trauma. The mechanical stress can convert the interchain configuration of the monoclinic form back into orthorhombic configuration. The process is very slow in room temperature. The rate can be increased by increasing the temperature. In this study, although there is no battering, the samples were subjected to sunlight and the heat of solar radiation as well as the natural wind action which may result in similar effects.

Both of orthorhombic and monoclinic reflections were observed in LDPE polymer where they are expected to be. The monoclinic reflections as 200, 201, and 400, observed in samples before aging, were transformed into orthorhombic reflections with time. The 001 monoclinic reflections were observed for each pipe sample after 24 months weathering. The degree of crystallinity may give a gist about the change in the structure. Therefore, the degree of crystallinity (DC) given in Table 4.4 was calculated by using eq. 4.4.

The DC of all the samples remained constant, with the aging time being different from what is normally expected. These values were compared with those found by calorimetric analysis as given in Figure 4.5. It was observed that the crystallinity values obtained from calorimetric investigation were lower than that found from the XRD analysis.

There is no considerable difference between XRD curves of samples with stabilizers before and after AW. However, the intensity of 200 orthorhombic reflection observed on a sample surface without stabilizer increased with time.

TABLE 4.3 XRD Pattern of LDPE Samples before and after Weathering.

2θ	d (A°)*	hkl	Crystal type	Aging period (day) and d (A°)											
				PE				PE/HALS				PE/CB			
				0	201**	420	720	0	201**	420	720	0	201**	420	720
19.45	4.563	001	Mono	—	4.75	—	4.75	—	4.53	—	4.71	—	4.74	—	4.71
21.56	4.11	110	Ortho	4.21	4.15	4.06	4.12	4.21	4.13	4.15	4.15	4.06	4.14	4.19	4.16
23.17	3.84	200	Mono	3.80	—	—	—	3.80	—	—	—	—	—	3.79	—
23.99	3.71	200	Ortho	—	3.74	3.68	3.76	—	3.77	3.73	3.74	3.68	3.6	—	3.76
30.12	2.96	210	Ortho	3.04	—	2.94	—	3.03	—	2.99	—	—	—	—	—
35.08	2.56	201	Mono	2.51	—	—	—	2.50	—	—	—	—	—	2.51	—
36.33	2.47	020	Ortho	2.51	—	2.46	—	2.50	—	2.49	—	2.46	—	2.51	—
40.80	2.21	310	Ortho	2.26	—	2.23	—	2.28	—	2.26	—	2.25	—	2.29	—
43.95	2.06	220	Ortho	—	—	—	—	—	—	—	—	—	—	2.09	—
47.43	1.91	400	Mono	—	—	—	—	—	—	—	—	1.93	—	—	—

*Literature data.
**AW.

TABLE 4.4   DC of Weathered PE Samples Estimated by Means of XRD Patterns.

Aging period, day	Weathering type	Degree of crystallinity, %		
		PE	**PE/HALS**	**PE/CB**
0	–	54.1	53.8	52.7
201	AW	65.5	52.6	50.0
420	NW	55.3	52.0	53.9
720	NW	51.6	53.6	56.5

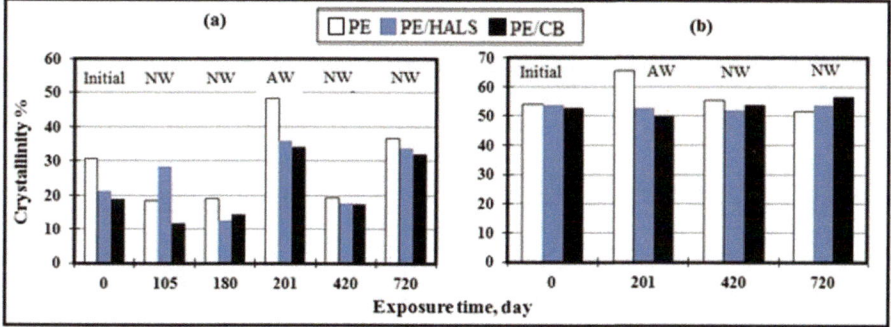

FIGURE 4.5   The crystallinity of PE samples as a function of aging time, determined by (a) calorimetry and (b) X-ray diffraction.

There was a 21% increase in degree of crystallinity of sample without stabilizer, while that of others still remained constant after 201 days AW. At longer exposure times, the increased crystallinity often results in the surface cracking which was also observed on the SEM micrograph that is given in a later section.

Although Joo et al.[33] observed the opposite, here the crystallinity values obtained from calorimetric investigation were lower than the values found from the XRD analyses. The difference between the two methods is thought to be that while the XRD experiments carried out at 25°C represent the crystallinity at the state of nonentangled conformation, the calorimetry may involve certain conformational changes during the measurement due to the heating process.

4.4.4   MECHANICAL STRENGTH OF POLYETHYLENE ON THE COURSE OF AGING

The mechanical properties such as tensile strength and elongation at the break of the PE pipes were measured by the tensile test. The tensile strength tests could be applied only to NW samples due to the sample size limitations. Tensile and yield strength values of PE pipes gave no considerable difference even after 1200 days of NW (Fig. 4.6).The smallest value was observed in the sample without stabilizer, while the PE/CB sample gave the highest mechanical performance.

As the elongation at break values are concerned, there is no significant difference between the pipes with and without stabilizer after 360 days NW. The samples without stabilizer were found to be the weakest after 1200 days NW (Fig. 4.6).

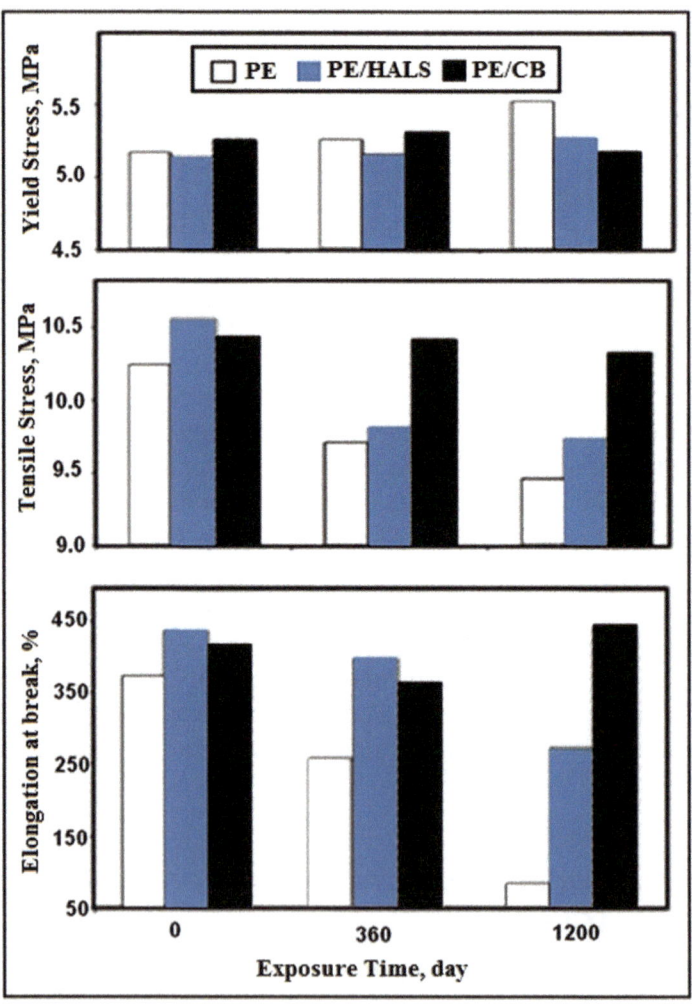

FIGURE 4.6 Strength and elongation of PE samples as a function of aging time.

The test conditions simulated 50 years lifetime, since the pipes have been designed for 50 years lifetime.[37]

The hydrostatic strength test results were given as the burst pressure of pipes versus aging time in Figure 4.7. The pipes could withstand not only to a minimum pressure stipulated in the related test standard, but also to quite high burst pressures, even after 720 day NW, suggesting that no severe degradation would happen during their service life.

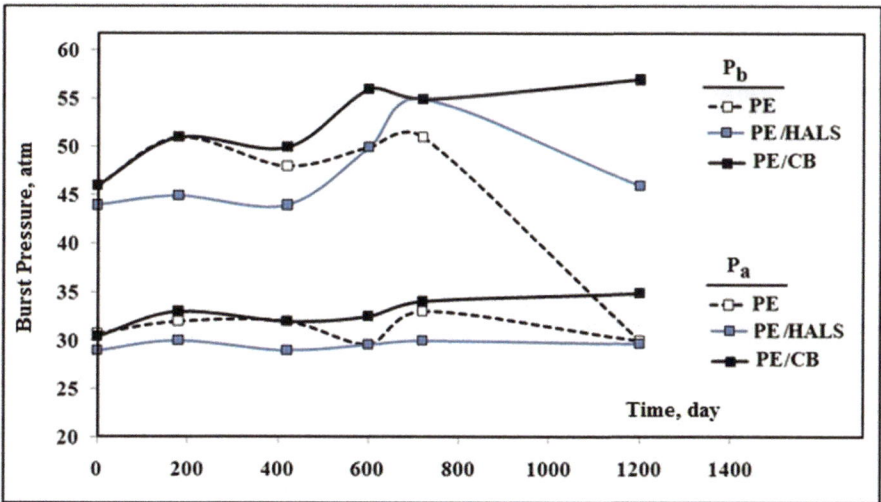

FIGURE 4.7 HPRT results as burst pressure with respect to NW time of the pipe samples and the applied pressure values.

The pipe samples subjected to hydrostatic tests after last two sampling as 600 and 720 days were depicted in Figure 4.8. The hydrostatic strength test revealed some structural differences between stabilized and unstabilized pipes after 600 and 720 days NW. While the unstabilized pipes cleaved longitudinally upon bursting, the stabilized pipes burst, enlarging their diameters. There were lines as sign of cracks all along the unstabilized pipe and it was cleaved from one of these cracks during bursting.

Another significant effect of the extrusion processing is the orientation of the long polymer chain, due to the applied shear forces in the machine direction. In this oriented state, the spherulites take the appearance of ellipsoids, regularly also called lamellae. In general, this orientation imparts changes in the mechanical properties of the plastics.[44,45]

Chain scissions that occur due to aging are caused tight packing of molecules on the macroscopic scale over time. This causes an increase in the density and hardness of the material. However, there is no apparent increase in the hardness of the pipe samples (Table 4.5), which proves that the aging had minute effect on the present pipe samples.

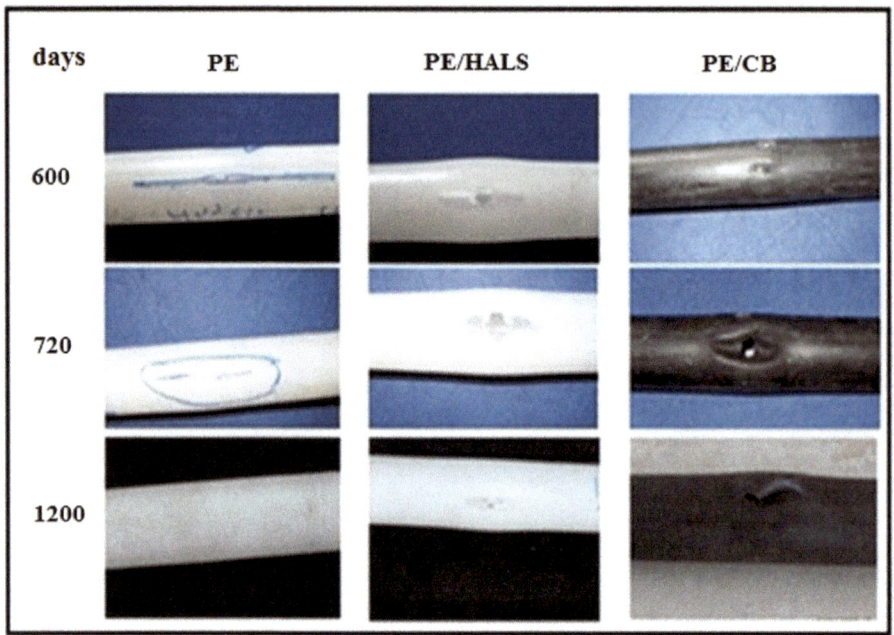

FIGURE 4.8   The NW pipe samples after subjecting to HPRT.

TABLE 4.5   Shore-D Hardness of LDPE Pipe Surfaces during Natural Weathering.

t_{dy}, day	Hardness, Shore D		
	PE	**PE/HALS**	**PE/CB**
0	50.0	44.0	46.0
120	45.0	46.0	50.0
360	42.0	50.0	50.0
1200	50.0	43.5	49.5

4.4.5 INVESTIGATION OF DEGRADATION BY MEANS OF SEM AND EDS

Surface features of LDPE samples subjected to NW and AW were presented in Figure 4.9. Although there were signs of flows and holes all over the sample surfaces, there is no comparable surface deformation on the samples during NW. On the other hand, the greater changes on accelerated weathered sample surfaces were expected because of a more pronounced effect of weathering conditions. The photomicrographs show that the surface of new and nondegraded pipes were smooth, without deep holes, cracks, and free of any kind of defects. In Figure 4.9, the surface features of the samples subjected to AW for 201 days were presented. The strong effect of weathering during exposure time on the sample without stabilizer is obvious. There is also a considerable surface deformation on the PE/CB sample. The results of FTIR and SEM analyses of samples exposed to AW are in agreement that the sample with HALS stabilizer showed the better performance under AW conditions.

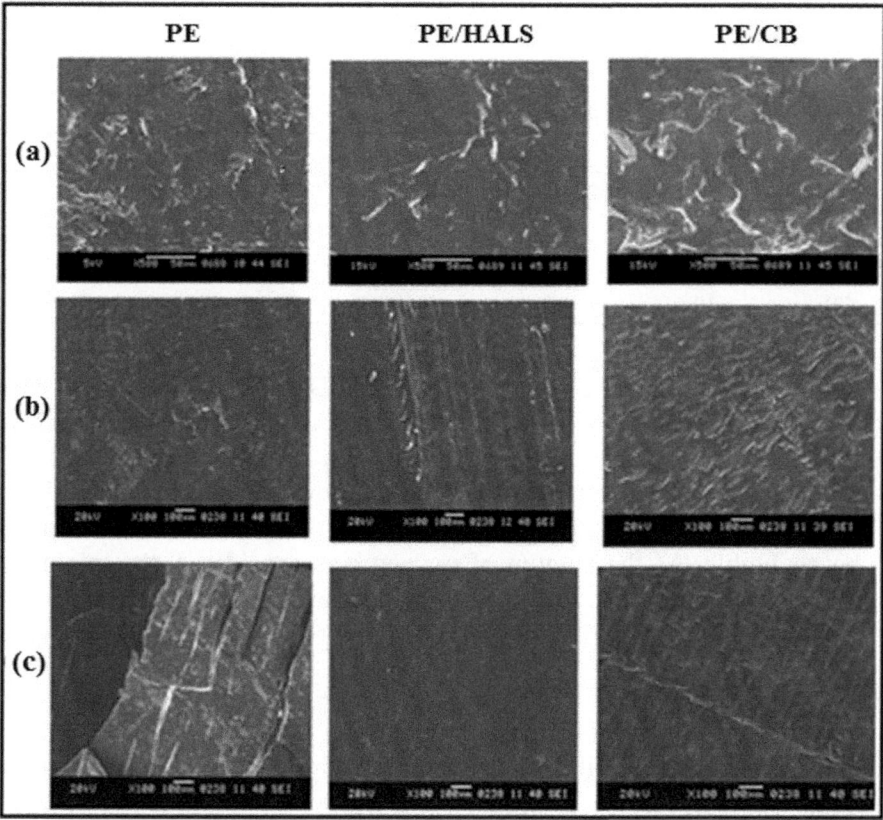

FIGURE 4.9 SEM micrographs of PE sample surfaces: (a) before weathering, (b) after NW for 720 day, and (c) after AW for 201 day.

The EDS analyses were done to obtain elemental information and identify foreign materials on the surface of pipes during natural and AW. Contaminants from rain and wind may accumulate on the surface and combine with condensation to create an environment which could be harmful to the stabilizing system in the polymer film.[22]

The 720 days NW sample surface EDS revealed the presence of oxygen, silica, and very small amounts of Na, Al, Ca, Fe, and K. On the sample surfaces, oxygen which originates from oxidative degradation and silica which comes by the winds were detected.

It is not possible to distinguish between the influences of UV radiation and other weathering conditions such as rain and wind, herein. Screening of the samples from the influences other than UV radiation would most likely preserve all the other weathering influences.[22] Yet, the oxygen concentration increased (Table 4.6) with time, in general.

TABLE 4.6   The Oxygen Amounts (%) on NW and AW Sample Surfaces.

Sample	Aging time, day				
	NW				AW
	0	180	435	720	201
PE	4.7	8.0	14.7	11.2	8.1
PE/HALS	5.6	8.5	11.2	11.9	5.5
PE/CB	10.4	6.0	11.4	12.9	8.4

In addition, it can be concluded from the EDS analyses that the HALS stabilizer has better stabilizing efficacy than CB, against UV light under NW conditions.

In this study, the amounts of silica on samples of PE, PE/HALS, and PE/CB were found to be 0.03, 0.06, and 0.1%wt, respectively. So they were thought to be negligible. As the amount of oxygen was considered, the minimum oxygen concentration was found on the sample with HALS stabilizer.

4.5   CONCLUSIONS

The pipes used on the soil for agricultural irrigation applications are prone to UV degradation under the particularly hot weather conditions of Izmir city of Turkey. Therefore, the effects of two traditional type stabilizers (CB and HALS) on the degradation of LDPE were compared under natural and accelerated weathering conditions.

Stabilizers play an important role to improve the life time of plastics and maintain the properties of finished products. In this study, it is found that the pipe with HALS stabilizer showed better performance against both natural and artificial weathering conditions. However, optimizing the performance of the product with an appropriate cost is also important during series production.

FTIR and SEM analyses revealed the higher resistance of PE/HALS samples to artificial weathering conditions when compared with the PE/CB ones. The change in melting temperature of sample without stabilizer is higher than that of samples with stabilizer and the melting temperatures are lower than that of samples with stabilizer for the same UV exposure time period. No considerable change was observed in melting temperature of PE/HALS. On the other hand, the melting temperature of PE and PE/CB samples showed a decreasing tendency under AW conditions. The determined degree of crystallinity of all samples, especially, the sample without stabilizer, exhibited a sharp increase with exposure time. These behaviors of samples indicate that the molecular mobility and chain scission started, and samples suffered from oxidative degradation. The obtained values from NW and AW are in agreement that the PE/HALS sample showed better performance when compared with PE and PE/CB samples. The lower crystallinity values obtained from calorimetric investigation than that found from the XRD analysis are thought to be due to the different molecular conformation of the polymers during DSC and XRD measurements.

Incorporation of HALS and CB had positive contribution to the mechanical performance of the pipes, while the 1200 days of NW did not result in decrement of strength.

ACKNOWLEDGMENTS

The authors express their deep gratitude to Göktepe Plastics Company of Turkey which supported the project financially, allowed using of their extruder, and performed the hydrostatic tests. The partial funding of Ege University (Project No: 06 MUH 027) is also acknowledged. The project was presented in two local congresses, namely, UPTS-2015 and UKMK-2016 organized by Yildiz Technical University and Izmir Institute of Technology, respectively.

KEYWORDS

- **LDPE**
- **stabilizer**
- **oxidation**
- **crystallinity**
- **weathering**

REFERENCES

1. Azwa, Z. N.; Yousif, B. F.; Manalo, A. C.; Karunasena W. A Review on the Degradability of Polymeric Composites Based on Natural Fibres. *Mater. Design.* **2013,** *47,* 424–442.
2. Boru İzmirli, G. UV Stabilization of Low Density Polyethylene, Ph.D. Thesis, Ege University, Izmir, 2007.
3. Moore, G. R.; Kline, D. E. *Properties and Processing of Polymers for Engineers;* Prentice-Hall: Englewood Cliffs, NJ, 1984.
4. Santos, A. S. F.; Agnelli, J. A. M.; Trevisan, D. W.; Manrich, S. Degradation and Stabilization of Polyolefins from Municipal Plastic Waste During Multiple Extrusions under Different Reprocessing Conditions. *Polym. Degrad. Stab.* **2002,** *77,* 441–447.
5. Al-Madfa, H.; Mohamed, Z.; Kassem, M. E. Weather Ageing Characterization of the Mechanical Properties of the Low Density Polyethylene. *Polym. Degrad. Stab.* 1998, *62,* 105–109.
6. Küpper, L.; Gulmine, J. V.; Janissek, P. R.; Heise, H. M. Attenuated Total Reflection Infrared Spectroscopy for Micro-Domain Analysis of Polyethylene Samples After Accelerated Ageing within Weathering Chambers. *Vib. Spectrosc.* 2004, *34,* 63–72.
7. Tidjani, A. Comparison of Formation of Oxidation Products During Photo-Oxidation of Linear Low Density Polyethylene Under Different Natural and Accelerated Weathering Conditions. *Polym. Degrad. Stab.* **2000,** *68,* 465–469.
8. Frank, A.; Pinter, G.; Lang, R. W. Prediction of the Remaining Lifetime of Polyethylene Pipes after up to 30 Years in Use. *Polym. Test.* **2009,** *28,* 737–745.
9. Rosa, D. S.; Sarti, J.; Mei, L. H. I.; Filho, M. M.; Silveira, S. Test Method, A Study of Parameters Interfering in Oxidative Induction Time(OIT) Results Obtained by Differential Scanning Calorimetry Inpolyolefin. *Polym. Test.* **2000,** *19,* 523–531.
10. Möller, K.; Gevert, T.; Holmström, A. Examination of a Low Density Polyethylene (LDPE) Film After 15 Years of Service as an Air and Water Vapour Barrier. *Polym. Degrad. Stab.* **2001,** *73,* 69–74.
11. Gugumus, F. Possibilities and Limits of Synergism with Light Stabilizers in Polyolefins 1. HALS in Polyolefins. *Polym. Degrad. Stab.* **2002,** *75,* 295–308.
12. Stepek, J.; Daoust, H. *Additives for Plastics;* Springer-Verlag Inc: New York, **1983**.
13. Ewais, A. M. R.; Rowe, R. K.; Scheirs, J. Degradation Behaviour of HDPE Geomembranes with High and Low Initial High-Pressure Oxidative Induction Time. *Geotext. Geomembr.* **2014,** *42,* 111–126.

14. Liu, M.; Horrocks, A. R. Effect of Carbon Black on UV Stability of LLDPE Films Under Artificial Weathering Conditions. *Polym. Degrad. Stab.* 2002, *75,* 485–499.

15. Pena, J. M.; Allen, N. S.; Edge, M.; Liauw, C. M.;Valange, B. Interactions between Carbon Black and Stabilisers in LDPE Thermal Oxidation. *Polym. Degrad. Stab.* 2001, *72,* 163–174.

16. Wong, W. K.; Hsuan, Y. G. Interaction of Antioxidants with Carbon Black in Polyethylene Using Oxidative Induction Time Methods. *Geotext. Geomembr.* **2014,** *42,* 641–647.

17. Blazso, M. Thermal Decomposition of Oligomeric and Polymeric Hindered Amine Light Stabilisers, *J. Anal. Appl. Pyrol.* **2001,** *58,* 29–47.

18. Hodgson J. L.; and Coote, M. L. Clarifying the Mechanism of the Denisov Cycle: How do Hindered Amine Light Stabilizers Protect Polymer Coatings from Photo-Oxidative Degradation? *Macromolecules* **2010,** *43,* 4573–4583.

19. Pfaendner, R. How Will Additives Shape the Future of Plastics, *Polym. Degrad. Stab.* **2006,** *91,* 2249–2256.

20. Gugumus, F.; Lelli, N. Light Stabilization of Metallocene Polyolefins. *Polym. Degrad. Stab.* 2001, *72,* 407–421.

21. Liauw, C. M.; Childs, A.; Allen, N. S.; Edge, M.; Franklin, K. R.; Collopy, D. G. Effect of Interactions between Stabilisers and Silica Used for Anti-Blocking Applications on UV and Thermal Stability of Polyolefin Film 2. Degradation Studies. *Polym. Degrad. Stab.* 1999, *65,* 207–215.

22. Sampers, J. Importance of Weathering Factors Other Than UV Radiation and Temperature in Outdoor Exposure. *Polym. Degrad. Stab.* **2002,** *76,* 455–465.

23. Luzuriaga, S.; Kovarova, J.; Fortelny, I. Degradation of Pre-Aged Polymers Exposed to Simulated Recycling: Properties and Thermal Stability. *Polym. Degrad. Stab.* **2006,** *91,* 1226–1232.

24. Mendes, L. C.; Rufino, E. S.; Paula, F. O. C.; Torres Jr., A. C. Mechanical, Thermal and Microstructure Evaluation of HDPE After Weathering in Rio De Janeiro City. *Polym. Degrad. Stab.* **2003,** *79,* 371–383.

25. Qin, H.; Zhao, C.; Zhang, S.; Chen, G.; Yang, M. Photo-Oxidative Degradation of Polyethylene/Montmorillonite Nanocomposite. *Polym. Degrad. Stab.* 2003, *81,* 497–500.

26. Dilara, P. A.;Briassoulis, D. Degradation and Stabilization of Low-density Polyethylene Films Used as Greenhouse Covering Materials, *J. Agric. Engng. Res.* **2000,** *76,* 309–321.

27. Mueller, W.; Jakob, I. Oxidative Resistance of High-Density Polyethylene Geomembranes. *Polym. Degrad. Stab.* **2003,** *79,* 161–172.

28. Schmid, M.; Affolter, S. Test Method Comparison, Interlaboratory Tests on Polymers by Differential Scanning Calorimetry (DSC): Determination and Comparison of Oxidation Induction Time (OIT) and Oxidation Induction Temperature (OIT*). *Polym. Test.* **2003,** *22,* 419–428.

29. Corrales, T.; Catalina, F.; Peinado, C.; Allen, N. S.; Fontan, E. Photooxidative and Thermal Degradation of Polyethylenes: Interrelationship by Chemiluminescence, Thermal Gravimetric Analysis and FTIR Data. *J. Photochem. Photobiol. A Chem.* **2002,** *147,* 213–224.

30. Dintcheva, N. T.; La Mantia, F. P.; Acierno, D.; DiMaio, L.; Camino, G.; Trotta, F.; Luda, M. P.; Paci, M. Characterization and Reprocessing of Greenhouse Films. *Polym. Degrad. Stab.* 2001, *72,* 141–146.

31. Mojumdar, S. C.; Sain, M.; Prasad, R. C.; Sun, L.; Venart, J. E. S. Selected Thermoanalytical Methods and Their Applications from Medicine to Construction, Part I. *J. Therm. Anal. Calorim.* **2007,** *90,* 653–662.

32. Buchanan, F. J.; White, J. R.; Sim, B.; Downes, S. The Influence of Gamma Irradiation and Aging on Degradation Mechanisms of Ultra-High Molecular Weight Polyethylene. *J. Mater. Sci. Mater. Med.* **2001,** *12,* 29–37.

33. Joo, Y. L.; Han, O. H.; Lee, H. K.; Song, J. K. Characterization of Ultrahigh Molecular Weight Polyethylene Nascent Reactor Powders by X-ray Diffraction and Solid State NMR. *Polymer.* **2000,** *41,* 1355–1368.

34. Russell, K. E.; Hunter, B. K.; Heyding, R. D. Monoclinic Polyethylene Revisited. *Polymer.* **1997,** *38,* 1409–1414.

35. Rodriguez, F. *Principles of Polymer Systems;* 2nd ed.; McGraw-Hill: Singapore, 1982.

36. Farshad, M. Two New Criteria for the Service Life Prediction of Plastics Pipes, Property Modeling. *Polym. Test.* **2004,** *23,* 967–972.

37. Maria, R.; Rode, K.; Schuster, T.; Geertz, G.; Malz, F.; Sanoria. A.; Oehler, H.; Brull, R.; Wenzel, M.; Engelsing, K.; Bastian, M.; Brendle, E. Ageing Study of Different Types of Long-Term Pressure Tested PE Pipes by IR-Microscopy. *Polymer.* **2015,** *61,* 131–139.

38. Meththananda, I. M.; Parker, S.; Patel, M. P.; Braden M. The Relationship between Shore Hardness of Elastomeric Dental Materials and Young's Modulus. *Dent. Mater.* **2009,** *25,* 956–959.

39. Vriend, N. M.; Kren, A. P. Determination of the Viscoelastic Properties of Elastomeric Materials by the Dynamic Indentation Method. *Polym. Test.* **2004,** *23,* 369–375.

40. Weon, J. I.; Effects of Thermal Ageing on Mechanical and Thermal Behaviors of Linear Low Density Polyethylene Pipe. *Polym. Degrad. Stab.* **2010,** *95,* 14–20.

41. Hassini, N.; Guenachi, K.; Hamou, A.; Saiter, J. M.; Marais, S.; Beucher E. Polyethylene Greenhouse Cover Aged Under Simulated Sub-Saharan Climatic Conditions. *Polym. Degrad. Stab.* **2002,** *75,* 247–254.

42. Gulmine, J. V.; Janissek, P. R.; Heise, H. M.; Akcelrud, L. Degradation Profile of Polyethylene After Artificial Accelerated Weathering. *Polym. Degrad. Stab.* **2003,** *79,* 385–397.

43. Yenigül, M.; Tuna, L. *Proceedings of III. International Packaging Congress and Exhibition;* Chamber of Chemical Engineers of Turkey; İzmir, 2003; Vol. 2, pp 415–422.

44. Briassoulis, D.; Aristopoulou, A.; Bonora, M.; Verlodt, I. Degradation Characterisation of Agricultural Low-Density Polyethylene Films. *Biosyst. Eng.* **2004,** *88,* 131–143. doi:10.1016/j.biosystemseng.

45. Patel, R. M.; Butler, T. I.; Walton, K. L.; Knight, G. W. Investigation of Processing-Structure-Properties Relationships in Polyethylene Blown Films. *Polym. Eng. Sci.* **2004,** *34,* 1506–1514.

CHAPTER 5

THEORETICAL CALCULATIONS ON AZA-SCORPIAND SYSTEMS

J. V. DE JULIÁN-ORTIZ[1,*], L. POGLIANI[1], E. BESALÚ[2], B. VERDEJO[3], and E. GARCÍA-ESPAÑA[3]

[1]*Departament de Química Física, Facultat de Farmàcia, Universitat de València, Av V. Andrés Estellés 0, 46100 Burjassot, València, Spain*

[2]*Institut de Química Computacional i Catàlisi, Universitat de Girona, 17003 Girona, Spain*

[3]*Institut de Ciència Molecular, Universitat de València, Paterna, Valencia, Spain*

Corresponding author. E-mail: jejuor@uv.es

CONTENTS

ABSTRACT

Rearrangements and their control are a hot topic in supramolecular chemistry due to the possibilities that these phenomena open in the design of synthetic receptors and molecular machines. Macrocycle aza-scorpiands constitute an interesting system that can reorganize their spatial structure depending on pH variations or the presence of metal cations. In this study, some simulations on these systems are reviewed, including the relative stabilities of their conformations predicted computationally.

5.1 INTRODUCTION

The possibility of controlling the conformation of chemical structures is interesting because it opens the possibility of creating molecular machines and synthetic receptors that react to the desired stimuli.[1] Among the molecular structures showing such properties, the aza-macrocycles stand out because they show a coordinating tail, known as a scorpiand.[2]

These systems merit attention due to their potential biological and pharmacological applications, since their ability to recognize hydrophilic and hydrophobic amino acids has been demonstrated.[3] Molecular dynamics (MD) studies have pointed out that aza-scorpiands block the access of substrates to the active site of iron superoxide dismutase and break the hydrogen bond pattern in it.[4] This could explain their ability as inhibitors of *Trypanosoma cruzi*.[4] Receptors for amino acid sensing[5], or drug delivery[6] are two additional fields for their potential applications.

Changes in the protonation state are able to induce conformational reorganizations in scorpiand-like ligands.[7,8] It has been observed by others that the presence of metal centers produces similar effects. Also, the metal coordination is influenced by the protonation state because of electrostatic repulsion.[2a]

Materials reviewed in the present article can be found in de Julián-Ortiz et al.[9]

The simplified structures presented in Figure 5.1 were taken from the X-ray geometries. Experimental values of the energy required to change the conformations from closed (A) to open (B) and vice versa, depending on the protonation state, are lacking. Conformation A is preferred for the neutral and monoprotonated species. H—bond and π–π stacking contribute to this stabilization. For instance, Figure 5.2 shows one of the possible H—bonds, depending on the orientation of the covalent N—H bonds, that stabilizes A;

That is, a N1H···N4 2.08 Å distance.[8] The π–π stacking is evidenced by the minimal distance from the pyridine ring to the naphthalene ring equal to 3.26 Å.[10]

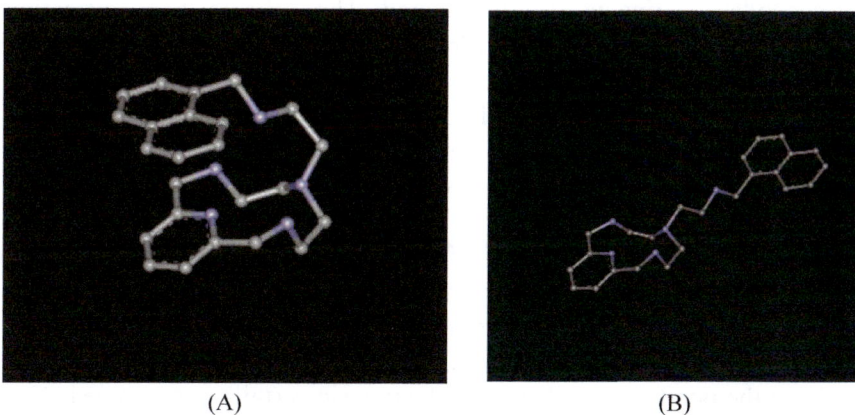

(A)   (B)

FIGURE 5.1 Conformations closed (A) and open (B) of the aza-scorpiand macrocycle studied.

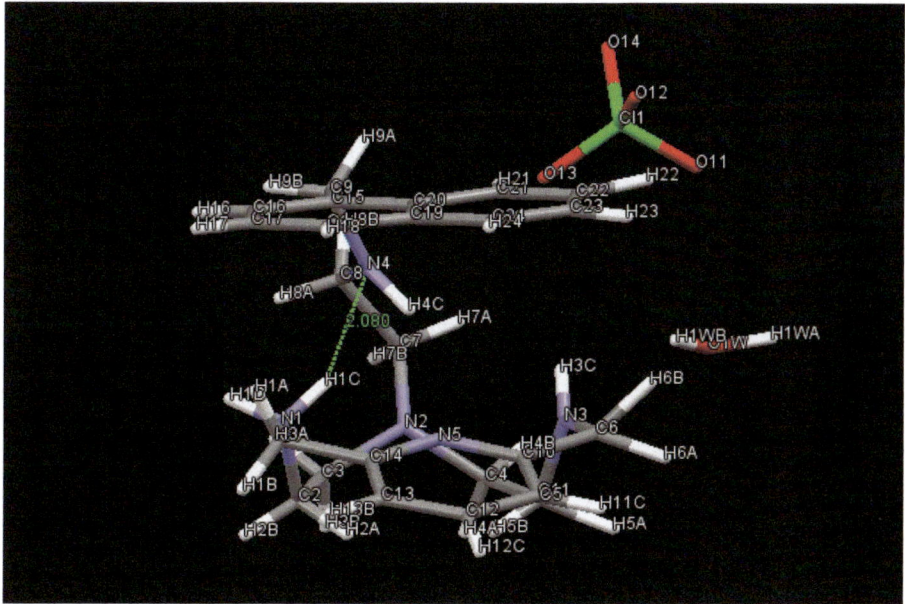

FIGURE 5.2 Distance between NH and N, which proves the presence of the H—bond in the monoprotonated A.

5.2   CONFORMATIONAL SEARCH

Experimentally, it is observed that as the pH in the solution decreases, the two macrocyclic secondary amines are first protonated, and then the pendant arm's secondary amine is also protonated. Thus, the H—bond attraction between the pendant arm and the macrocyclic amines and the attraction with the pyridine nitrogen in A are progressively strengthening as the protonation increases, but the electrostatic repulsion increases more sharply. This makes conformation B more stable for the triprotonated species. In order to have an understanding of the conformational stabilities involved in this system and to model its changes, the following simulation has been undertaken.[9]

Setting the conformation A as the initial one, the molecule structure was passed from monoprotonated to triprotonated. The conformational search was performed by the Monte Carlo multiple minimum (MCMM) method[11–13] with the MM+ force field.[14] The free rotation of the dihedral angles comprising the pendant arm that joins the two rings (tail) was allowed. This method finds the lowest energy conformations of a molecule by randomly varying specified dihedral angles to generate new starting conformations. These were then minimized.

MCMM was preferred to MD trajectories because the conformational changes pursued were mainly torsions of the pendant arm atoms, and only MCMM gives these in a straightforward manner. Furthermore, the groups attached to the extremes of the pendant arm were too large to switch efficiently under MD. This last method is useful to obtain all the possible conformations in a molecule much smaller than the one studied here, or to study thermodynamic properties in molecular ensembles. An example can be the study of chemical structures surrounded by water.[15,16]

Figure 5.3 shows the neutral chemical structure with the nomenclature used for the different dihedral angles that have been rotated. The labels point out the central bond that undergoes torsion for each dihedron.

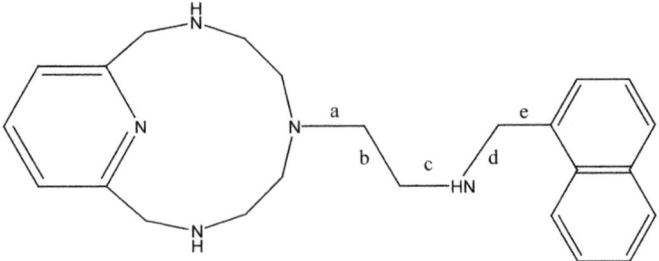

FIGURE 5.3   Labels for the dihedral angles in the pendant arm.

The conformation reorganization between A and B could be done in several ways. It can be done by the rotation of the bond between the two carbon atoms in the pendant arm, "b," and the consecutive bond that joins to the 12-ring amine "a." Alternatively, it can be done by inverting the 12-ring amine and rotating "b" and "e." Thus, if we track the changes, we see that the sequence of the torsion angles is not linear, but is even corkscrewing in some cases. These are the changes that must overcome the main rotational barriers, since all the dihedral angles suffer minor changes to obtain B, due to the energy minimization.

As pointed out, starting from the conformation A, it was triprotonated, and MCMM simulations were performed by allowing different possible rotations in different runs.

These calculations were first attempted with the molecule in a water constant-density periodic box (standard water molecules TIP3P, equilibrated at 300 K and 1 atm, minimum distance between solvent and solute atoms: 2.3 Å), but the results failed to converge. This can be due to the fact that triprotonated conformation A is too far from the equilibrium, the total system with solvation water is too complex, and the dihedral torsions are too sudden to quickly balance the hydration sphere.

For this reason, MCMM simulations were run in vacuum and the final conformations achieved were minimized in the referred periodic box boundary conditions.

It is possible to order the resulting conformations obtained and figure a move in which B is obtained from A in successive steps, which represent relative energy minima. Table 5.1 and Figure 5.4 describe different conformations obtained as local minima, all of them starting from A. Their respective values for the torsion angles are shown in Table 5.1. As said, these conformations result from MCMM and minimization within their

TABLE 5.1 Torsion Angles (°) in the Tail for Different Conformations Obtained with MCMM, Starting with Conformation A.

Conformation	Torsions allowed	a	b	c	d	d
A	–	−146.8	74.4	−170.9	80.7	77.0
1	12 ring, d	−148.4	59.1	176.8	144.9	66.3
2	12 ring, b, d	−68.2	179.2	−91.2	71.3	74.7
3	a, b	−67.9	−175.5	94.2	−67.3	105.3
4	d	−62.6	−167.8	−177.3	−176.9	−814
5	12 ring, a	−154.9	170.8	177.6	172.9	−87.1
B	–	69.4	−176.5	−177.2	−179.2	84.6

corresponding water periodic boxes. For clarity, water molecules have been removed from Figure 5.4. Table 5.2 shows the parameters involved in the periodic box boundary minimization. These five structures are computationally accessible examples of intermediates, representing local minima. There is no evidence that the route shown is the most probable, and there is no experimental confirmation for each structure. The simulation methods allow us to figure out these kinds of processes for which we have only the starting and final points.

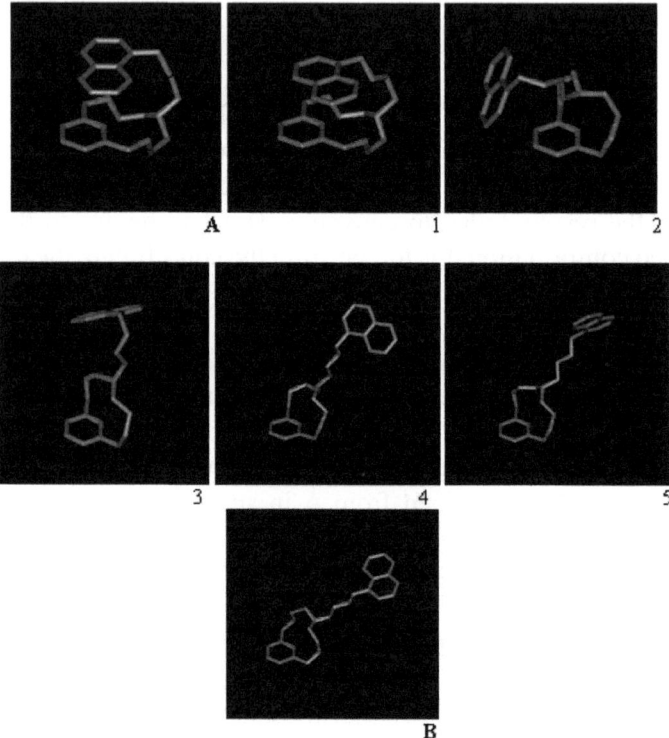

FIGURE 5.4 (A,B) Conformations from X-ray minimized in the periodic box. (1–5) Conformations obtained from MCMM in vacuum and further minimization in standard water density-constant TI3P box until their respective local minima.

The conformations for the 12 ring obtained with MM+ are distorted with respect to the X-ray diffraction model. Furthermore, it seems that ring torsion is not well achieved for macrocycles, with the algorithms currently implemented in MCMM for ring inversion. It can be due to deficiencies in the MM+ parameters, which were not optimized for macrocycles. In spite of these drawbacks, conformation 5, near B, is obtained and the sequence can be illustrative of the conformation rearrangement.

TABLE 5.2 Geometric Parameters Involved in the Water Box Calculations for Each Conformation Obtained from MCMM.

Conformation	Smallest box enclosing solute/Å			Cubic periodic box edge/Å	Maximum number of water molecules
	X	Y	Z		
A	6.90	6.19	7.73	18.10	216
1	7.65	5.87	7.80	18.10	216
2	8.08	6.91	9.07	18.10	216
3	7.76	5.71	10.75	21.51	329
4	7.49	5.42	15.48	30.96	980
5	7.30	5.17	19.91	31.81	1064
B	6.90	3.29	17.62	35.25	1447

Unfortunately, it was not possible to obtain a conformation near A starting from B with MCMM, probably due to the huge number of freedom degrees involved and the underestimation of H bond interaction in MM+ which implied calculation times beyond our possibilities. (The entropy of B must be the greatest.)

5.3 COMPARISON OF TOTAL ENERGIES

Preliminary calculations to compare the total energy differences between **A** and **B** were performed by the semiempirical methods CNDO, INDO, MINDO3, MNDO, AM1, RM1, PM3, TNDO, and PM7.[17]

For this, a similar procedure to the followed previously was performed. Thus, for the conformations closed A and open B, four protonation states were depicted. Then, semiempirical minimizations were run to obtain the total energy for each one of the eight resulting structures, to compare their respective stabilities.

All the methods at a level lower than AM1, and also TNDO, gave at least one final distorted conformation that could not be classified as closed or open, thus giving the idea that such methods were not applicable to the system and the total energies obtained with them were not reliable. It must be pointed out here that TNDO has been parameterized specifically to reproduce NMR chemical shifts.

The results of "total energy" with the reliable methods confirm (Table 5.3) that mono- and diprotonated species are more stable in the closed A conformation, while the triprotonated is more stable in the open B conformation. The method RM1 gave wrong relative stabilities for the nonprotonated and monoprotonated structures. The method PM3 performs better than PM7 in this case, since the difference of the energies for the monoprotonated species calculated by PM7 should not justify the observed predominance of A in solution at a less acidic pH. This can be due to the adjustment of the parameters in PM7, which has been optimized for bio molecules.

The values of "total energy" are not absolute values. They depend on the contributions that each method uses in its determination. Values obtained with different methods cannot be compared. Values obtained with one given method for n-protonated structure cannot be compared with those obtained for another protonation state, because they have different numbers of atoms. Only energies obtained with exactly the same method for isomers or "conformers" can be directly compared. Thus, the absolute value of the energy is not important. The key is the difference in such values between species with the same composition. An experimental value related to this "total energy" is the heat of formation. These have not been determined experimentally for scorpiands at present.

It is noteworthy that, for all the methods that gave acceptable results, which are shown in Table 5.3, the difference in energy between the conformations A and B for the diprotonated species is greater, in absolute value, than this difference for the other protonation states. This trend is independent from the method used. A priori, this seems logic that the difference A − B should approach zero from the negative values of the less protonated states, and being positive for the triprotonated species. Thus, the A − B difference for diprotonated species "should be" lower than for the monoprotonated ones. By contrast, for every method accounted in Table 5.3, the result obtained is the opposite.

When plotting the total energy versus the number of H^+ on the molecule, it is apparent a nonlinear relationship that justifies the result obtained. Figures 5.5 and 5.6 show these values obtained with AM1 and PM3, respectively.

Density functional theory (DFT) calculations[18,19] were tried by using several exchange functionals and basis sets.[9] The water environment was simulated by the COSMO method.[20,21] Relativistic contributions were not considered significant because no heavy nucleus was involved. It was considered the atomization energy, this is, the stabilization of the molecule relative to the free constitutive atoms. The determinant importance of hydrogen

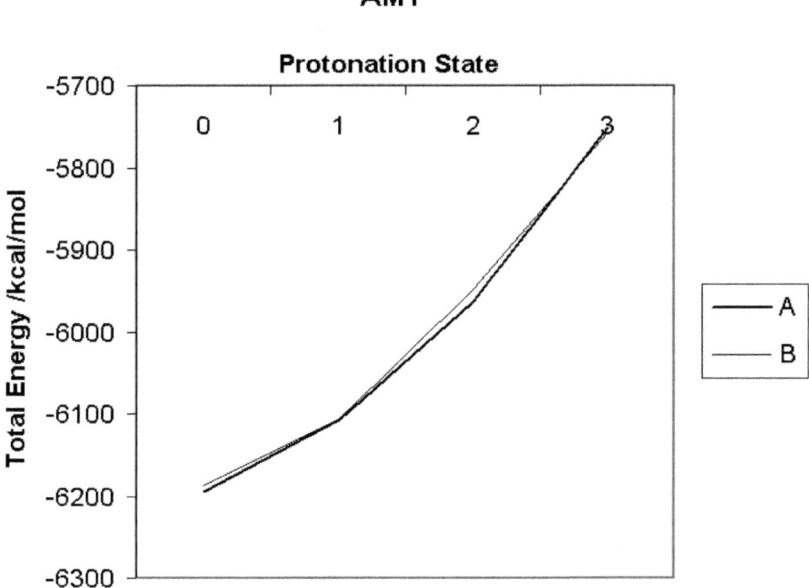

FIGURE 5.5 Total energy versus protonation state for energies obtained with the method AM1.

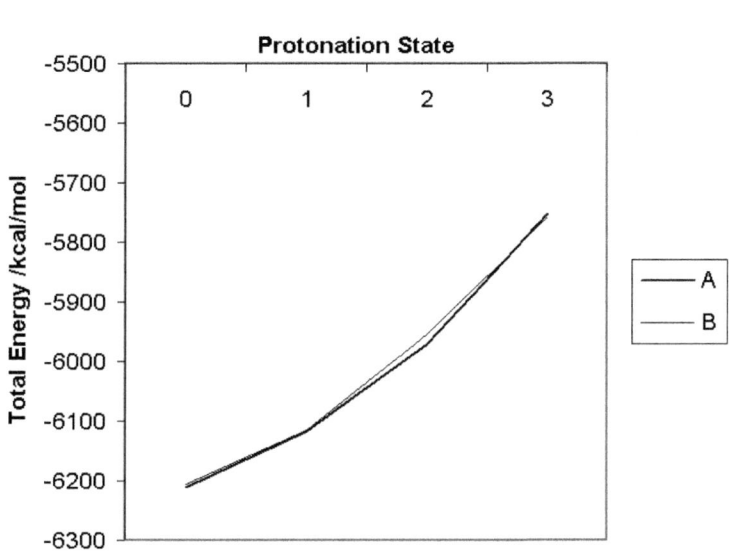

FIGURE 5.6 Total energy versus protonation state for energies obtained with the method PM3.

bonds in the conformational equilibrium made necessary using several polarization functions. For this reason, BP86/TZP and GGA-BP/TZ3P did not give good results, and the all-electron QZ4P basis was necessary. All the calculations performed assigned the correct energy order for the triprotonated species: The open conformation B was found more stable. However, this result was also found with the monoprotonated stage when polarization functions were not used. For computational details, see Julián-Ortiz et al.[9]

The local density approximation (LDA) can be useful to estimate properties in big molecules, and it was the only method that allowed good results within a reasonable calculation time (ca. 8 h with a Pentium 4 @ 3.2 GHz running on Windows XP, 1 GB RAM). Although the predicted values for atomization energies estimated are maybe not accurate, the method qualitatively predicted the experimental results. Table 5.3 shows the results obtained. The integration accuracy was four decimal places. Three decimal places were also tried to see the influence in the final result. It was seen that the dispersion of the result was lower than the differences between the values to be compared. The conclusions were, thus, unchanged.

TABLE 5.3 Calculated Energies for Each Protonation State and Conformation, Closed (A) and Open (B).

Starting conformation	Number of H⁺	Method	Environment simulation method	Total energy (kcal/mol)*	Difference A – B (kcal/mol)
A	0	AM1	Vacuum	**−6194.40**	−6.55
B	0	AM1	Vacuum	−6187.85	
A	1	AM1	Vacuum	**−6107.66**	−1.99
B	1	AM1	Vacuum	−6105.67	
A	2	AM1	Vacuum	**−5962.60**	−14.61
B	2	AM1	Vacuum	−5947.99	
A	3	AM1	Vacuum	−5750.75	7.10
B	3	AM1	Vacuum	−5757.85	
A	0	RM1	Vacuum	−6196.46	4.61
B	0	RM1	Vacuum	−6201.08	
A	1	RM1	Vacuum	−6122.17	0.39
B	1	RM1	Vacuum	−6122.56	
A	2	RM1	Vacuum	**−5993.12**	−18.80
B	2	RM1	Vacuum	−5974.32	
A	3	RM1	Vacuum	−5786.49	7.41
B	3	RM1	Vacuum	**−5793.90**	

TABLE 5.3   *(Continued)*

Starting conformation	Number of H+	Method	Environment simulation method	Total energy (kcal/mol)*	Difference A − B (kcal/mol)
A	0	PM3	Vacuum	**−6210.98**	−6.32
B	0	PM3	Vacuum	−6204.66	
A	1	PM3	Vacuum	**−6124.84**	−9.14
B	1	PM3	Vacuum	−6115.7	
A	2	PM3	Vacuum	**−5967.66**	−14.13
B	2	PM3	Vacuum	−5953.53	
A	3	PM3	Vacuum	−5752.39	−5.49
B	3	PM3	Vacuum	**−5757.88**	
A	1	PM7	Vacuum	**−99257.16**	−0.16
B	1	PM7	Vacuum	−99257	
A	2	PM7	Vacuum	**−99368.16**	−6.42
B	2	PM7	Vacuum	−99361.74	
A	3	PM7	Vacuum	−99411.69	12.63
B	3	PM7	Vacuum	**−99424.32**	
A	1	LDA/QZ4P	COSMO	**−9171.05**	−8.21
B	1	LDA/QZ4P	COSMO	−9162.84	
A	2	LDA/QZ4P	COSMO	**−9178.82**	−26.1
B	2	LDA/QZ4P	COSMO	−9152.72	
A	3	LDA/QZ4P	COSMO	−9133.36	1.04
B	3	LDA/QZ4P	COSMO	**−9134.4**	
A	3	GGA-BP/TZ2P	Vacuum	−8396.54	11.47
B	3	GGA-BP/TZ2P	Vacuum	**−8408.01**	

*The minimum energy for each pair A − B is indicated in bold.

The difference of energy for the two conformations of the triprotonated species obtained with the method LDA/QZ4P was not high enough to justify the predominance of B in solution at a more acid pH.

In order to have a confirmation of the trends obtained with a more complex functional, GGA-BP/TZ2P was used for a simulation without solvent.[19]

All calculations were performed to compare the atomization energies of the triprotonated species in its crystal conformation, B open, and in A closed. The integration accuracy was fixed in three decimal places. The calculation time for each process was approximately 8 days using the same computer as before. These gave a more reliable stability prediction at the greatest protonation state. The results are displayed in Table 5.3.

The optimized conformations obtained were compared with the crystal ones. Using the Carbó index measured through Coulomb integrals and the local Newton–Raphson as a superposition algorithm, the similarity was quantified.[22] Thus, monoprotonated A and its minimized conformation obtained by LDA/QZ4P showed a Carbó index equal to 0.932, and triprotonated B and its respective optimization gave 0.964.

5.4 CONCLUSIONS

The MCMM/MM+ method was useful in the reviewed study, but some drawbacks must be pointed out. Thus, ring torsions in macrocycles do not seem well simulated. The pH-dependent opening of the scorpiand was easy to predict, but the reverse effect was not possible to simulate.

Among the semiempirical methods, AM1 and its reparametrization, PM3, are the best to give account of the experimental behavior. Less sophisticated methods were inapplicable to the studied systems.

DFT was a good method to give account of the respective experimental stabilities, although the all-electron basis was necessary. The LDA with the COSMO method was enough to obtain the relative stabilities clearly according to the experiments at lower protonation. At more acidic pH, the more complex functional GGA-BP with a somewhat simple basis gave more reliable results. The optimized conformations obtained by DFT were in accordance with the crystal ones, as verified by quantum molecular similarity. However, these methods are computationally expensive, and the results can be qualitatively modeled by using advanced semiempirical Hamiltonians.

Due to the nature of the studied molecules, the balance among intramolecular forces that determines a given conformation is due to several counteracting interactions. These make it so that pH changes induce conformation rearrangements in these species. The key aspect is that the relatively weak intramolecular forces that induce a given folded conformation, such as H—bonds and π–π stacking, are balanced and eventually overcome by electrostatic forces that can induce a conformation change that minimizes the intramolecular charge repulsion by extending out the structure. This process can be consistently modeled by the computational methods used here. Since the intramolecular forces involved are common to other kinds of molecular systems, the methods disclosed here could be applied to other structures, particularly to molecular systems in which electrostatic forces are balanced with staking and H—bonding.

This study points out that molecular modeling at minimum AM1 level could be used to design new molecules able to reorganize their conformations due to pH variations. This opens interesting possibilities in the design of molecular systems with controllable motion.

ACKNOWLEDGMENT

Financial support by the Spanish Ministerio de Economía y Competitividad (Projects CONSOLIDER INGENIO CSD-2010-00065 and CTQ2013-48917-C3-1-P), Generalitat Valenciana (Project PROMETEOII2015-002) is gratefully acknowledged. Emili Besalú acknowledges the Generalitat de Catalunya (Departament d'Innovació, Universitats i Empresa) for the financial support given to the QTMEM (Química Teòrica i Modelatge i Enginyeria Molecular) research group of the University of Girona (code 2014-SGR-1202).

KEYWORDS

- **pH controlled**
- **supramolecular chemistry**
- **synthetic receptors**
- **aza-scorpiands**
- **molecular mechanics**
- **semiempirical**
- **density functional**
- **Monte Carlo multiple minimum**
- **COSMO**

REFERENCES

1. Kinbara, K.; Aida, T. Toward Intelligent Molecular Machines: Directed Motions of Biological and Artificial Molecules and Assemblies. *Chem. Rev.* **2005,** *105,* 1377–1400.

2. (a) Pallavicini, P. S.; Perotti, A.; Poggi, A.; Seghi, B.; Fabbrizzi, L. *N*-(aminoethyl) Cyclam: A Tetraaza Macrocycle with a Coordinating Tail (Scorpiand). Acidity Controlled Coordination of the Side Chain to Nickel(II) and Nickel(III) Cations. *J. Am. Chem. Soc.* **1987,** *109,* 5139–5144. (b) Bencini, A.; Berni, E.; Bianchi, A.; Fornasari, P.; Giorgi, C.; Lima, J. C.; Pina, F. A Fluorescent Chemosensor for Zn(II). Exciplex Formation in Solution and the Solid State. *Dalton Trans.* **2004,** *14,* 2180–2187.

3. Blasco, S.; Verdejo, B.; Bazzicalupi, C.; Bianchi, A.; Giorgi, C.; Soriano, C.; García-España, E. A Thermodynamic Insight into the Recognition of Hydrophilic and Hydrophobic Amino Acids in Pure Water by Aza-scorpiand Type Receptors. *Org. Biomol. Chem.* **2015,** *13,* 843–850.

4. Olmo, F.; Clares, M. P.; Marín, C.; González, J.; Inclán, M.; Soriano, C.; Sánchez-Moreno, M. Synthetic Single and Double Aza-scorpiand Macrocycles Acting as Inhibitors of the Antioxidant Enzymes Iron Superoxide Dismutase and Trypanothione Reductase in *Trypanosoma Cruzi* with Promising Results in a Murine Model. *RSC Adv.* **2014,** *4,* 65108–65120.

5. Bernier, N.; Esteves, C. V.; Delgado, R. Heteroditopic Receptor Based on Crown Ether and Cyclen Units for the Recognition of Zwitterionic Amino Acids. *Tetrahedron.* **2012,** *68,* 4860–4868.

6. Aydın, I.; Aral, T.; Karakaplan, M.; Hoşgören, H. Chiral Lariat Ethers as Membrane Carriers for Chiral Amino Acids and Their Sodium and Potassium Salts. *Tetrahedron Asymmetry.* **2009,** *20,* 179–183.

7. Verdejo, B.; Acosta-Rueda, L.; Clares, M. P.; Aguinaco, A.; Basallote, M. G.; Soriano, C.; García-España, E. Equilibrium, Kinetic, and Computational Studies on the Formation of Cu^{2+} and Zn^{2+} Complexes with an Indazole-containing Azamacrocyclic Scorpiand: Evidence for Metal-induced Tautomerism. *Inorg. Chem.* **2015,** *54,* 1983–1991.

8. Verdejo, B.; Ferrer, A.; Blasco, S.; Castillo, C. E.; González, J.; Latorre, J.; García-España, E. Hydrogen and Copper Ion-induced Molecular Reorganizations in Scorpionand-like Ligands. A Potentiometric, Mechanistic, and Solid-state Study. *Inorg. Chem.* **2007,** *46,* 5707–5719.

9. Julián-Ortiz, J.; Verdejo, B.; Polo, V.; Besalú, E.; García-España, E. Molecular Rearrangement of an Aza-Scorpiand Macrocycle Induced by pH. A Computational Study. *Int. J. Mol. Sci.* **2016,** *17,* 1131.

10. Valencia, L.; Bastida, R.; García-España, E.; de Julián-Ortiz, J. V.; Llinares, J. M.; Macías, A.; Pérez Lourido, P. Nitrate Encapsulation Within the Cavity of Polyazapyridinophane. Considerations on Nitrate–Pyridine Interactions. *Cryst. Growth Des.* **2010,** *10,* 3418–3423.

11. Chang, G.; Guida, W. C.; Still, W. C. An Internal-coordinate Monte Carlo Method for Searching Conformational Space. *J. Am. Chem. Soc.* **1989,** *111,* 4379–4386.

12. Saunders, M.; Houk, K. N.; Wu, Y. D.; Still, W. C.; Lipton, M.; Chang, G.; Guida, W. C. Conformations of Cycloheptadecane. A Comparison of Methods for Conformational Searching. *J. Am. Chem. Soc.* **1990,** *112,* 1419–1427.

13. Kolossváry, I.; Guida, W. C. Torsional Flexing: Conformational Searching of Cyclic Molecules in Biased Internal Coordinate Space. *J. Comput. Chem.* **1993,** *14,* 691–698.

14. Allinger, N. L. Conformational Analysis. 130. MM2. A Hydrocarbon Force Field Utilizing V1 and V2. *J. Am. Chem. Soc.* **1977,** *99,* 8127–8134.

15. Bushuev, Y. G.; Sastre, G.; de Julian-Ortiz, J. V. The Structural Directing Role of Water and Hydroxyl Groups in the Synthesis of Beta Zeolite Polymorphs. *J. Phys. Chem. C.* **2009,** *114,* 345–356.

16. Bushuev, Y. G.; Sastre, G.; de Julian-Ortiz, J. V.; Gálvez, J. Water–Hydrophobic Zeolite Systems. *J. Phys. Chem. C.* **2012,** *116,* 24916–24929.

17. Thiel, W. Semiempirical Quantum–Chemical Methods. *Wiley Interdiscip. Rev.: Comput. Molec. Sci.* **2014,** *4,* 145–157.

18. Versluis, L.; Ziegler, T. The Determination of Molecular Structures by Density Functional Theory. The Evaluation of Analytical Energy Gradients by Numerical Integration. *J. Chem. Phys.* **1988,** *88,* 322–328.

19. Perdew, J. P.; Burke, K. Comparison Shopping for a Gradient-corrected Density Functional. *Int. J. Quantum Chem.* **1996,** *57,* 309–319.

20. Klamt, A.; Schüürmann, G. COSMO: A New Approach to Dielectric Screening in Solvents with Explicit Expressions for the Screening Energy and Its Gradient. *J. Chem. Soc. Perkin Trans.* **1993,** *2,* 799–805.

21. Pye, C. C.; Ziegler, T. An Implementation of the Conductor-like Screening Model of Solvation Within the Amsterdam Density Functional Package. *Theor. Chem. Acc.* **1999,** *101,* 396–408.

22. Bultinck, P.; Gironés, X.; Carbo-Dorca, R. Molecular Quantum Similarity: Theory and Applications. *Rev. Comput. Chem.* **2005,** *21,* 127.

CHAPTER 6

GLOBAL WATER CRISIS, GROUNDWATER REMEDIATION, AND FUTURISTIC VISION OF ENVIRONMENTAL ENGINEERING TECHNIQUES: A FAR-REACHING REVIEW

SUKANCHAN PALIT[1,2,*]

[1]*Department of Chemical Engineering, University of Petroleum and Energy Studies, Energy Acres, Post Office Bidholi via Premnagar, Dehradun 248007, India*

[2]*43, Judges Bagan, Post Office Haridevpur, Kolkata 700082, India*

Corresponding author. E-mail: sukanchan68@gmail.com, sukanchan92@gmail.com

CONTENTS

ABSTRACT

Environmental engineering science is today witnessing one paradigmatic shift over another. Science, technology, and engineering science are taking giant steps in this century. Environmental engineering needs to be re-envisioned and reshaped to its utmost as human endeavor and scientific research pursuit enters into a new era of wide scientific regeneration.

The world of membrane science and chemical process engineering are in a similar manner undergoing drastic changes and surpassing visionary frontiers. Man's vision, civilization's prowess, and the march of science are emboldened with each step of human scientific endeavor. Membrane separation processes is moving toward a newer visionary era. Environmental sustainability is the need of the hour. The author with cogent insight and deep comprehension brings to the scientific forefront the scientific doctrine of membrane science particularly ultrafiltration (UF) and microfiltration (MF), its vast and versatile challenges, and the recent advances in scientific endeavor in advanced oxidation processes (AOPs). Human scientific endeavor in today's world is in the path of new scientific regeneration. The progress of science and technology and the world of scientific validation in the field of membrane science and novel separation processes need to be re-envisioned at each step of research pursuit. In this treatise, the author delves deeply into the intricate details in the vision and aim of UF and MF, its immense applications, the barriers of fouling, the vision of concentration polarization, and the world of process design in the area of membrane science. Diverse branches of science and engineering need to be re-envisioned with each step of new and visionary scientific awakening. Environmental engineering applications and chemical process engineering pursuits are covered in this chapter. Scientific vision and man's versatile prowess are emboldened with each step of endeavor in novel separation processes such as membrane science. This detailed treatise opens up new windows of environmental engineering science in particular, in the years to come. The author pinpoints the wide world of wastewater treatment and the applications of membrane science with a wide review on groundwater remediation. Vision of science and engineering are rapidly changing. Global water shortage and membrane science in today's world have an umbilical cord. The author with deep comprehension and cogent insight brings to the scientific forefront the technological vision of membrane science, especially UF and MF. The treatise also provides wide glimpses of groundwater remediation and the future of global water initiatives. Water and wastewater treatment is witnessing drastic challenges and immense barriers. Membrane science is the only answer. This treatise also brings to the reader the wide world of AOPs and the recent endeavor behind it.

6.1 INTRODUCTION

The vast and versatile world of membrane science and novel separation processes is undergoing immense metamorphosis, in spite of innumerable hindrances. Environmental engineering science is the backbone of application of membrane separation phenomenon. Progress of science and advancement of technology is gearing for newer scientific horizon. Technology and science are surpassing visionary boundaries. The important need of the hour of humankind is the true realization of environmental sustainability and preservation of ecology. UF and MF are membrane separation processes. The author delves deeply with immense comprehension into the doctrine of UF and MF with a broad and visionary scientific perspective. The author pointedly focuses on the future of global water initiatives and groundwater heavy metal remediation. Innovations and remediation technologies are the other wide facets of this well observed study. Man's immense vision and a scientist's wide prowess and comprehension needs to be reshaped with respect to the realization of environmental sustainability. The effectivity of a membrane separation process such as UF and MF needs to be readdressed at each step of the vision of science. The fruits of membrane science and technology are opening up new windows of scientific innovation and challenges. The broad and versatile vision of application of UF and MF are discussed with minute details. Scientific cognizance, scientific rigor, and the progress of technology are the torchbearers toward a greater visionary future in the field of membrane separation phenomenon.[1] AOPs are changing the face of scientific research pursuit. Thus, the author also reviews AOPs. AOPs are the pillars and visionary supports of today's integrated environmental engineering techniques. The paradigm of science and engineering is slowly changing. This treatise widely observes the immense potential, the wide success, and the vast and versatile scientific rigor in the field of nontraditional environmental engineering techniques.

6.2 VISION OF THE PRESENT TREATISE

Technology and engineering needs to be readdressed and re-envisioned in today's world at every step of scientific research pursuit. The vision of this treatise is to bring forward to the scientific forefront the challenges, difficulties, and barriers of application of UF and MF and its futuristic vision. The author dives deep into the concept of fouling and cleaning of the membranes and the domain of process design. Chemical process modeling and application of chemical process engineering in modeling are at the helm of scientific vision in today's world. Application of UF and MF and the recent advances in the world of membrane separation phenomenon lies as a major backbone to this treatise.[1] This treatise fundamentally addresses the wide world of global water initiatives, the vast challenges behind it, and the success of groundwater remediation tools. Membrane science, nontraditional environmental engineering techniques such as AOPs and the new technologies of groundwater heavy metal remediation stands as a major backbone of this study.

Science and engineering of membrane separation processes are breaking vast and versatile scientific frontiers. Novel separation phenomenon and its scientific sagacity are the hallmarks of a new generation of scientific understanding and vision. The author pinpoints with cogent insight and deep introspection the present and future scientific understanding of membrane science particularly MF and UF in details and with precision. Human scientific research pursuit, scientific validation, and scientific innovation are leading human civilization to newer heights and newer realms. In such a critical juxtaposition, vision of science and technology needs to be targeted toward realization of basic human needs such as water. In such a similar vein, global water crisis and its alleviation need to be re-envisioned. Membrane separation phenomenon and novel separation processes are the only definite answer.[1-4]

The author in this study pointedly focuses on the immense scientific potential and wide scientific vision in the application domain of membrane science and other novel separation processes. The challenge and vision of scientific research pursuit is unimaginable and paving the path toward zero-discharge norms in the field of environmental engineering.

6.3 SCOPE OF THE STUDY

Science and engineering in today's world are pillars and colossus without a definite visionary will of its own. Environmental engineering science and membrane science are linked in today's scientific horizon by an unsevered umbilical cord. Ecological balance and application of tools of technologies such as membrane are showing us the way toward a newer visionary area.

The scope of this study is to evolve new technologies and membrane science paradigm. The author treads a weary and definite path toward the true realization of membrane separation processes.[1] The author also delves into the depths of heavy metal remediation technologies and the success of many arsenic remediation technologies in many parts of the world.

An area which needs to be veritably explored is the burning issue of arsenic groundwater remediation in developed and developing countries of the world. The raging issue of global water shortage due to various environmental catastrophes, such as arsenic groundwater contamination, needs to be readdressed and re-envisioned. Vision of science and progress of technology are the pallbearers of a greater visionary human society of tomorrow. Scientific validation, scientific vision, and scientific fortitude need to be addressed at every step of endeavor in membrane science and environmental engineering science. Ecological misbalance is degrading the environmental engineering scenario. A deep introspection and deep scientific understanding is needed in the implementation of environmental science policies. The author with definite understanding and vision elucidates upon the subject of application of membrane science such as UF and MF in details. Environmental engineering science is moving toward a new era and a new realm. The grave concern of environmental degradation is the serious concern of the day. The vision of humankind is veritably leaving no stone unturned as regards to ecological balance and environmental pollution control.[1,4–7]

6.4 THE NEED AND THE RATIONALE OF THE STUDY

Human civilization is moving toward one visionary frontier over another. Environmental engineering paradigm in a similar vein is moving toward a newer visionary era. The concern for ecological imbalance and the imminent target toward environmental sustainability are gearing human mankind toward novel environmental engineering separation processes and newer innovations. The other grave concern is global water shortage and groundwater contamination. Here rises the immediate need of novel environmental engineering tools such as membrane science. This study widely observes the pros and cons of the application scenario of membrane science, particularly MF and UF. The challenge and the success of novel separation phenomenon are far-reaching and in today's world targeted toward global water crisis. The need and the rationale of this study surpass wide and versatile frontiers. The importance, the vision and the efficacy of membrane separation processes are changing the true face of global water shortage and the future of environmental engineering science. The author skillfully delineates the futuristic vision of membrane technology, particularly MF and UF.[1–6,28]

6.5   ENVIRONMENTAL SUSTAINABILITY, ENVIRONMENTAL ENGINEERING SCIENCE AND THE WIDE VISION FOR THE FUTURE

Environmental sustainability is moving from one visionary frontier over another. Effective environmental pollution control stands at the helm of effective environmental sustainability. Countries, whether developing as well as developed, are moving toward new direction of true realization of environmental sustainability. Energy sustainability is changing the face of human civilization. Sustainable development, the wide progress of human civilization and the futuristic vision of environmental engineering are the torchbearers of a holistic sustainable development of tomorrow, especially with respect to environment. Man's improved vision in today's scientific scenario, the human civilization's prowess and the futuristic progress of human mankind will all lead a long way in the true emancipation of the application of novel separation processes, especially membrane science. In the present discussion, the author widely informs the scientific audience the scientific doctrine of membrane science especially UF and MF and the true realization of successful environmental sustainability.[1–4]

The vision for the future in the true application of membrane science in chemical process engineering and environmental engineering science is far-reaching in today's path of human civilization and improved scientific rigor. Successful sustainable development is the coin word of the future. Environmental sustainability stands as a major component in the research and development in the field of membrane science. In such a critical juncture, vision of science and technology is of immediate and utmost importance. Environmental sustainability is the coin word of the newer scientific generation and the vision for the future. Science in today's world needs to be re-envisioned. Membrane separation processes and applications in the domain of UF and MF will surely and veritably realize the concept of environmental sustainability.[1,28]

Scientific vision is of immense importance in today's world of environmental sustainability and environmental engineering science. Human mankind's prowess, man's future progress, and the future challenges are the pallbearers toward a greater visionary tomorrow in the scientific world of alleviation of global water crisis.[1–4,28]

6.5.1   *GLOBAL WATER CRISIS AND GROUNDWATER REMEDIATION*

Water science and technology today are witnessing drastic challenges. Scientific vision and scientific truth are the pillars of research and development initiatives in water technology. Groundwater heavy metal contamination is a major environmental disaster today. Remediation and decontamination of polluted drinking water are the utmost need of the hour. Technological objectives need to be re-envisioned as scientific research pursuit enters into a newer far-reaching domain. Arsenic groundwater contamination is a major disaster in many developing countries. The author in this treatise delves deep into the murky depths of groundwater remediation with a view toward solving drinking water crisis. Civilization's immense vision, man's prowess, and the wide scientific rigor will go a long way in true realization of sustainable development and of environmental engineering science.

6.6   MEMBRANE SCIENCE: ITS TECHNOLOGICAL DOCTRINE AND WIDE SCIENTIFIC VISION

Membrane science and its doctrine and scientific vision are in today's human civilization, in the path of immense reshaping and re-envisioning. Chemical process engineering and bioengineering are emerging as major pallbearers of the future scientific society. The scientific doctrine, the scientific sagacity and the deep scientific understanding of membrane science still needs to be uncovered. Progress of membrane science today stands in the midst of immense scientific optimism and deep hope. Technological vision of environmental engineering science in the present decade is in the path of definite glory and scientific fortitude. Global water crisis, water and wastewater treatment and the progress of human civilization will all lead a long way in the true realization of the application of environmental engineering tools.[1-4]

Membrane science, global water shortage and the vision of environmental sustainability are surpassing vast and versatile scientific and engineering frontiers. Human scientific endurance and scientific rigor needs to be reshaped with each step of human life. Environmental catastrophes and recurrent disasters are challenging the face of environmental engineering tools such as membrane science. Membrane separation phenomenon today stands in the midst of immense scientific challenges and deep scientific optimism (Table 6.1). The question of environmental sustainability needs to be answered at each step of human life.[1-4]

TABLE 6.1   Characteristics of Membrane Processes.

Process	Driving force	Retentate	Permeate
Osmosis	Chemical potential	Solutes/water	Water
Dialysis	Concentration difference	Large molecules/water	Small molecules/water
Microfiltration	Pressure	Suspended particles/water	Dissolved solutes/water
Ultrafiltration	Pressure	Large molecules/water	Small molecules/water
Nanofiltration	Pressure	Small molecules/divalent salts/dissociated acids/water	Monovalent ions/undissociated acids/water
Reverse osmosis	Pressure	All solutes/water	Water
Electrodialysis	Voltage/current	Nonionic solutes/water	Ionized solutes/water
Pervaporation	Pressure	Nonvolatile molecules/water	Volatile small molecules/water

6.7   MEMBRANE SCIENCE: A VISIONARY SOLUTION FOR MULTITUDE OF GLOBAL WATER ISSUES

Clean water, clean energy, global warming, and affordable global health-care are the four major global concerns resulting from clean-water short-ages, widely fluctuating oil prices, rapid climate change, and high cost of healthcare. Chemical process engineering and bioengineering are changing the scientific horizon of membrane science applications. Current and future trends in membrane science aim to tackle these four major issues stated earlier by means of innovative and groundbreaking membrane technologies for the separation and purification of highly valuable products or the removal of unwanted species. In today's world, the visionary future of membrane science and technology stands in major crossroads of human scientific history and time. Today, the whole world is concerned with climate change and global warming. The successes of realization of energy and environmental sustain-ability today are linked by an umbilical cord with environmental engineering tools such as membrane science. In such a crucial situation, scientific vision and scientific forbearance are of utmost importance.[1]

Global water shortages are the vexing issues in the path of human civili-zation today. Scientific vision, scientific rigor and the wide path of scientific progress are the veritable torchbearers toward a greater realization of envi-ronmental engineering today. Innovations, scientific intuitions, and the wide world of scientific forbearance will definitely lead a long way in the true emancipation of global research and development initiatives. Water science and technology are ushering in a new pragmatic paradigm as science and engineering treads a difficult path in its tryst with scientific destiny. The vast and versatile world of scientific profundity is reshaping the domain of water science and technology as humankind moves toward a newer eon.

6.7.1 MEMBRANE CHEMISTRY, STRUCTURE AND FUNCTION, AND THE WIDE VISIONARY APPLICATIONS

Membrane chemistry and its structure stand as a major backbone to the futuristic vision of membrane separation phenomenon. Vision of science and scientific validation are the major components of membrane function and effective process design. Chemical process engineering and environmental engineering science in today's human scientific endeavor are in the path of vision, glory, and emancipation. The structure and function of membranes needs to be re-envisioned and restructured with each step of research pursuit. Membrane chemistry and function are the bedrock of the immense application domain of UF and MF. Chemical process technology and its world of vision are entering a new phase with respect to the application of membrane science, especially UF and MF. The major thrust area in scientific rigor in this field is to obliterate membrane fouling. Scientific research pursuit, the question of scientific validation, and the wide world of scientific vision will all lead a long way in effective emancipation of the application of membrane separation processes.

Filters are manufactured from a variety of materials using several methods, but they can all be classified into two general categories: depth filters or screen filters. Depth filters derive their name from the fact that filtration or particle removal occurs within the depths of the filter material. A screen filter, in contrast, separates by retaining particles on its surface, in much the same manner as a sieve. The structure is usually more rigid, uniform, and continuous, with pore size more accurately controlled during manufacture. Science of membrane chemistry, structure and function, is attaining new heights with visionary innovations and path-breaking discoveries. The vision of science stands in the midst of intense introspection and groundbreaking future.[1–6]

6.7.2 MICROPOROUS VERSUS ASYMMETRIC MEMBRANES

Screen filters can be further classified according to their ultrastructure as either microporous or asymmetric (the latter are also referred to as skinned membranes). Microporous membranes are sometimes further classified as isotropic (with pores of uniform size throughout the body of the membrane) or anisotropic (where the pores change in size from one surface of the membrane to the other). Frequently, the terms anisotropy and asymmetry are used interchangeably.[1]

6.7.3 GENERAL AND STANDARD METHODS OF MEMBRANE MANUFACTURE

There are several methods for manufacturing membranes. Some of the methods are applicable to a variety of polymers and others are material-specific, for example, the heating-stretching method used to make pores in microporous polytetrafluoroethylene (PTFE) membranes. Each of these methods results in different ultrastructures, porosity, and pore size distribution. For example, track-etched membranes have a narrow pore size distribution but a low porosity (the number of pores per unit surface area). On the other hand, the phase-inversion process is a good way to form the asymmetric skin structure on a membrane and can result in fairly high porosity in certain cases.[1,28]

6.7.4 MEMBRANE PROPERTIES

Membrane properties need to be readdressed with serious and immense challenges. Vision of science and progress of engineering are the hallmarks of human scientific endeavor today. Membrane properties are the genre of scientific achievements and scientific vision of membrane science today. The fundamental question of membrane science lies in its intricate properties that are still yet not explored. Human scientific research pursuit is still latent in the first decades of 21st century. Chemical process engineering is in the path of new glory with the evolution of new techniques such as membrane science.[1,28]

Science, technology, and engineering are moving toward a newer realm and a newer visionary direction. In the similar vein, membrane science and technology is ushering in a new era of scientific vision and scientific cognizance. Global water crisis is the cornerstone of membrane science research in today's visionary path of human civilization. Environmental sustainability needs to be re-envisioned at this need of the hour.[1]

Of special importance for UF and MF membranes are the pore statistics, for example, pore "size" (usually expressed as pore diameter), pore density (number of pores per unit membrane surface area), and bulk porosity (or void volume) which is the fraction of the membrane volume occupied by the pores.[1]

The most common methods of determining pore size are (1) bubble point; (2) direct observation, for example, with electron microscopes; (3) function measurement or the challenge test. It should be recognized of course that polymeric membranes used for UF and MF are essentially screen or surface filters (as opposed to depth filters) and that very rarely do we observe pores of just one diameter but rather a distribution of pore sizes.[1]

6.7.5   MEMBRANE FOULING, BIOFOULING, AND CLEANING

Technological vision of membrane separation phenomenon has immense barriers and impediments such as membrane fouling. This is a major limiting step. Fouling manifests itself as a decline in flux with time of operation. In its strictness sense, the flux decline should occur when all operating parameters, such as pressure, flow rate, temperature, and feed concentration, are kept constant. Membrane fouling led to slow acceptance of membrane technology in its early days. The challenge and then success of separation phenomenon and research and development initiatives have catapulted the whole domain of membrane science toward a newer height.[1]

6.7.5.1   CHARACTERISTICS OF FOULING

Flux of a real world feed stream is usually much lower than the flux of the pure solvent (e.g., water) for several feasible reasons:

- Changes in membrane properties: These can occur as a result of physical or chemical deterioration of the membrane. Since membrane processing is a pressure-independent process, it is possible that under high pressures, the membrane can undergo a "creep" or "compaction" phenomenon, which may change the permeability of the membrane. This generally occurs when pressures are in the hundreds of pounds per square inch and is usually not of concern in MF and UF where pressures are typically 15–100 psi (1–7 bars).[1]
- Change in feed properties: Solvent transport through porous MF and UF membranes is usually considered a viscous flow phenomenon governed by the Hagen–Poiseuille equation or supported by mass transfer relationships.[1]

- Concentration polarization: Flux-depressing effects due to membrane fouling are frequently confused with flux-lowering phenomena associated with concentration polarization. In theory, concentration polarization effects should be reversible by decreasing the transmembrane pressure, lowering the feed concentration, or increasing the cross-flow velocity.[1]

6.7.6 PERFORMANCE AND ENGINEERING MODELS

Performance of a membrane and engineering models stands as a major backbone and a hallmark in the success of a separation process phenomenon. The challenge, the vision and the prowess will lead a long way in the true realization of environmental sustainability and novel separation processes. The immense challenges will veritably mitigate the difficulties and barriers to membrane separation phenomenon. Engineering models and performance models are the focal point in the design of a membrane separation process and its subsequent applications. Scientific rigor, the progress of chemical process engineering, and the wide vision of environmental engineering science are all the hallmarks of greater realization of membrane separation phenomenon.[1]

Several mathematical models are available in the literature that attempt to describe the mechanism of transport through the membranes. Although the operating techniques of MF, UF, nanofiltration (NF), and reverse osmosis (RO) are similar, the latter two are almost certainly not separation merely by size alone. MF and UF, on the other hand, due to their relatively large pores have most frequently been visualized as sieve filtration. Science, technology, and engineering are moving at a rapid pace in today's world of chemical process engineering and environmental engineering science. Engineering modeling and simulation of membrane separation processes stands as a major issue in the success of separation and purification. The scientific challenge needs to be redrawn and re-envisioned. Global water crisis, the deep scientific understanding of membrane science and the barriers of fouling will all lead a long and visionary way in the true emancipation of chemical engineering science.[1]

6.7.7 PROCESS DESIGN AND THE VAST AND VERSATILE WORLD OF CHEMICAL PROCESS ENGINEERING

Process design and its immense capability are the forerunners to the novel separation phenomenon. Novel separation processes includes membrane science. Human scientific research pursuit, the immense vision of membrane science and the future path of excellence will all lead a long, arduous visionary in the true emancipation of membrane separation phenomenon. Chemical process engineering today stands in the midst of tremendous introspection and technological vision. Science and technology of chemical process design is at the helm of humankind's progress. Process design and the future of its scientific vision needs to be addressed at the utmost. Global water crisis and desalination will witness a new scientific revolution if proper process design is addressed. Process design and application of membrane science in today's scientific arena are linked by an unsevered umbilical cord. The author with deep introspection pointedly deals with the contribution of process design in the vast and versatile applications of membrane science and technology. The ingenuity of this treatise surpasses visionary frontiers and leads the chemical process engineers toward a newer realm and an innovative scientific horizon.[1]

6.8   RECENT SIGNIFICANT SCIENTIFIC ENDEAVOR IN THE FIELD OF MEMBRANE SCIENCE

Science and engineering of membrane science are moving at a rapid pace. Human scientific endeavor, the challenges of global water initiatives, and the unmitigated barriers of science will lead a long way in the true emancipation of chemical process engineering and environmental engineering science.

Wenten[43] delineated with deep and cogent insight recent development in membrane science and its industrial applications. The word "membrane" comes from the Latin word "membrane" that means a skin. Today's word "membrane" has been extended to describe a thin flexible sheet or film, acting as a selective boundary between two phases because of its semi permeable properties. Synthetic membrane history began in 1748 when French Abble Nollet demonstrated semipermeability for the first time, that animal bladder was more semipermeable to water than to wine. One century later, Fick published his phenomenological law of diffusion, which we still use today as a first-order description of diffusion through membranes. Sartorius Werke GmbH manufactured industrial scale membranes, micro-filtration membranes, for the first time in 1950. The principal advantage of membrane processes compared to other separation processes are low energy consumption, simplicity, and environmental friendliness. Flux and selectivity problems arise as an increase in flux and are usually followed by a decrease in selectivity. Therefore, membrane processes are suitable for selective separation in which flux is not concerned such as that carried out in pharmaceutical industries. Other major problems are material sensitivity and fouling. Emerging processes are development of improved membranes and membrane materials, high-performance RO membranes, stabilization of supported liquid membranes, preparation of composite hollow fiber membranes, membrane reactors/contactors, and development of novel membrane processes.

Zhao et al.[44] reviewed recent developments in forward osmosis (FO) with its opportunities and challenges. FO is a recent scientific endeavor of membrane science. So it needs immense attention. Membrane science is slowly moving toward a newer scientific regeneration and revamping. Recently, forward FO has attracted growing attention in many potential applications such as power generation, desalination, wastewater treatment, and food processing. However, there are many viable challenges such as concentration polarization, membrane fouling, reverse solute diffusion, and the need for new membrane development. These critical and vital challenges are discussed in versatile details in this treatise.

Bae et al.[45] delineated with deep precision the concept of a high-performance gas-separation membrane containing submicrometer-sized metal-organic framework (MOF) crystals. MOFs are an emerging class of nanoporous materials consisting of metal centers connected by various organic linkers to create one-, two- and three-dimensional porous structures with tunable pore volumes, surface areas, and chemical properties.

Nagarale et al.[46] discussed recent developments on ion-exchange membranes and electromembrane processes. Technological vision of membrane science is moving toward a newer frontier with each step of this innovative endeavor. Ion-exchange membrane technologies are nonhazardous in nature and are being widely used for not only separation and purification but their application also extended to energy conversion devices, storage batteries and sensors, etc. In this well-informed and well-observed treatise, the authors have reviewed the preparation of various types of ion-exchange membranes, their characterization, and applications for different electromembrane processes.

Baker[47] discussed in details future directions of membrane gas separation technology. More than 90% of gas separation equipment business involves the separation of noncondensable gases: nitrogen from air, carbon dioxide from methane, and hydrogen from nitrogen, argon, and methane. However, a much larger potential market involves separation of condensable gases. The authors with cogent and deep insight delineate the veritable success of separation phenomenon of condensable gases.

6.9   THE VISION OF INDUSTRIAL WASTEWATER AND DRINKING WATER TREATMENT TODAY AND THE FUTURE TRENDS OF RESEARCH AND DEVELOPMENT INITIATIVES

The vision and purpose of science and technology are ever-growing. Industrial wastewater treatment and application of membrane science in today's scientific horizon have an unsevered umbilical cord. Ecological misbalance, the growing concerns for environmental process safety and the challenges of research and development initiatives will all lead a long way in the true emancipation of science and engineering. Desalination and global water initiatives are paving the way of science and engineering endeavor. The first historic discovery of Loeb–Sourirajan asymmetric RO membrane enabled seawater desalination on an industrial scale. The challenge of human civilization and the interlinked scientific endeavor are opening up new vistas in membrane science and industrial wastewater treatment. The author repeatedly focuses on different areas of industrial wastewater treatment and widely delineates the immense potential of novel separation processes.

6.10   BASIC FUNDAMENTALS AND SCIENTIFIC UNDERSTANDING OF UF AND MF

Scientific research pursuit and scientific endeavor is moving toward a visionary domain of novel separation processes and membrane science. Scientific progress, scientific rigor, and deep scientific understanding are the immediate need of the hour. Scientific validation is on the other side of the coin of science and technology. Without effective and proper scientific validation, human scientific research is defunct. The world of vicious challenges, the instinct of introspection and a deep comprehension of science will lead a long way in the proper scientific emancipation of UF and NF. Global water crisis is in a state of distress. Today's scientific advancement is veritably linked to the provision of basic human needs. Science and engineering are being groomed toward that direction. Scientific rigor should be re-envisioned in such a situation such as global water shortage. In today's world, global water crisis has an unsevered umbilical cord with environmental engineering rigor. Scientific doctrine of UF and MF needs to be re-envisioned at each step of environmental engineering science rigor.

6.11   RECENT SCIENTIFIC ENDEAVOR AND THE WORLD OF CHALLENGES AND VISION IN THE DOMAIN OF UF

Scientific endeavor and scientific validation today stands in the midst of immense introspection. The challenges, the vision and the subsequent progress of engineering and technology all depends on the vast and versatile world of scientific validation. Man's wide scientific vision, the world of difficult challenges and the path toward a new future of science are the pallbearers of the futuristic domain of membrane separation phenomenon. After the discovery of Loeb–Sourirajan model, membrane science witnessed drastic challenges and visionary roads for the future. UF and NF are moving toward a new world order in environmental engineering science and chemical process engineering. Science and technological vision in today's world stands in the midst of scientific optimism and immense hope. The author in every step of scientific creation reveals a wondrous fact that is to alleviate global water shortage. UF, MF, and other branches of membrane science are the hallmarks of scientific rigor in tackling global water shortage.

6.11.1   RECENT SCIENTIFIC ADVANCES IN THE DOMAIN OF UF AND MF

Recent scientific endeavor in MF touches upon vast and varied world of process design. Today's remarkable worlds of membrane separation processes are surpassing wide and visionary frontiers. Pursuit of science in today's human civilization has no definitive barriers yet in the midst of deep instinct and cogent comprehension. MF has immense applications in environmental engineering. The targets and vision of science in the field of MF are in the domain of water and wastewater treatment and the wide world of environmental engineering. The author in this section elaborates on the recent advances in the field of MF in deep and visionary details.[1]

Ghosh[2] lucidly discussed in a well-informed review protein separation using membrane chromatography, its opportunities and challenges. This chapter reviews the current state of development in the area of membrane chromatographic separation of proteins. The transport phenomenon of membrane chromatography and the recent scientific endeavor are discussed in lucid details.

Zularisam et al.[3] in a well-detailed treatise elucidated upon behaviors of natural organic matter (NOM) in membrane filtration for surface water treatment. This review encompasses on membrane application in surface

water treatment. The authors delineate the advantages of membrane application over conventional treatment. The vision of science is upheld at each
step of scientific rigor in this review. Fouling issue stands as a major issue
in the successful operation of membrane separation phenomenon. It is a
burning and vexing problem since it restricts widespread application due
to increases in hydraulic resistances, operational and maintenance costs,
deterioration of productivity, and frequency of membrane regeneration. This
treatise discusses NOM and its components as the major membrane foulants
that occur during the water filtration process, possible fouling mechanisms
relating to reversible and irreversible of NOM fouling, current techniques
used to characterize fouling mechanisms and methods to control fouling.

Merin and Daufin[4] discussed cross flow microfiltration (CFMF) in the
dairy industry in a state of the art review. CFMF is a relatively new membrane
process that was introduced to the dairy industry in the last decade. In this
review, the authors with precision and cogent insight relate the problems
associated with operation and cleaning of membranes. MF is one of the
first filtration processes, commercially developed by Sartorius-Werke in
Germany in 1929. In the beginning, it was only used for research but during
World War II, it was adapted for bacteriological analysis of water supplies.
Until 1963, microfilters were predominantly nitrocellulose or a mixture of
cellulose esters. The main uses of microfilters are water purification and sterilization and microbiological and related analytical applications, such as the
direct epifluorescent fiber technique (DEFT). The author delineates tangen-
tial or CFMF which is a pressure driven membrane process, similar to UF.
MF could be used for coarse filtration of particulates and bacteria as well as
to finely separate soluble protein, small molecular solutes, and water.

Mohammad et al.[5] detailed in a far-sighted review UF in food processing
industry. In this review, the authors touch upon application, membrane
fouling, and fouling control. Vision of science needs to be readdressed at
each step of scientific and academic rigor in the field of UF. This review work
informs the vast and varied world of UF and its versatile scientific rigor. The
challenge, the vision and the purpose is unimaginable. UF process has been
applied widely in food processing industry for the last 20 years due to its
advantages over conventional separation processes such as gentle product
treatment, high selectivity, and lower energy consumption. Recent forays in
this domain show the various intensive studies carried out to improve UF,
focusing on membrane fouling control and cleaning of fouled membranes.
A subtle review and an informed narrative gains higher vision as the author
gleans through the fundamental concept of UF.

Howe and Clark[6] elucidated on the fouling of MF and UF membranes by natural waters. Membrane filtration (MF and UF) has become an accepted process for drinking water treatment but membrane fouling remains a significant problem. The objective of this study was to systematically investigate the mechanisms and components in natural waters that contribute to fouling. Natural waters from five sources were filtered in a benchtop filtration system. The vision of science is emboldened at each step of this treatise with immense precision and deep comprehension. Membrane filtration has become an accepted process in water treatment. Many full-scale facilities are operating with either MF or UF membranes. One of the significant and vexing issues affecting the development of membrane filtration is fouling. Interactions between the membrane and components in the raw water cause a rapid and often irreversible loss of flux through the membrane. The authors stresses on the concept of fouling of MF and UF membranes.

Foley[7] lucidly delineated in a phenomenal review paper factors affecting filter cake properties in dead-end MF of microbial suspensions. Scientific vision and scientific understanding are at its helm in the futuristic research endeavor in the field of MF. Dead-end MF of microbial suspensions is reviewed with particular emphasis on the factors affecting the specific cake resistance. The effects of cell size and shape, cell surface properties (including charge), ionic environment, fermentation medium components, and ageing effects are reviewed in details. The separation of cells from the fermentation broth is a crucial step in the recovery of many chemicals of biological origin and dead-end filtration remains an useful tool for performing the separation.

Cuoto and Herrera[8] reviewed in a research paper industrial and biotechnological applications of laccases. Laccases have received much attention from researchers in last decades due to their ability to oxidize both phenolic and nonphenolic lignin related compounds as well as highly recalcitrant environmental pollutants which makes them very useful for their application to several biotechnological processes. Biotechnological applications are vast and versatile. The authors delineates various applications which include the detoxification of industrial effluents, mostly from the paper and pulp, textile, and petrochemical industries, use as a tool for medical diagnostics and as a bioremediation agent to clean up herbicides, pesticides, and certain explosives in soils. Today's world of biotechnological applications is surpassing wide scientific frontiers. Cuoto and Herrera[8] deeply delineates with cogent insight the nanobiotechnology area of scientific endeavor.

Elimelech and Phillip[9] discussed deeply with cogent insight the scientific domain of desalination and its tremendous potential. The challenge of research pursuit, the grave concern for global water crisis and the application of membranes in desalination has urged the environmentalists to devise effective technologies. Desalination is more energy intensive and there are potential environmental impacts. With this strong focus in mind, the authors target energy efficiencies of desalination plants and its global impact. This lucid treatise trudges a weary mile in the wide world of global water crisis. Desalination and membrane science are the two opposite sides of a coin in today's scientific regeneration. Elimelech and Phillip[9] describe with scientific precision the sustainability aspect of desalination as a probable solution to global water issues. Technological vision in today's scientific scenario is deeply retrogressive and in deep crisis as civilization confronts with climate change and environmental sustainability. In such a crucial juxtaposition, the author establishes the future of desalination, the energy efficiency, and the massive challenges that lie forward in successful realization of environmental sustainability.

Howell[10] discusses future of membranes and membrane reactors in green technology and green science and its immense importance in water reuse. Water shortage problems stands in extremely difficult times today. The author skillfully delineates membrane technology that can make a great contribution in providing clean water to common mass. Science and engineering of membranes are moving at a drastic pace toward a newer realm. Water reuse, application of membrane bioreactors and the success of separation phenomenon are today all targeted toward a newer eon of water sustainability.

Geise et al.[11] describe water purification by membranes and the wide application of membrane science. Today, as move drastically forward in the 21st century, two major impacts on human society are energy and environmental sustainability. Two highly interrelated resources of provision of clean water and energy self-sufficiency are shaping and re-envisioning the future of environment and future of humankind. Depletion of conventional energy resources and water stresses in human society has urged humankind to yearn for new technologies. This article describes the current state of polymeric membranes for water purification and the forays into areas where application of polymer science can be envisaged. The scientific challenge, the scientific vision and the purpose needs to be augmented as human society moves toward a newer world of application of polymer technology.

Shrimali[12] discusses in a review on RO technology. Development of RO technology and RO membranes of very high rejection, while maintaining high permeability has the potential to reduce energy consumption. This review re-envisions the overall targets of RO technology as an alternate method for wastewater treatment. Water problems involve the following:

- Sea water desalination in coastal areas
- Brackish water desalination
- Water purification
- Water reuse
- Rainwater harvesting
- Water supply schemes

This review justifies the immense importance of RO technology in tackling global water issues. The challenge to move forward in scientific research pursuit is opening up new vistas of innovation and new avenues of vision in years to come.

A white paper[13] discusses sustainable solutions to the global freshwater crisis and lessons from a developing economy such as India. The unique feature of this paper targets the success of a water treatment plant on the Yamuna river that supplies drinking water to the Delhi population, a community based fluoride treatment plant in Jaipur, India, and a sewage treatment plant in Jaipur, India. Practical challenges to efficient and effective water-quality management are envisioned in this study (Table 6.2).

TABLE 6.2 Visionary Scientific Endeavor in the Field of Membrane Science.

Authors and researchers	Scientific research pursuit
Ghosh(2002)[2]	Review on membrane chromatographic separation of proteins.
Zularisam et al(2006)[3]	A review on behaviors of natural organic matter in membrane filtration for surface water treatment.
Merin et al(1990)[4]	A state of the art review on crossflow microfiltration in the dairy industry.
Mohammad et al(2012)[5]	A review on ultrafiltration in food processing industry—its application, membrane fouling and fouling control.
Howe et al(2002)[6]	Fouling of microfiltration and ultrafiltration membranes by natural waters.
Foley(2006)[7]	A review of important parameters affecting filter cake properties in dead-end microfiltration.
Cuoto(2006)[8]	A review on industrial and biotechnological applications of laccases.
Elimelech et al(2011)[9]	A review on seawater desalination
Howell(2004)[10]	Future of membranes and membrane bioreactors.

TABLE 6.2   *(Continued)*

Authors and researchers	Scientific research pursuit
Giese et al(2010)[11]	Water purification by membranes and the role of polymer science.
Shrimali et al(2015)[12]	A short review on reverse osmosis technology.
International workshop on sustainability and water quality (2011)[13]	A review on drinking water scenario in India and water quality remediation.
Palit(2016)[20]	A review on frontiers of nanofiltration.

6.12   DIFFICULTIES AND BARRIERS OF MEMBRANE SEPARATION PHENOMENON

Membrane separation phenomena in today's world have an unsevered umbilical cord with the barriers of fouling. The efficiency and effectiveness of membrane separation processes and novel separation phenomenon are attached to the immense barrier of fouling. Membrane separation processes has an unfinished and fascinating journey in the avenues of environmental engineering science. Difficulties, barriers, and subsequent success in membrane separation processes are the hallmarks of separation phenomenon.

A typical example of UF in food processing industry is elucidated in details by the Mohammad et al.[5] Food processing industry is a glorious example of application domain of UF. Mohammad et al.[5] dealt with lucid details the application of UF and stressed on membrane fouling and fouling control.

Membrane filtration processes have gained popularity in the food processing industry over the last 25 years. It is estimated that 20–30% of the current 250 million Euros turnover of membrane used in the manufacturing industry worldwide was from food processing industry. To date, this market is still undergoing rapid growth, approximately 7.5% per year, particularly in dairy industry, followed by beverages and egg products. The total membrane market for the food and beverages industry has been estimated to be worth US$ 1182 billion in 2008. In the dairy industry, it is estimated that over 75% of membrane usage is dedicated to whey processing, while 25% of UF membranes is accounted for milk processing.

Fouling refers to the irreversible alteration in membrane properties, resulting from several interactions of feed stream components and membrane. In food application, membrane is usually fouled by biofoulants such as protein and polysaccharide.

6.13 FOULING, CLEANING, AND THE SUCCESS OF MEMBRANE SEPARATION PHENOMENON

Science and technology of membrane science is ushering in a new era of scientific understanding. Fouling stands as a major obstacle to membrane separation processes. The success of membrane separation lies on the domain of fouling and cleaning. Scientific advancements in the field UF and MF are vast, versatile, and visionary. Scientific research output and scientific validation of fouling and cleaning needs to be highlighted and re-envisioned. The barriers to membrane separation regarding fouling and cleaning needs to scale visionary heights in the near future. Science and technology of fouling is ushering in a new era of scientific forbearance and scientific fortitude.

Formation of concentration boundary layer and the phenomenon of fouling and cleaning stands as a major imperative in the process design of a membrane separation process. The challenge, the vision and the motivation of research pursuit remains unparalleled. The visionary aura of science and technology needs to be restructured and re-envisioned at each step of scientific life.

6.14 CONCENTRATION POLARIZATION AND THE VISION OF MEMBRANE SEPARATION PROCESSES

Concentration polarization is a major impediment to successful membrane separation process. The author delineates in details the future of successful membrane separation phenomenon. The authors stress on the field of application of membrane science in environmental engineering. Concentration polarization and fouling are the pallbearers to a larger extent toward the true realization of membrane science and technology. Scientific rigor, scientific candor, and scientific forbearance are the major backbones of the success of the application of membrane technology such as UF and MF. The path of scientific glory is arduous and vicious. Advancement of science needs to be re-envisioned at each step of technological and engineering rigor.

6.15 MEMBRANE SCIENCE, ENVIRONMENTAL ENGINEERING, AND THE WIDE VISION OF SCIENCE AND ENGINEERING

Membrane science in today's visionary generation has an unsevered umbilical cord with environmental engineering science. Human scientific struggle is ushering a new era of scientific achievement and a holistic scientific advancement. Environmental engineering science is gearing toward newer challenges and a holistic sustainable development. Environmental and energy sustainability are the hallmark of a new scientific generation and the immediate need of the hour of our present day human civilization. Membrane science is ushering in a new eon of holistic research pursuit. Parallely, human scientific generation should veritably target toward alleviating global water crisis. Membrane science, in this crucial juxtaposition of science and engineering, should target toward the vexing issue of groundwater remediation.

6.16   SCIENTIFIC ACUITY AND RECENT ADVANCES IN MEMBRANE SCIENCE

In today's human civilization and human scientific pursuit, membrane science is connected with environmental engineering science by an umbilical cord. Global water shortage and the crisis linked with it is changing the face of human civilization. Membrane science needs to be envisaged at each step of environmental engineering pursuit. The vision and the challenge lies in the hands of future scientists and engineers. The scientific acuity and the scientific wisdom needs to be re-envisioned at each step of human life. Membrane science in today's world is linked by an umbilical cord with global water scenario. The separation phenomenon needs to be re-envisaged at each step of scientific endeavor. UF and MF are witnessing a new dawn of human scientific research pursuit. The present status of global water shortage has urged the scientific generation to devise new and frontier technologies. Scientific grandiloquence and immense scientific vision are in the path of newer regeneration. Membrane science is truly showing the path of immense scientific vision and understanding. The immediate need of the hour for the scientific domain is to delve deep into the unknown world of membrane separation phenomenon and unfold the hidden characteristics of different membrane separation processes such as UF and MF.

6.17   SCIENTIFIC VISION, SUSTAINABLE DEVELOPMENT, AND THE FUTURE PROGRESS IN THE FIELD OF MEMBRANE SEPARATION PROCESSES

Environmental and energy sustainability are the two pillars of humankind today. Progress in the field of MF and UF is changing the face of environmental engineering science today. The immediate need of the hour is the visionary scientific endeavor in the field of ecological balance and industrial pollution control. Sustainable development today stands in the midst of unimaginable crisis and in the similar vein deep introspection. Developed as well as developing economies in today's human civilization are in deep distress. The grave concern for ecological imbalance, the environmental disasters, and the future progress of science are all the visionary parameters toward a greater emancipation of environmental sustainability in years to come.

6.18   GLOBAL WATER CRISIS AND THE STATUS OF MEMBRANE SCIENCE

Global water crisis and the world of global water shortage are changing the face of environmental engineering applications of membrane science. UF and MF today stand in the midst of immense challenge and deep comprehension. The status of membrane applications is inspiring as well as thought provoking. Scientific hindsight and scientific challenges in today's world will lead a long way in the true realization of environmental sustainability. Global water catastrophe and its alleviation are the path-breakers and hallmarks of a newer generation of scientific realm and deep scientific understanding. Membrane science has surpassed many expectations with the passage of time and eon. Water crisis is surely and veritably linked with an umbilical cord with the successful application of membrane science. The status of membrane separation processes is veritably visionary today. Environmental catastrophes, the blunders of ecological diversity, and the wide scientific vision and rigor are reframing humankind's scientific framework. Global water crisis will in future be a strong support for global environmental emancipation.

6.18.1   GLOBAL WATER CHALLENGES, WATER RESEARCH AND DEVELOPMENT INITIATIVES, AND THE VISION FOR THE FUTURE

The undeniable seriousness of the global water situation was first brought to the attention of the international fraternity at the 1992 United Nations Conference on Environment and Development in Rio de Janiero, at what came to be known as the Rio Earth summit.[51] The vision of the future global water initiatives turned out to be clear and far-reaching. Environmental engineering science witnessed an overwhelming challenge after that summit. After two decades after the Rio summit, the global situation with respect of water has improved in some areas but still has long ways to go in other avenues because of the needs and effects of rapidly growing population in this decade. Water scarcity has become a major issue in our planet. The important issues are well known. These include: a rapidly growing population; competition between sectors such as industry, agriculture, and energy for precious land and water resources; inadequate access to water supply and sanitation services; the failure to address the issue of indigenous water rights; matters related to environmental protection; and finally growing tension over transboundary water challenges.[51]

The author stresses on the effective needs of environmental engineering techniques such as novel separation processes and nontraditional environmental engineering techniques in view of this critical situation. Desalination and membrane science, in such a crucial juxtaposition, widely in the scientific horizon, need re-envisioning. Advanced oxidation techniques also will unearth global water issues and the veritable challenges. The success of human scientific endeavor will open new chapters tackling global water issues.[51]

6.19 THE FUTURE OF SEA WATER DESALINATION AND THE WIDE VISIONARY WORLD OF GLOBAL WATER INITIATIVES

In recent years, numerous large-scale seawater desalination plants have been built in water-stressed countries. Desalination today stands as a vibrant example in the application domain of membrane technology. In today's scientific arena global water shortage, the visionary world of desalination and the future progress of human research pursuit are changing the face of scientific vision and deep scientific understanding. The fate of energy and environment stands in the midst of unimaginable crisis today. Developed and developing economies are gearing toward a new eon of scientific fortitude. Human life's challenges has become immeasurable with the progress of environmental engineering science.[48–50]

Water scarcity is the most severe global challenges of our times. Scientific vision and deep scientific introspection has no answers. Presently, over one-third of the world's population lives in water-stressed countries and by 2025, this figure is predicted to rise to nearly two-thirds. The challenge and the scientific struggle of providing ample and safe drinking water is further complicated by population growth, industrialization, contamination of available freshwater resources, and climate change. The fate of human civilization and the future of science is at a state of immense distress. In recent years, numerous large-scale seawater desalination plants have been built in water-scarce countries to enhance water resources and construction of new desalination plants needs to be envisioned with each step of scientific endeavor. Elimelech and Phillip[9] widely reviewed to a larger scientific audience the decisive challenges and the future of desalination plants throughout the world and its immense scientific potential.

6.20   THE WIDE VISION OF AOPS AND ITS SCIENTIFIC DOCTRINE

Technology and science of AOPs are moving toward newer visionary domains of environmental engineering science. Scientific and academic rigors in the field of AOPs are gaining new heights. Advances in chemical and wastewater treatment have led to a range of novel and nontraditional techniques such as AOPs. These processes has shown immense potential in degrading and destroying pollutants of low or high concentrations and have found deep introspection and effective applications in groundwater treatment, industrial wastewater treatment, municipal wastewater sludge destruction, and volatile organic compounds treatment (VOCs). Broadly speaking, the AOPs have proceeded along the two routes:

- Wet air oxidation (WAO) techniques—oxidation with O_2 in temperature ranges between ambient conditions and those found in incinerators in the region of 1–2 MPa and 200–300°C.
- The use of high-energy oxidants such as ozone and hydrogen peroxide and/or photons those are able to generate highly reactive intermediates —OH (hydroxyl) radicals.

However, specifically AOPs have been broadly defined as "near ambient temperature and pressure water treatment processes which involve the generation of hydroxyl radicals in sufficient quantity to effect water purification." The hydroxyl radical (OH) is a powerful (redox potential = 2.80 V), nonselective chemical oxidant, which acts very rapidly with most organic compounds. Technology of AOPs is today crossing visionary boundaries. The scientific imagination and deep scientific vision behind AOPs are the pallbearers toward a greater emancipation of green science and green technology.

Scientific endeavor in the field of nontraditional environmental engineering techniques such as AOPs are being widely challenged and needs to be re-envisioned and rebuilt.

Yalfeni[52] discussed in a widely observed research project new catalytic AOPs for wastewater treatment. This project is aimed to present new catalytic AOPs (specifically Fenton process and catalytic ozonation) for the degradation of organic pollutants in industrial wastewater. A special area of research was the in situ generated hydrogen peroxide. A new innovation was the appropriate substitute of for H_2 in the relevant pathway. The immense potential of integrated advanced oxidation techniques is being re-envisioned and re-envisaged at each step of scientific research pursuit.

Gilmour[53] delineated in deep details in a visionary research work application perspectives in water treatment using AOPs. AOPs using hydroxyl radicals and other oxidative radical species are being studied extensively for removal of organic compounds from various wastewater streams. This study focuses on the evaluation of the upstream processing and downstream posttreatment analysis of selected AOPs. A pilot scale immobilized photocatalytic reactor was used in this study.

Goi[54] in a research thesis AOPs for water purification and soil remediation. The aim and objective of this study was to enlarge the existing knowledge in AOPs applications for water and soil contamination. Technological objectives and scientific motivation are immensely required for the advancement of environmental engineering science.

AOPs are the challenging and the visionary goals of environmental engineering today. The challenge and the immense vision of environmental protection needs to reshaped in the midst of immense industrial disasters and wastewater treatment issues. The vision of science, the scientific and academic rigor is opening up new chapters in scientific history.

6.20.1 AOPS, NONTRADITIONAL ENVIRONMENTAL ENGINEERING TECHNIQUES, AND DEEP COMPREHENSION

AOPs are the next generation environmental engineering techniques. The efficiency and the effectivity of the processes are opening a new chapter in the field of environmental engineering science. In many avenues of scientific endeavor, primary and secondary treatments are not helpful to degrade recalcitrant substances. Here comes the need for tertiary treatment such as advanced oxidation techniques such as ozonation. Gogate and Pandit[15] deeply related with cogent insight in a far-reaching review imperative technologies for wastewater treatment such as oxidation technologies at ambient conditions. Their visionary pursuit involves five different oxidation processes operating at ambient conditions that are cavitation, photocatalytic oxidation, Fenton's chemistry and ozonation, use of hydrogen peroxide (belonging to the class of chemical oxidation technologies). AOPs are defined as the processes that generate hydroxyl radicals in sufficient quantities to be able to oxidize majority of the complex chemicals present in the effluent water. Gogate et al.[15] elucidated upon cavitation, acoustic cavitation and reactors used for the generation of acoustic cavitation. The other facets of their research involve optimum parameters for sonochemical reactors. Hydrodynamic cavitation and photocatalysis are the other visionary pursuits of this study. Stasinakis[16] deeply comprehends in a well-informed review use of AOPs for wastewater treatment. The aim and purpose of this study encompasses the use of titanium dioxide/UV light process, hydrogen peroxide/UV light process and Fenton's reactions in wastewater treatment (Table 6.3).

TABLE 6.3   Scientific Research Pursuit in Advanced Oxidation Processes.

Authors	Scientific endeavor in advanced oxidation processes
Gogate et al.(2004)[15]	A review of appropriate technologies for wastewater treatment with special emphasis on chemical oxidation technologies.
Stasinakis(2008)[16]	A review of use of advanced oxidation processes (AOPs) for wastewater treatment.
Oller et al.(2011)[17]	A review on combination of advanced oxidation processes and biological treatments for wastewater decontamination.
Chakinala et al.(2008)[18]	Treatment of industrial wastewater effluents using hydrodynamic cavitation and advanced Fenton processes.
Comninellis et al.(2008)[19]	Perspective, Advanced oxidation processes for water treatment: advances and trends for R&D
Palit(2015)[21]	A review on advanced oxidation processes, nanofiltration and application of bubble column reactor.
Palit(2016)[22]	A review on advanced oxidation processes and bioremediation.

6.21   GROUNDWATER REMEDIATION AND THE WIDE SCIENTIFIC PROGRESS

Groundwater remediation globally is the immediate need of the hour. Different parts of the world especially South Asia are witnessing this difficult water challenge. Science has a definitive vision of its own. The colossus of science and technology needs re-envisioning and revamping at each step of scientific endeavor. Developing and developed nations in the world are moving toward the murky depths of groundwater contamination. Techno-logical and scientific validation stands today in the midst of an environ-mental catastrophe. The motivation and objective of this treatise is crystal clear which is toward tackling global water calamity. The author rigorously targets the success of novel separation processes such as membrane science. Global sustainability vision needs to be re-emphasized and re-envisioned at this stage of human civilization.[14,24–26,40–42]

6.22 HEAVY METAL CONTAMINATION, ARSENIC GROUNDWATER CONTAMINATION, AND THE FUTURE OF SCIENCE AND TECHNOLOGY

The contamination of groundwater by heavy metal, originating either from natural soil sources or from anthropogenic sources is a matter of grave concern to human society's public health. Remediation of contaminated groundwater is of highest priority since billions of people all over the world use it for drinking water purpose. Selection of appropriate and innovative technology is of utmost challenge to the environmental engineer and scien-tist. Hashim et al.[14] discussed in a well-researched paper remediation tech-nologies for heavy metal contaminated groundwater. Scientific vision is at its helm at every step of this innovative scientific endeavor. Selection of a suitable technology assumes immense importance due to complex soil chemistry and aquifer characteristics and no thumb rule can designate this complex scientific understanding.[14,27–30,34–36]

6.23 HEAVY METALS IN GROUNDWATER: SOURCES, CHEMICAL PROPERTY, SPECIATION, AND WIDE VISION FOR THE FUTURE

Heavy metals occur in the earth's crust and may be solubilized in ground water through natural processes or by change in soil pH. Hashim et al.[14] delineates with extreme clarity heavy metal status in groundwater. Ground-water can be contaminated with heavy metals from landfill leachate, sewage, leachate from mine tailings, deep-well disposal of liquid wastes, seepage from industrial waste lagoons or from industrial spills and leaks. A variety of reactions in soil environment, for example, acid/base, precipitation/

dissolution, oxidation/reduction, sorption or ion-exchange processes can influence the speciation and mobility of metal contaminants.[14] The rate and extent of these reactions will depend on factors on pH, Eh, complexation with other dissolved constituents, sorption and ion exchange capacity of the geologic materials and organic matter content. Groundwater flow characteristics are vital in influencing the transport of metal contaminants.[14,30–33]

6.24   TECHNOLOGIES FOR TREATMENT OF HEAVY METAL CONTAMINATED GROUNDWATER

Several technologies exist for the remediation of heavy metals contaminated groundwater and soil and they have some definite outcomes such as Hashim et al.[14] (1) complete or substantial destruction/degradation of the pollutants, (2) extraction of pollutants for further treatment or disposal, (3) stabilization of pollutants in forms less mobile or toxic, (4) separation of noncontaminated materials and their recycling from polluted materials that require further treatment, and (5) contaminant of the polluted material to restrict exposure of the wider environment.[14] Technological validation, the wide avenues of science and the technology drivers will lead a long way in the true realization and effective emancipation of groundwater remediation.[14,34–36]

The treatment technologies can be classified as: (1) chemical treatment technologies, (2) biological/biochemical/biosorptive treatment technologies, and (3) physicochemical treatment technologies.[14]

1. Chemical treatment technologies:

(a) In situ treatment by using reductants, (b) reduction by diothinite, (c) reduction by gaseous hydrogen sulfide, (d) reduction by iron-based technologies, (e) removal of chromium by ferrous salts, and (f) soil flushing.[14]

2. Biological treatment technologies:

(a) Biological activity on the subsurface, (b) natural biological activity, (c) enhanced biorestoration, (d) in situ bioprecipitation process (ISBP), (e) biosorption of heavy metals, (f) metal removal by bio surfactants, (g) metal uptake by organisms, and (h) biosorption of heavy metals by cellulosic materials and agricultural wastes.[14]

3. Physicochemical treatment technologies:

(a) Permeable reactive barriers (PRB), (b) sorption process in PRB, (c) sorption within red mud in PRB, (d) activated carbon and peat in PRB, and (e) filtration and absorption mechanisms.[14]

6.25   THE CHALLENGE AND THE SCIENTIFIC FORESIGHT IN R&D INITIATIVES IN WATER TECHNOLOGY

Scientific foresight in today's world is moving toward a newer visionary eon. The challenge and the vision need to be restructured at every step of human civilization. The struggles of science, the scientific vision to excel and immense scientific foresight are leading a long way in the path toward greater sustainable development of humankind. Scientific foresight in today's human civilization is gearing toward immense and unimaginable challenges. This treatise pointedly delineates the success of application of UF and MF in tackling global water crisis.[48-50] The recurrent catastrophes, the immense environmental calamities and the scientific potential of membrane separation phenomenon are the precursors toward a greater scientific foresight and a greater scientific vision. Global water shortage is restructuring and re-envisioning the domain of membrane science.[26-29,37-39,51]

6.26   MOTIVATION AND OBJECTIVES OF THE FUTURE OF APPLICATION OF MEMBRANE SCIENCE

Membrane science and technology in today's human civilization are moving at a rapid pace. Motivation and objectives of the application of membrane science needs to be re-envisioned with the progress of environmental engineering science. Global water issues are changing the face of human scientific endeavor. The challenge of science, the global environmental concerns and the fire and yearning of scientific imagination are the torchbearers of greater visionary tomorrow.

Scientific imagination and the future challenges are changing the face of future scientific research pursuit. In today's world, global water crisis is re-envisioning the scientific domain of membrane science. The immediate and the imminent need of research pursuit should be toward provision of basic needs such as water.[20,22,23,40-42,51]

6.27   FUTURE SCIENTIFIC DOCTRINE AND THE FUTURISTIC VISION

Future scientific doctrine and the futuristic vision of engineering science are in today's world challenged with the passage of history and time. Scientific research pursuit in the field of water science and water technology are changing the path of environmental sustainability and the aisles of holistic sustainable development. The global water crisis in developed and developing world has shaken the scientific domain and scientific paradigm. The immediate concern, the imminent scientific vision and the innovative scientific understanding are the pallbearers toward a newer age of membrane science and water technology. Groundwater remediation is another avenue of scientific research pursuit. The future of water technology stands in the midst of global water imbroglio and immense calamities. Science and engineering have few answers to it. This treatise forcefully focuses with deep insight the futuristic vision of application of membrane science especially UF and MF. Its impact on global water shortage and the success of membrane science research pursuit are the ultimate issues facing humankind today. The author deeply observes the global water crisis with deep scientific conscience.[20–23]

6.28   FUTURE RECOMMENDATIONS FOR THE STUDY

The challenge of science is awe-inspiring. Human scientific endeavor and the development of environmental engineering tools are the torchbearers toward a newer generation of holistic sustainable development. Human society's challenges and civilization's prowess are witnessing in today's world a new generation of scientific ideas and scientific innovations. The future recommendations of this study should be toward more membrane science applications toward groundwater remediation. Arsenic groundwater contamination is devastating the scientific domain and social scenario of many developed and developing countries of the world. Future thoughts, future trends and future recommendations of the study in the field of membrane science should be toward tackling global water crisis. Future recommendations of this study aims at the academic and scientific rigor in the domain of membrane science and water and wastewater treatment.[22,23,28,35]

6.29 FUTURE TRENDS IN RESEARCH AND FUTURISTIC CHALLENGES

Futuristic challenges in the field of membrane science and environmental engineering science should be targeted toward fighting global water crisis phenomenon. Human scientific research pursuit is ushering in a new world of scientific imagination and scientific forbearance. Global water shortage today stands in the midst of immense scientific fortitude and scientific sagacity. Man's prowess and a scientist's directed vision stands as a major backbone of this century's scientific research pursuit. Arsenic groundwater remediation and heavy metal remediation is opening up a new era of scientific understanding and scientific fortitude.[24,25,36]

Trends in scientific research and immense scientific rigor are the torchbearers to a greater vision of environmental engineering tools and membrane technology. The challenge in UF, MF, and membrane science lies in removal of fouling and effective cleaning of the membranes. Scientific adjudication, scientific understanding, and scientific sagacity are in the futuristic road of immense vision.[28,35,36]

Green chemistry in today's scientific horizon has an unsevered umbilical cord with application of nanotechnology. Ecological damage, environmental catastrophes and the grave concern for climate change has urged the scientific domain to target green technologies. Futuristic challenges should be targeted toward this vision. Extensive applications of nanotechnology in environmental engineering are the other broad scientific perspective. The challenge for the future needs to be re-envisioned at each step of environmental engineering forays.

6.30 CONCLUSION

The challenges, the visionary future of science and the wide roads toward progress are opening up new avenues of scientific emancipation in this century. Membrane science, particularly UF and MF, is ushering in a new eon of scientific rigor and immense understanding. The grave concerns of environmental pollution and the enhanced vision of water and wastewater treatment are gearing for new exploration and new avenues. Science and technology needs to be re-envisioned at every step of human life. The scientific rigor of membrane science and water and wastewater treatment needs to be reshaped. In this century, the success of realization of environmental sustainability is immense and far-reaching. This is the largest question of environmental engineering science. Sustainable development and realization of environmental sustainability will be the coin words of the future at every step of human life. The crisis of human civilization lies in the fact that global water shortage and drinking water treatment are at a state of deep distress. Wastewater treatment and industrial water pollution control needs to be re-envisioned and reshaped at each step of scientific rigor. Science and engineering in today's human civilization are in the path of new glory and new scientific understanding. The challenge, the vision and the progress of science needs to be overhauled with each step of human life. Water and wastewater treatment and application of membrane science are in the road toward newer scientific innovation and newer scientific instinct in years to come.

ACKNOWLEDGMENT

The author wishes to acknowledge the contribution of Chancellor, Vice-Chancellor, Head of Department (Chemical Engineering), all faculty, and students of University of Petroleum and Energy Studies, Dehradun, India without whom this writing project would not had been complete. The author with great respect acknowledges the contribution of Shri Subimal Palit, the author's father and an eminent textile engineer from India who taught the rudiments of chemical engineering to the author.

KEYWORDS

- **water**
- **membrane**
- **science**
- **ultrafiltration**
- **microfiltration**
- **sustainability**

REFERENCES

1. Cheryan, M. *Ultrafiltration and Microfiltration Handbook;* Technomic Publishing Company Inc.: Lancaster, PA, 1998; pp 1–920.

2. Ghosh, R. Protein Separation Using Membrane Chromatography: Opportunities and Challenges. *J. Chromatogr. A.* **2002,** *952,* 13–27.
3. Zularisam, A. W.; Ismail, A. F.; Salim, R. Behaviours of Natural Organic Matter in Membrane Filtration for Surface Water Treatment: A Review. *Desalination.* **2006,** *194,* 211–231.
4. Merin, U.; Daufin, G. Crossflow Microfiltration in the Dairy Industry: State-of-the-art. *Le Lait* **1990,** *70,* 281–291.
5. Mohammad, A. W.; Ng, C. Y.; Lim, Y. P.; Ng, G. H. Ultrafiltration in Food Processing Industry: Review on Application, Membrane Fouling and Fouling Control. *Food Bioprocess Technol.* **2012,** *5,* 1143–1156.
6. Howe, K. J.; Clark, M. M. Fouling of Microfiltration and Ultrafiltration Membranes by Natural Waters. *Environ. Sci. Technol.* **2002,** *36,* 3571–3576.
7. Foley, G. A Review of Factors Affecting Filter Cake Properties in Dead-end Microfiltration of Microbial Suspensions. *J. Membr. Sci.* **2006,** *274,* 38–46.
8. Cuoto, S. R.; Herrera, J. L. T. Industrial and Biotechnological Applications of Laccases: A Review. *Biotechnol. Adv.* **2006,** *24,* 500–513.
9. Elimelech, M.; Phillip, W. A. The Future of Seawater Desalination: Energy, Technology and the Environment. *Science.* **2011,** *333,* 712–717.
10. Howell, J. A. Future of Membranes and Membrane Reactors in Green Technologies and for Water Re-use. *Desalination.* **2004,** *162,* 1–11.
11. Geise, G. M.; Lee, H. S.; Miller, D. J.; Freeman, B. D.; McGrath, J. E.; Paul, D. R. Water Purification by Membranes: The Role of Polymer Science. *J. Polym. Sci. Part B Polym. Phys.* **2010,** *48,* 1685–1710.
12. Shrimali, H. V. A Brief Review on Reverse Osmosis Technology. *IJRAT.* **2015,** *3*(5), 93–97.
13. *Finding Sustainable Solutions to the Global Freshwater Crisis: Lessons from India*; A White Paper from the International Workshop on Sustainability and Water Quality (IWSWQ): Delhi, India, January 2011.
14. Hashim, M. A.; Mukhopadhyay, S.; Sahu, J. N.; Sengupta, B. Remediation Technologies for Heavy Metal Contaminated Groundwater. *J. Environ. Manage.* **2011,** *92,* 2355–2388.
15. Gogate, P. R.; Pandit, A. B. A Review of Imperative Technologies for Wastewater Treatment I: Oxidation Technologies at Ambient Conditions. *Adv. Environ. Res.* **2004,** *8,* 501–551.
16. Stasinakis, A. S. Use of Selected Advanced Oxidation Processes (AOPs) for Wastewater Treatment: A Mini Review. *Global NEST J.* **2008,** *10*(3), 376–385.
17. Oller, I.; Malato, S.; Sanchez-Perez, J. A. Combination of Advanced Oxidation Processes and Biological Treatments for Wastewater Decontamination: A Review. *Sci. Total Environ.* **2011,** *409,* 4141–4166.
18. Chakinala, A. G.; Gogate, P. R.; Burgess, A. E.; Bremner, D. H. Treatment of Industrial Wastewater Effluents Using Hydrodynamic Cavitation and the Advanced Fenton Process. *Ultrason. Sonochem.* **2008,** *15,* 49–54.
19. Comninellis, C.; Kapalka, A.; Malato, S.; Parsons, S. A.; Poulis, I.; Mantzavinos, D. Perspective Advanced Oxidation Processes for Water Treatment: Advances and Trends for R&D. *J. Chem. Technol. Biotechnol.* **2008,** *83,* 769–776.
20. Palit, S. Filtration: Frontiers of the Engineering and Science of Nanofiltration—a Far-reaching Review. In *CRC Concise Encyclopedia of Nanotechnology;* Ubaldo Ortiz-Mendez, Kharissova. O. V., Kharisov. B. I., Eds.; Taylor and Francis: Boca Raton, FL, 2016; pp 205–214.

21. Palit, S. Advanced Oxidation Processes, Nanofiltration, and Application of Bubble Column Reactor. In *Nanomaterials for Environmental Protection*; Boris, I., Kharisov, Oxana, V., Kharissova, Rasika Dias, H. V., Eds.; Wiley: Hoboken, NJ, 2015; pp 207–215.
22. Palit, S. Advanced Oxidation Processes, Bioremediation and Global Water Shortage: A Vision for the Future. *Int. J. Pharm. Bio. Sci.* **2016,** *7*(1), 349–358.
23. Palit, S. Microfiltration, Groundwater Remediation and Environmental Engineering Science: A Scientific Perspective and a Far-reaching Review. *Nat. Environ. Pollut. Technol.* **2015,** *14*(4), 817–825.
24. Qu, X.; Alvarez, J. J.; Li, Q. Application of Nanotechnology in Water and Wastewater Treatment. *Water Res.* **2013,** *47*, 3931–3946.
25. Tansel, B. New Technologies for Water and Wastewater Treatment: A Survey of Recent Patents. *Recent Pat. Chem. Eng.* **2008,** *1*, 17–26.
26. Byrne, J.; Shen, Bo. The Challenge of Sustainability, Balancing China's Energy, Economic and Environmental Goals. *Energy Pol.* **1996,** *24*(5), 455–462.
27. Hanley, N.; Mcgregor, P. S.; Swales, J. K.; Turner, K. Do Increases in Energy Efficiency Improve Environmental Quality? *Ecol. Econ.* **2009,** *68*(3), 692–709.
28. Masters, G. M.; Wendell, P. E. *Environmental Engineering and Science;* Prentice Hall India Learning Private Limited: India, 2013.
29. Banerjee, A.; Solomon, B. Eco-labeling for Energy Efficiency and Sustainability: A Meta Evaluation of US Programs. *Energy Pol.* **2003,** *31,* 109–123.
30. Goodland, R. The Concept of Environmental Sustainability. *Annu. Rev. Ecol. Syst.* **1995,** *26,* 1–24.
31. Jenkins, D. *Renewable Energy Systems: The Earthscan Expert Guide to Renewable Energy Technologies for Home and Business;* Routledge-Taylor and Francis Group: London, 2013.
32. Nair, J. *Impending Global Water Crisis*; Pentagon Press: India, 2009.
33. Newell, P.; Phillips, J.; Mulvaney, D. *Human Development Research Papers, Pursuing Clean Energy Equitably*, United Nations Development Programme, Plaza NY, November 2011.
34. Palit, S. Concept of Sustainability and Development in Indian Perspective: A Vision for the Future. *J. Environ. Res. Develop.* **2013,** *8*(1), 189–192.
35. Kalam, Abdul, A. P. J.; Singh, S. P. *Target 3 Billion PURA: Innovative Solutions towards Sustainable Development;* Penguin Books: London, 2011.
36. Mukherjee, A.; Sengupta, M. K.; Amir Hossain, M.; Ahamed, S.; Das, B.; Nayak, B.; Lodh, D.; Rahman, M. M.; Chakrabarti, D. Arsenic Contamination of Groundwater: A Global Perspective with Emphasis on Asian Scenario. *J. Health Popul. Nutr.* **2006,** *24*(2), 142–163.
37. McCutcheon, J. R.; McGinnis, R. L.; Elimelech, M. A Novel Ammonia–Carbon dioxide Forward (direct) Osmosis Desalination Process. *Desalination.* **2005,** *174,* 1–11.
38. Fritzmann, C.; Lowenberg, J.; Wintgens, T.; Melin, T. State-of-the-art of Reverse Osmosis Desalination. *Desalination.* **2007,** *216,* 1–76.
39. Lattemann, S.; Hopner, T. Environmental Impact and Impact Assessment of Seawater Desalination. *Desalination.* **2008,** *220,* 1–15.
40. Kumar, A.; Bisht, B. S.; Joshi, V. D.; Dhewa, T. Review on Bioremediation of Polluted Environment: A Management Tool. *Int. J. Environ. Sci.* **2011,** *1*(6), 1079–1093.
41. Sikdar, S. K.; Grosse, D.; Rogut, I. Membrane Technologies for Remediating Contaminated Soils: A Critical Review. *J. Membr. Sci.* **1998,** *151,* 75–85.

42. Thornton, E. C. Jackson, R. L. *Laboratory and Field Evaluation of the Gas Treatment Approach for In-situ Remediation of Chromate Contaminated Soil*, Prepared for the Department of Energy, Pasco, WA, 1994.
43. Wenten, I. G. Recent Developments in Membrane Science and its Industrial Applications. *Songlanakarin J. Sci. Technol.* **2002,** *24,* 1009–1024.
44. Zhao, S.; Zou, L.; Tang, C. Y.; Mulcahy, D. Recent Developments in Forward Osmosis: Opportunities and Challenges. *J. Membr. Sci.* **2012,** *396,* 1–21.
45. Bae, T. H.; Lee, J. S.; Qiu, W.; Koros, W. J.; Jones, C. W.; Nair, S. A High-performance Gas-separation Membrane Containing Submicrometer-Sized Metal – Organic framework Crystals. *Angew. Chem. Int. Ed.* **2010,** *49,* 9863–9866.
46. Nagarale, R. K.; Gohil, G. S.; Shahi, V. K. Recent Developments on Ion-exchange Membranes and Electro-membrane Processes. *Adv. Colloid Interface Sci.* **2006,** *119,* 97–130.
47. Baker, R. W. Future Directions of Membrane Gas Separation Technology. *Ind. Eng. Chem. Res.* **2002,** *41,* 1393–1411.
48. Chaudhery, M. H. Carbon Nanomaterials as Adsorbents for Environmental Analysis. In *Nanomaterials for Environmental Protection;* Boris, I., Kharisov, Oxana, V., Kharissova, Rasika Dias. H. V., Eds.; Wiley: Hoboken, NJ, 2015; pp 217–236.
49. Bhatnagar, A.; Sillanpaa, M. Application of Nanoadsorbents in Water Treatment. In *Nanomaterials for Environmental Protection;* Boris, I., Kharisov, Oxana, V., Kharissova, Rasika Dias, H. V., Eds.; Wiley: Hoboken, NJ, 2015; pp 237–247.
50. Liu, P. Organoclay Nanohybrid Absorbents in the Removal of Toxic Metal Ions. In *Nanomaterials for Environmental Protection;* Boris, I., Kharisov, Oxana, V., Kharissova, Rasika Dias, H. V., Eds.; Wiley: Hoboken, NJ, 2015; pp 249–268.
51. Axworthy, T. S. In *The Global Water Crisis: Addressing an Urgent Security Issue*, Papers for the Inter Action Council, 2011–2012.
52. Yalfeni, M. S. New Catalytic Advanced Oxidation Processes for Wastewater Treatment. Doctoral Thesis, Universitat Rovira I Virgili, March 2011.
53. Gilmour, C. *Water Treatment Using Advanced Oxidation Processes: Application Perspectives*, Master of Engineering Science, University of Western Ontario, September 2012.
54. Goi, A. Advanced Oxidation Processes for Water Purification and Soil Remediation. Doctoral thesis, Department of Chemical Engineering, Tallinn University of Technology, March 2005.

PART II
Biochemistry, Bioproducts and Bioprocessing Technology

CHAPTER 7

WHEY PROTEIN-BASED EDIBLE FILMS: PROGRESS AND PROSPECTS

OLGA B. ALVAREZ-PÉREZ[1], RAÚL RODRÍGUEZ-HERRERA[1], ROSA M. RODRÍGUEZ-JASSO[1], ROMEO ROJAS[2], MIGUEL A. AGUILAR-GONZÁLEZ[3], and CRISTÓBAL N. AGUILAR[1,*]

[1]*Department of Food Research, School of Chemistry, Universidad Autónoma de Coahuila, Saltillo, 25280 Coahuila, Mexico*

[2]*Research Center and Development for Food Industry, School of Agronomy, Universidad Autónoma de Nuevo León, General Escobedo, 66050 Nuevo León, Mexico*

[3]*CINVESTAV, Center for Research and Advanced Studies, IPN Unit Ramos Arizpe, Coahuila, Mexico*

Corresponding author. E-mail: cristobal.aguilar@uadec.edu.mx

CONTENTS

ABSTRACT

The dairy industry generates large amounts of serum that are not used for generating high amounts of pollutants being disposed of improperly causing environmental damage. In Mexico, whey is a bioproduct of the dairy industry that has no added value and is produced in large quantities. This product has high potential to be reused as a raw material for other processes. The use of this product currently discarded, solves a clear problem of pollution and is a potential residue for integrating products into production chains, so source that promotes good sustainable management of this resource, with the use of products that are currently discarded and that cause pollution and could be used as feedstock for other processes, avoiding ecological imbalance and reinstating materials to the production chain in the dairy sector. Manufacturing of edible films from whey protein products might represent an effective means of utilization of excess whey. The formation of the heat induced gel structure involves a complex series of chemical reactions involving dissociation, denaturation, and exposure of hydrophobic amino acid residues. These reactions are influenced by experimental conditions such as protein concentration, pH, heating temperature, and ionic strength. The aim of this study was to determine the progress in the use of whey and interactions that are generated when mixed in a food matrix.

7.1   INTRODUCTION

Most foods are highly perishable and subject to alterations and modifications caused by various factors (chemical, physical, and biological) that are primarily responsible for its deterioration. It has been said that a food is rot when it loses its normal characteristics. To avoid or delay senescence, a lot of methods have been developed for preserving and processing foods taking as a fundamental principle prevent the alteration or decomposition, mainly in their organoleptic properties (taste, odor, color, texture, among others) or avoiding the process of putrefaction, typical of the breakdown of protein foods of animal origin.

In the last 50 years, new technologies have been developed. These techniques enable a great food distribution worldwide. Some methods have been used over the years such as heat, cooling, added sugar, acidification, fermentation, drying or dehydration, modified atmospheres, among others.[1] One of the conservation methods used recently is modified atmospheres, using inert gases that reduce the maturation process using specialized materials for containing a product to generate a micro atmosphere.[2] This arises due to the trend toward production and/or consumption of minimally processed food or organic food that is forcing to the food industry, research centers, and regulatory agencies, among others to transform their technologies and find different ways to processing foods.

A viable alternative is the formulation of edible films that help in controlling the deterioration occurred in foods, without neglecting that the requirements to support the control shall be based on the nature of food in which will be applied. Some of desirable characteristics in coatings and films are controlling the water vapor permeability to gases and volatile compounds, among others. Currently, there are a variety of films or edible coatings based on various polymers as pectin,[3] Arabic gum,[4] galactomannans,[5] chitosan,[6] starch,[7,8] alginate,[9,10] xanthan gum,[11,12] zein,[13,14] pullulan,[15,16] hydroxypropyl methylcellulose,[17,18] locus bean gum,[19] etc. However, protein-based films are characterized by its important functional properties, among which is the delay or decrease of the mass transfer through it, because it possesses a complex structure. Also, serve as an alternative to synthetic materials used as packings.[20]

In Mexico, the whey is a by-product of the dairy industry, which has no added value although is produced in large quantities and in spite of the diversity of applications and products being developed, wastage has become the main source of pollution in the industry, making it a serious environmental problem, and this is mainly due to its high biochemical oxygen demand (BOD). For example, when the serum is pouring in a water body, the microorganisms need a large amount of oxygen to degrade it and consequently reduce the concentration of dissolved oxygen killing the fauna that exist in these ecosystems.[21]

The application or use of biodegradable resources coming from renewable sources is a strategy that reduces environmental problems and adds value. In recent years, the recovery of proteins from renewable agricultural or industrial waste, as is the case of some effluents of the dairy industry (mainly whey), are an important part of the most severe contaminants that exist, that despite of its many uses, it discharged into the soil, drains, and water bodies, becoming a serious problem for the environment. Its use as a source of conservation and recycling becomes an excellent choice for innovation to develop new biodegradables products.[22] Applications include the alcoholic fermentation, demineralization, hydrolyzed, forming edible films, among others. This latest packaging technology enables the development of products with specific barrier, mechanical, and thermal characteristics in certain packaging as films.

Natural biopolymers come from four main sources: animal origin (e.g., collagen and gelatin), seafood (e.g., chiton and chitosan), agricultural origin (e.g., lipids for instance triglycerids and hydrocolloids such as protein and polysaccharides), and microbial origin [such as polylactic acid (PLA) and polyhydroxyalkanoates (PHA)].[22,23] Proteins provide the opportunity to be used as raw material for the production of films. The films are formed by different types of amino acids that allow the development of intermolecular interactions (as ionic interactions), that can be combined with other compounds and carried with different temperatures, allows the obtention of coatings with different chemical and physical properties. Optimizing interactions between amino acids, the formation of polymers with improved stability, barrier, mechanical, and solubility properties are favored.[20,24]

The coatings prepared from whey protein have advantages over coatings made from other biopolymers due to the excellent nutritional value, imperceptible taste, and carrying capacity of food additives.[25] The objective of this review is to have knowledge in progress of the use of whey proteins in the manufacture of packaging materials and/or edible coatings as well as the interactions that occur in the matrix formed from various materials as lipids.

7.2 THE MILK WHEY: A BY-PRODUCT OF THE DAIRY INDUSTRY

Each sector of the food industry generates waste in different amounts depending on the type of product that is produced. In Mexico, the food industry has become one of the productive sectors of higher socioeconomic and environmental impacts from their production chains to their degree of pollution. In the dairy industry, it is necessary to subject the raw material to various processes to obtain the desired product with prolonged periods of storage and good preservation, which generates a large volume of air pollutants, solid waste, hazardous toxic waste, and liquid effluents.[26–28]

Only in 2013, Mexico generated about 10,926,771 L of milk leaving as residue an estimated of 4,964,099 L of serum. From January to March 2014, milk production generate an estimate of 2,595,134 L that will generate 848,427 L of serum.[29] Within the dairy industry, cheese is a primary product, which uses about 25% of total world production in its preparation. Undoubtedly, the main product of the dairy industry is the serum used in the preparation of cheese, which retains about 55% of the components of milk.

The whey is defined as one liquid translucent green substance obtained by removal of clot milk in the elaboration of cheese after precipitation of the protein.[30–32] There are several types depending on whey casein removal. The first is sweet whey, which is based on coagulation of renin at pH 6.5. The second is the acid whey that results from the fermentation process in which organic or mineral acids to coagulate casein are added. The nutritional composition of both types may vary slightly, as the content of lactose and protein that are prevalent in sweet whey[31,32] (Table 7.1). It is estimated that for every kg of cheese are produced 9 kg of whey, which represents about 85–90% of the volume of milk.[28,33–35]

TABLE 7.1   Composition of Whey.

Component	Sweet whey (g/L)	Whey (g/L)
Total solids	63.0–70.0	63.0–70.0
Lactose	46.0–52.0	44.0–46.0
Protein	6.0–10.0	6.0–8.0
Calcium	0.4–0.6	1.2–1.6
Phosphates	1.0–3.0	2.0–4.5
Lactate	2.0	6.4
Chlorides	1.1	1.1

Within the chemical composition of whey are: lactose (4.5–5% w/v), soluble protein (0.6–0.8% w/v), lipids (0.4–0.5% w/v), and minerals (8–10% dry matter) as potassium, calcium, phosphorus, sodium, and magnesium. It also has B vitamins (thiamin, pantothenic acid, riboflavin, pyridoxine, nicotinic acid, and cobalamin) and ascorbic acid.[32,34,36,37] The high nutrient content generates BOD and chemical oxygen demand (COD), 3.5 and 6.8 kg per 100 kg of liquid whey, being the lactose the main component contributing to the high BOD and COD.[32,34,38–41]

Although proteins are not the most abundant component in whey, it is the most important economically and nutritionally.[32,42] It has about 20% of the proteins in bovine milk, being the main component in the beta-lactoglobulin (β-Lg, 10%) and alpha-lactalbumin (α-La, 4%) of whole milk protein, also contains other proteins such as lactoferrin, lactoperoxidase, immunoglobulins, and glycomacropeptide. Its nutritional property and high biological value is attributed to the amount of essential amino acids (leucine, lysine, tryptophan, threonine, cysteine, methionine, histidine, valine, isoleucine, and phenylalanine) present in about 26%.[32,43,44]

However, despite their composition, they are discarded into rivers, sewage, and industrial water recollection centers without given or provide them any benefit. In Mexico, the penalties provided in the General Law of Ecological Equilibrium and Environmental Protection, National Water Law and other applicable breach of the Official Mexican Standard NOM-PA-CCA-009/93 that establishes the maximum permissible limits of entities of pollutants in wastewater discharges into receiving bodies of the processing industry of milk and its derivatives. This leads to increased demand for innovation in technological processes for treating effluents that due to the constant increase in dairy industry will not cover. Unlike developed countries (e.g., USA, Germany, China, etc.), serum is dehydrated for use in making beverages, dairy products and meat extenders.[45–47]

In order to reduce the environmental problem caused by the deposition of whey, some alternatives have been proposed to transform this problem of waste generation in a potential economic resource that could give to this waste source a high value-added. Traditionally, whey is used to feed the population of cattle and pigs to provide them energy, protein and minerals; however, there are some techniques used the serum.

7.2.1 ALTERNATIVES TO THE USE OF SERUM

7.2.1.1 ETHANOL PRODUCTION

Ethanol production from serum has been widely studied and some industrial processes have been implemented in developed countries in milk production. Industrial plants are operating in Ireland, USA, New Zealand, among other countries.[48,49] Generally, deproteinized serum is used before the ultrafiltration. The first studies were conducted in the thirties, using yeast capable of fermenting lactose.[50] The most used species can ferment this disaccharide are *Kluyveromyces marxianus* (*Kluyveromyces fragilis* before), *Kluyveromyces lactis,* and *Candida kefyr* (formerly *Candida pseudotropicalis*). The main limitation of this process is the low concentration of ethanol obtained by intolerance of some strains and the low lactose level that generates between 2% and 3% as maximum of ethanol at the end of fermentation.[51]

7.2.1.2 DEMINERALIZATION

Serum has a high content of salt and other minerals, calculating the dry weight; this is ~8–12%, that makes its application, as a food additive is limited. By the serum demineralization, it is possible to open new avenues for use as partially demineralized whey (25–30%) or much demineralized whey (90–95%).[52] The partially demineralized whey concentrate can be used, for example, in the manufacture of ice cream and bakery products, while much demineralized whey concentrate or powder may be used in infant formulas and a large degree of products. Demineralization involves removing certain organic salts with organic ions reduction of citrates and lactates using techniques as nano- and/or ultrafiltration.[53] Ion exchange resins or electrodialysis can perform this procedure.[54]

7.2.1.3 PROTEIN CONCENTRATES WHEY

Protein concentrates whey is prepared by ultrafiltration, consisting of a semipermeable membrane, which selectively allows passage of materials of low molecular weight as water, ions, and lactose, while retaining high molecular weight materials such as protein. The retentate is concentrated by evaporation and lyophilized.[32,34,55] These concentrates are prepared as substitutes of skim milk and are used in the production of yogurt, processed cheese, in several applications of drinks, sauces, noodles, cookies, ice cream, cakes, dairy, bakery, acrne, beverages, and products of infant formulas due to their excellent functional properties of the proteins and their nutritional benefits.[30,32,34]

7.2.1.4 HYDROLYSATES

The preparations of enzymatic hydrolysates rich in oligopeptides represent a way to improve protein utilization. These preparations have been used in countries such as dietary supplements or physiological needs, to senior age, premature babies, athletes who control the weight through dieting, and children with diarrhea. This is because the amino acids provided by protein hydrolysates are completely absorbed in the digestive system compared with the intact protein without solubilizing.[32,56,57]

7.2.1.5 ISOLATED

Isolated whey protein is one that contains no fat, sugar, and lactose includes major bovine proteins such as beta-lactalbumin, lactoglobulin, and lactoferrin. Protein supplements based protein isolated from whey provides a pure source of high-quality protein with minimal amounts of fat, carbohydrates, and lactose. This protein also has the characteristic of being fast digestion and increase levels of amino acids available in the tissues needed to build muscle,[58–60] and has a high antioxidant capacity.[61,62] In this way, several technologies have been used. Thus, concentration of whey may be realized by heating and drying (evaporation, spray-drying, and freeze-drying) or by reverse osmosis. Membrane separation technologies have been used for obtain proteins ingredients from whey.[63]

7.2.1.6   INFANT FORMULAS

The interest in improving the biological and nutritional milk yield that is modified to resemble human and thus be used in infant formulas has increased in recent years. For this, the ingredients are isolated from bovine milk and are adapted using the human milk as suitable reference. The composition of the bovine milk, despite differs in many aspects relating to human milk, such as to the content of casein, lactose, mineral salts, and specifically in the proportion of proteins found (β-Lg is not in the breast milk), is still the main source of nutrition for infant formulas.[32]

7.2.1.7   EDIBLE FILMS

Environmental pollution caused by packaging made of polyethylene has been a major reason for the development of research to increase the interest in the use of biodegradable polymers forming food packaging. One of the most studied polymers was the whey proteins due to the excellent nutritional value, taste, and soft capacity to serve as a means to add color, flavor, and functional food ingredients.[25]

7.3   FEATURES OF PROTEINS AS A STARTING MATERIAL

Proteins are biomolecules consisting of carbon, nitrogen, hydrogen, and oxygen, which may also contain sulfur, phosphorus, iron, magnesium, copper, among others. Consist of 20–500 amino acids, in addition to the alpha amino and carboxyl alpha groups involved in the peptide bonds have a lateral chain with different functional groups, which give a distinctive characteristic are able to form three-dimensional amorphous networks mainly stabilized by noncovalent interactions and the functional properties of these materials are very dependent on the structural heterogeneity (due to amino acids that form the primary structure), thermal sensitivity, and the hydrophilic behavior of proteins.[64]

Depending on their own-primary sequence of each polypeptide chain structure, assumes different organizations on its axis—secondary structure—stabilized by hydrogen bonds and the tertiary structure—the three-dimensional organization reflects the poly peptide chain, based on the hydrogen bonding, Van der Waals forces, electrostatic and hydrophobic interactions, and disulfide bonds, to form globular protein or random fibrous structures. Finally, quaternary structure occurs as a consequence of the association of different polypeptide, equal or not to each other, which interact through noncovalent bonds resulting in single molecules. The functionality of proteins it is known as the expression of their physicochemical properties, which affect systems or food matrices.[65] These properties can be classified based on the ability of proteins to interact with water molecules to establish interactions with other proteins and with its molecular surface characteristics.[66] From materials, technology is important to recognize properties related to protein–protein interactions, which give rise to the formation of matrices with specific characteristics. With outstanding properties such as gel forming ability and the ability of forming materials such as films, coatings, fibers, among others.[64,67]

The inherent properties of these biopolymers make them excellent starting materials for films and/or biodegradable coatings. Within amino acids, polar, and nonpolar, the load distribution along the chain creates a chemical potential. In β-Lg, the domains of the polar and nonpolar areas can generate a matrix in a protein-based film through the interactive forces. These arrays or systems can be stabilized from electrostatic interactions, hydrogen bonding, Van der Waals force, covalent, and disulfide bonds.[68,69]

The functional properties of composite films, which take advantage of each component and decrease its disadvantages, depend on their composition and formation process. Knowledge of how each component interacts each other (physically or chemically) facilitates the design of the films or coatings with structural features and properties specific barrier.[70] Among the side benefits for using proteins to form films and coatings is the existence of multiple sites for its chemical interactions as function of its various functional groups of amino acid.

7.3.1 WHEY PROTEIN

The Food and Drug Administration (FDA), mentions that milk proteins must have all the proteins found naturally in milk, and must be in the same relations, while that isolates and concentrates of whey protein, are those obtained by removing nonprotein components and are free of casein. These milk proteins have properties that can provide desirable texture and other attributes to the final product. They have multiple applications in traditional foodstuffs. Various types of milk protein, similar to whey protein concentrates (WPC), whey protein isolates (WPI), casein and caseinates, among others, can be obtained from the waste generated by industrial complexes. To concentrate and separate the protein, various techniques have been used as ultrafiltration or ion exchange technology. Then subjected to a drying process to obtain the WPC and WPI, which are highly soluble, probably because in water, acquire colloidal dimensions, are amphoteric and the complete hydrolysis produces a mixture of amino acids. Depending on the influence of different pH and ionizable groups to which they are, they can develop loads, either positive or negative, or reach a net charge of zero to reach the isoelectric point (PI).[65]

The whey proteins differ from the caseins in their net negative charge that is uniformly distributed along the chain. Hydrophobic, polar, and charged amino acids are likewise uniformly distributed. Consequently, proteins fold and so most hydrophobic groups are enclosed within the protein molecule. Protein interactions that occur between chains determine the formation of the network of the films and their properties.[71]

Whey is a significant source of functional proteins such as β-Lg and α-La mainly obtained as a byproduct of the cheese industry and casein.[72] These functional proteins confer to whey a distinctive characteristic such as solubility, stability of systems interphase, and protein thermal stability.[73] The whey proteins possess an excellent nutritional value, variable solubility in water, and aptitude as emulsifier agent.[74] There is research on the use of milk proteins as edible films.[75,76] For their high-protein content, both the concentrate (WPC, 80% protein approx.) and isolated (WPI, >90% protein) of whey proteins, are ideal for the formation of edible films.[77] However, the prior denaturation of β-Lg and α-La, is necessary to expose the hidden sulfhydryl (–SH) and disulfide (SS) groups in the hydrophobic center of the native globular tertiary structure of these proteins. The subsequent formation of intermolecular disulfide bonds, mainly in the nanomeric units of β-Lg[78], promotes the generation of a stable three-dimensional network. It is therefore necessary to incorporate a plasticizer (e.g., glycerol or sorbitol) to decrease the density and reversibility of intermolecular interactions and increase chain mobility and thus flexibility of the film.[79] Whey-based Films plasticized with glycerol are excellent barriers to O_2, CO_2, and C_2H_4.[75,80] Unfortunately, its highly hydrophilic characteristic make it turns into not good barriers to water vapor.[77] Using this type of film in conjunction with hydrophobic membranes makes this type of coatings the ideal ones for the study of postharvest fruits and vegetables, especially in those highly perishable fruit.

7.4 INTERFACIAL AND ELECTROSTATIC INTERACTIONS

Homogenization is a technique widely used in the food industry that obtains consistent uniformly mixing, in order to obtain a dispersion of oil in these products, that is, a system having a dispersed phase and a continuous phase. In the process, aggregates are formed that are absorbed into the surface with the oil droplets newly formed. Proteins being the amphipathic molecules, have polar and nonpolar parts that are oriented in the interphase so that a portion of nonpolar amino acids are in contact with the oil phase and the polar groups with the aqueous phase.[81–83]

The bilayer film formation is formed by the homogenization of lipid in a concentrated protein solution to form an emulsion, which will allow dehydration.[84] The result is a continuous protein matrix covered by lipid droplets embedded on film. Although edible films are not effective as barriers bilayers films, these are superior to those of a single layer in the mechanical properties. The presence of the emulsion droplets in the film increases the distance traversed by water molecules, which diffuse through the film, therefore the water vapor permeability decreases. This is called tortuosity effect; part of the protein in the film is partially immobilized at the interface of the immobilized lipid droplets. This interfacial protein can adopt few configurations than the protein in the film mass. Biopolymer segments are less mobile, so the diffusion of water through the interfacial protein is reduced. Therefore, the water vapor permeability of the interfacial protein is lower than that of the mass of protein. This effect has been termed as interfacial interaction.[75] Tortuosity has been thought to depend only on the volume fraction of the lipid film, while the interfacial interaction is a function of the surface area of emulsified lipid, and this depends on the volume fraction and particle size.[75] Many researchers have attempted to obtain the lowest possible particle size in the emulsions in order to maximize stability and interfacial interactions. The phases of dispersed solids should provide a good effect of interfacial interaction because the absorbed protein cannot move in the plane of the interface. However, production of very small droplets of dispersed solid takes energy.[85]

A microstructural study of mixtures of isolated whey protein and mesquite gum stressed the importance of taking into account the electrostatic interactions of the components.[70] The electrostatic interactions of the polymer are determined by the physicochemical characteristics of each polymer (charge density, molecular weight, etc.), their concentrations, and solution conditions (pH, ionic strength, ion type, etc.).[86] For this study, the mixture of an anionic polysaccharide and a protein, electrostatic interaction has to be based on the net charge of each molecule that is pH dependent. In the study conditions, isolate whey protein is negatively charged, similar to mesquite gum, favoring the electrostatic repulsion between the two molecules, and consequently, the formation of aggregates.

Furthermore, the mixture of two different polymers often results in a phase separation of two domains in only a biopolymer. This happens due to the tendency of these molecules to be associated with other similar structures.[87] As a rule, the mixtures of biopolymers tend to segregate[88] and thermodynamically, protein and polysaccharides can be compatible or incompatible in aqueous solution. Incompatibility occurs when the repulsion between the biopolymers (e.g., when both are negatively charged) are in solution. In this case, the solvent/biopolymer interactions are favored as opposed to those biopolymer/biopolymer and solvent/solvent interactions. Finally becomes the system in two phases, with each phase becoming rich in one of the biopolymers.[89]

In a pH acid range from 1 to 3, protein molecules and k-carrageenan have opposite charges, thereby leaving a strong attraction there between. Under these conditions, it is difficult for the molecules of whey protein to form a three-dimensional protein matrix. The pH 6 provides the optimum conditions to obtain gels of mixture. A pH value of 6 is relatively close to the PI of the whey protein (Ip 5.2) where these have a low net charge and therefore tend to aggregate.[90]

The pH plays an important role in the formation of protein networks. The behavior of WPI is highly related to the PI, which is 5.2 to WPI. Below this, the total charge of WPI is mainly positive, while if the PI is greater, the overall charge is negative. WPI gels formed are transparent after heating, indicating protein denaturation in a pattern of fine wire. This behavior can be related to electrostatic repulsion between the protein molecules. In this case, proteins are negatively charged, when cations are added the bridges are enabled and the gelation is allowed.[91]

Complex formation of beta-casein and guar rubber causes a disruption of a small proportion of the secondary structure of the protein between galactomannans of *Mimosa scabrella* and milk proteins. Doublier[92] suggested that the mechanism of interaction and compatibility involve creating weak electrostatic complexes soluble in water, which can be destroyed in increasing ionic strength. Their rheological studies on dynamic and static systems showed that the fraction of whey protein (α-La and β-Lg) do not have interaction with the galactomannan of *M. scabrella*, shown by the absence of variation in the values of viscosity.[93]

7.4.1 HEATING

The application of heating in protein solutions improves its properties as a matrix, which is due to the connections between the protein chains. One of modifications frequently used are disulfide bonds or unions, which occur with heat treatment, followed by polymerization of protein chains.[94] This allows expose chains, sulfhydryl and hydrophobic groups. Hydrogen bonds and nonpolar hydrophobic groups in the molecule of the protein are disrupted, which expose amino groups solvent a more open structure.[95,96] The presence of thiol groups (β-Lg) is of importance to changes occurring in the solution during heating because engage in reactions with other proteins.[97] Polymerization occurs through the exchange of intermolecular disulfide bonds groups.[69,98] After heating, there may be noncovalent-type interactions for exposure of new groups in the whey proteins, these interactions may be ionic, hydrophobic, and van der Waals force.[99]

7.5 PROTEIN–LIPID INTERACTIONS

A study of diversity and versatility of lipid–protein interactions in biological systems demonstrated associations between these two models may be only in the surface of the hydrophobic membrane (located in biological systems) or involve penetration in a hydrophobic membrane. The regulation occurs through conformational changes resulting from phosphorylation, ligand binding, etc. Proteins utilize amphipathic helices to test the lipid composition of the membrane, particularly the overall content of anionic lipid in a specific anionic lipid.[100]

Biological studies of lipids in the membrane protein structures show that there are three general principles of lipid binding proteins that are distinguished by their lipid interactions with proteins. First, there is an annular cover of lipids bound to the surface of the protein, which resembles the bilayer structure. Second, the lipid molecules are embedded in cavities and crevices of the surface of the protein, frequently by multisubunit complexes and multimeric ensembles. These surface lipids not override are typically present in oligomeric subunits or interfaces. Finally, few examples represent lipids, which reside in the membrane protein or protein complex membrane and these are in unusual positions.[101]

7.6 PROTEIN–POLYSACCHARIDE INTERACTIONS

Proteins may present some modifications to their functional properties after they have been conjugated with some polysaccharides. These complexes require covalent bond formation, usually after a heat treatment and conditions of low water activity.[102,103]

Over time, various techniques have been used for better understanding of the possible interactions between proteins and polysaccharides present. Depending on aqueous environmental conditions such as pH, ionic strength, among others, it is possible to describe four different behaviors for interactions of milk proteins and anionic polysaccharides in the aqueous phase. (1) At neutral pH and low ionic strength, both proteins and polysaccharides are negatively charged and although electrostatic attractive interactions might exist between protein parts of positive and negative charges of the polysaccharide, these biopolymers are cosoluble in low concentrations, (2) pH values near to the PI or relatively low values, electrostatic protein–polysaccharide complexes are formed, (3) in a high reduction of aqueous phase, pH allows the aggregation of soluble complexes and complex coacervation, and finally (4) to less than pH 2.5, electrostatic complexes of biopolymers are generally removed due to the protonation of acidic functional groups of polysaccharide.[104,105]

Interactions exist thanks to the presence of attractive electrostatic and covalent bonding forces, which occur in the positively charged proteins and anionic polysaccharides with low energy that may result in an insoluble precipitate of both polymers.[103] The polysaccharide interactions with β-Lg, show that when β-Lg adsorbed onto the air–water interface in the presence of polysaccharides three phenomena can occur. First is that the polysaccharides are adsorbed on the interface competing with the protein for the interface (competitive adsorption). Second, the complexes of polysaccharides with adsorbed protein, are linked mainly by electrostatic interactions or hydrogen bonds[106,107] and finally the existence of limited thermodynamic compatibility between protein and polysaccharide. The polysaccharide adsorbed protein concentrate.

The performance of mixtures of polysaccharides is determined by the pH and the concentration and also by the ionic strength of the system under study. The pH generally strongly affects the net charge of the protein and plays an important role in interactions between protein and anionic polysaccharides.[86] Protein at the interface is partially deployed forming intermolecular hydrophobic associations or disulfide bounds.[108] Murray[109] mentions that some properties of multilayer films are enhanced by the mixture of proteins and polysaccharides. Protein–polysaccharide interactions usually are formed by the binding of the protein to the amino groups to reduce the carboxyl end groups of polysaccharides via a controlled Maillard reaction.[110,111] Frequently protein complexes–polysaccharide arise from noncovalent associations mainly by electrostatic attractive interactions. In a study of interfacial rheology measurements using the expansion method drop, an elastic behavior for the whey protein was found. Showed that at low concentrations of protein, where the first biopolymer molecules reach the interface at optimum conditions can completely absorb any locus and occupy a high interface area because the interface is still free of molecules. However, at high concentrations, abundant protein–protein interactions, side to side, at the interface, induce a more compact conformation of the protein.[112]

7.7 CONCLUDING REMARKS

Numerous studies have been conducted to elucidate the interactions of proteins with other polysaccharides, as well as for changes and/or modifications that occur in the protein structure when subjected to different processes to produce a final product. These studies have generated different models that allow predictions of what happens within a matrix made of proteins, and even further, theoretically elucidate what interactions with other components are distorted when the denaturalization is reached. It is important to note that all data shown were carried out under defined conditions of work; any changes or modifications can directly influence the behavior of interactions. Studies remain to be done to learn more about this topic and thus know if there are any changes or if there are new functional properties that can take advantage in proteins.

7.8   LOOKING AHEAD

Industrial and environmental practices require alternatives to the reutilization of serum for constructive purposes, since the degree of pollution caused by pouring the drain is high. The dairy industry it is constantly growing each year and it will so far, so it is necessary to further research and technological development to take new technological achievement of this waste, and thereby extend the diversification of products to reach the international market with innovative products and be able to handle large scale production benefiting both the environmental and economic sector. The dairy sector has suggested the implementation of a plant receiving serum from various companies to conduct dewatering processes thereof, so that represent a solution to the problem of disposal, which in turn would generate jobs and increased revenue.

KEYWORDS

- **dairy industry**
- **isolates**
- **reutilization**
- **edible films**
- **interactions**

REFERENCES

1. FAO. Manual de Capacitación en Nutrición y Alimentación de Peses y Camarones Cultivados, [Online] 2004, http://www.fao.org/docrep/field/003/ab492s/AB492S01.htm (accessed Nov 15, 2014).
2. Ruiz-Martínez, J. Cubiertas Comestibles óleo Proteicas Para Prolongar la Vida de Anaquel del tomate. M.S. Thesis, Universidad Autónoma de Coahuila, 2013.
3. Maftoonazad, N.; Ramaswamy, H. S.; Marcotte, M. Evaluation of Factors Affecting Barrier, Mechanical and Optical Properties of Pectin-based Films Using Response Surface Methodology. *J. Food Process. Eng.* **2007,** *30*(5), 539–563.
4. Ali, A.; Maqbool, M.; Ramachandran, S.; Alderson, P. G. Gum Arabic as a Novel Edible Coating for Enhancing Shelf Life and Improving Postharvest Quality of Tomato (*Solanum lycopersicum* L.) Fruit. *Postharvest Biol. Technol.* **2010,** *58,* 42–47.

5. Cerqueira, M. A.; Lima, A. M.; Texeira, J. A.; Moreira, R. A.; Vicente, A. A. Suitability of Novel Galactomannans as Edible Coatings for Tropical Fruits. *J. Food Eng.* **2009,** *94,* 372–378.

6. Bourbon, A. I.; Pinheiro, A. C.; Cerqueira, M. A.; Rocha, C.; Avides, M. C.; Quintas, M.; Vicente, A. A. Physico Chemical Characterization of Chitosan Based Edible Films Incorporating Bioactive Compounds of Different Molecular Weight. *J. Food Eng.* **2011,** *106,* 111–118.

7. De Aquino, A. B.; Blank, A. F.; De Aquino Santana, L. C. L.; Impact of Edible Chitosan–Cassava Starch Coatings Enriched with Lippia Gracilis Schauer Genotype Mixtures on the Shelf Life of Guavas (*Psidium guajava* L.) During Storage at Room Temperature. *Food Chem.* **2015,** *171,* 108–116.

8. Slavutsky, A. M.; Bertuzzi, M. A. Formulation and Characterization of Nanolaminated Starch Based Film. *LWT Food Sci. Technol.* **2015,** *61*(2), 407–413.

9. Guerreiro, A. C.; Gago, C. M. L.; Faleiro, M. L.; Miguel, M. G. C.; Antunes, M. D. C. The Effect of Alginate-based Edible Coatings Enriched with Essential Oils Constituents on *Arbutus Unedo* L. Fresh Fruit Storage. *Postharvest Biol. Tec.* **2015,** *100,* 226–233.

10. Narsaiah, K.; Wilson, R. A.; Gokul, K.; Mandge, H. M.; Jha, S. N.; Bhadwal, S.; Anurag, R. K.; Malik, R. K.; Vij, S. Effect of Bacteriocin-incorporated Alginate Coating on Shelf-life of Minimally Processed Papaya (Carica papaya L.). *Postharvest Biol. Technol.* **2015,** *100,* 212–218.

11. Arismendi, C.; Chillo, S.; Conte, A.; Del Nobile, M. A.; Flores, S.; Gerschenson, L. N. Optimization of Physical Properties of Xanthan Gum/Tapioca Starch Edible Matrices Containing Potassium Sorbate and Evaluation of Its Antimicrobial Effectiveness. *LWT Food Sci. Technol.* **2013,** *53*(1), 290–296.

12. Zambrano-Zaragoza, M. L.; Mercado Silva, E.; Del Real, L. A.; Gutiérrez-Cortez, E.; Cornejo Villegas, M. A.; Quintanar-Guerrero, D. The Effect of Nano-coatings with α-Tocopherol and Xanthan Gum on Shelf-life and Browning Index of Fresh-cut "Red Delicious" Apples. *Innov. Food Sci. Emerg. Technol.* **2014,** *22,* 188–196.

13. Ünalan, I. U.; Arcan, I.; Korel, F.; Yemenicioğlu, A. Application of Active Zein-based Films with Controlled Release Properties to Control Listeria Monocytogenes Growth and Lipid Oxidation in Fresh Kashar Cheese. *Innov. Food Sci. Emerg. Technol.* **2013,** *20,* 208–214.

14. Yin, Y. C.; Yin, S. W.; Yang, X. Q.; Tang, C. H.; Wen, S. H.; Chen, Z.; Xiao, B. J.; Wu, L. Y. Surface Modification of Sodium Caseinate Films by Zein Coatings. *Food Hydrocolloid.* **2014,** *36,* 1–8.

15. Khanzadi, M.; Jafari, S. M.; Mirzaei, H.; Chegin, F. K; Maghsoudlou, Y.; Dehnad, D. Physical and Mechanical Properties in Biodegradable Films of Whey Protein Concentrate–Pullulan by Application of Beeswax. *Carbohyd. Polym.* **2015,** *118,* 24–29.

16. Synowiec, A.; Gniewosz, M.; Kraśniewska, K.; Przybył, J. L.; Bączek.; Węglarz, Z. Antimicrobial and Antioxidant Properties of Pullulan Film Containing Sweet Basil Extract and an Evaluation of Coating Effectiveness in the Prolongation of the Shelf Life of Apples Stored in Refrigeration Conditions. *Innov. Food Sci. Emerg. Technol.* **2014,** *23,* 171–181.

17. Ding, C.; Zhang, M.; Li, G. Preparation and Characterization of Collagen/Hydroxypropyl Methylcellulose (HPMC) Blend Film. *Carbohyd. Polym.* **2015,** *119,* 194–201.

18. Fagundes, C.; Palou, L.; Monteiro, A. R.; Pérez-Gago, M. B. Effect of Antifungal Hydroxypropyl Methylcellulose-beeswax Edible Coatings on Gray Mold Development

and Quality Attributes of Cold-stored Cherry Tomato Fruit. *Postharvest Biol. Tec.* **2014,** *92,* 1–8.

19. Martins, J. T.; Cerqueira, M. A.; Bourbon, A. I.; Pinheiro, A. C.; Souza, B. W. S.; Vicente, A. A. Synergistic Effects Between κ-Carrageenan and Locust Bean Gum on Physicochemical Properties of Edible Films Made Thereof. *Food Hydrocolloid.* **2012,** *29*(2), 280–289.

20. Montalvo, C.; López, A.; Palou, E. Películas Comestibles De Proteína: Características, Propiedades Y Aplicaciones. *TSIA.* **2012,** *6–2,* 32–46.

21. Carrillo Aguado, J. L. Tratamiento Y Reutilización Del Suero De Leche. In *Mundo Lácteo y Cámico;* 2006; pp 27–30.

22. Villada, H. S.; Acosta, H. A.;Velasco, R. J. Biopolymers Naturals Used in Biodegradable Packaging. *Temas Agrarios.* **2007,** *12*(2), 5–13.

23. Tharanathan, R. N. Biodegradable Films and Composite Coatings: Past, Present and Future. *Trends. Food Sci. Tech.* **2003,** *14,* 71–78.

24. Vroman, I.; Tighzert, L. Biodegradable Polymers. *Materials.* **2009,** *2,* 307–344.

25. Li, C.; Chen, H. Biodegradation of Whey Protein-based Edible Films. *J. Polym. Environ.* **2000,** *8,* 135–143.

26. Comisión Nacional del Medio Ambiente Región Metropolitana (CONAMA/RM). Guía Para el Control y Prevención de la Contaminación Industrial. CRC Press; Santiago, Chile, **1998**; 58.

27. Restrepo, M. Producción más Limpia en la Industria Alimentaria. *Rev. Producción Limpia.* **2006,** *1*(1), 88–101.

28. González Cáceres, M. J. Aspectos Medio Ambientales Asociados a Los Procesos de la Industria Láctea. *Mundo Pecuario.* **2012,** *8*(1), 16–32.

29. SIAP. Resumen Nacional de Producción Agrícola. [Online]2014, http://www.siap.gob. mx/index.php?option=com_wrapperandview=wrapperandItemid=258 (accessed Oct 23, 2014).

30. Foegeding, S.; Luck, P. Whey Protein Products. In *Encyclopedia of Dairy Sciences;* Academic Press: Cambridge, MA, **2002**; 1957–1960.

31. Jelen, P. Whey Processing. In *Encyclopedia of Diary Sciences*; Roginski, H., Fuquay, J. F., Fox, P. F., Eds.; Academic Press (an Imprint of Elsevier Amsterdam): Boston, London, 2003; p 2740.

32. Parra, R. Lactosuero; Importancia en la Industria De Alimentos. *Rev. Fac. Nal. Agr. Medellin.* **2009,** *62*(1), 4967–4982.

33. Berruga, M. I. Desarrollos de Procedimientos Para El Tratamiento de Efluentes de Quesería. Ph.D. Thesis, Universidad Complutense de Madrid, Facultad de Veterinaria, 1999.

34. Muñi, A.; Paez, G.; Faria, J.; Ferrer, J.; Ramones, E. *Eficiencia De Un Sistema De Ultrafiltración/Nanofiltración Tangencial En Serie Para El Fraccionamiento Y Concentración Del Lactosuero. Rev. Científ. FCV-LUZ.* **2012,** *15*(4), 361–367.

35. Liu, X.; Powers, J. R.; Swanson, B. G.; Hill, H. H.; Clark, S. High Hydrostatic Pressure Affects Flavor-binding Properties of Whey Protein Concentrate. *J. Food Sci.* **2005,** *70,* C581–C585.

36. Londoño M. Aprovechamiento Del Suero Ácido De Queso Doble Crema Para La Elaboración De Quesillo Utilizando Tres Métodos De Complementación De Acidez Con Tres Ácidos Orgánicos. In *Revista Perspectivas en Nutrición Humana-Escuela de Nutrición y Dietética;* Universidad de Antioquia: Medallin, Colombia, **2006;** Vol.16, 11–20.

37. Panesar, P.; Kennedy, J.; Gandhi, D.; Bunko K. Bioutilisation of Whey for Lactic Acid Production. *Food Chem.* **2007**, *105,* 1–14.
38. Ghaly, A.; Kamal M. Submerged Yeast Fermentation of Acid Cheese Whey for Protein Production and Pollution Potential Reduction. *Water Res.* **2004**, *38*(3), 631–644.
39. Mukhopadhyay, R.; Chatterjee, S.; Chatterjee, B. P.; Banerjee, P.; Guha, A. Production of Gluconic Acid From Whey by Free and Immobilized Aspergillus Niger. *Int. Dairy J.* **2005**, *15*(3), 299–303.
40. Koutinas, A.; Papapostolou, H.; Dimitrellou, D.; Kopsahelis, N.; Katechaki, E.; Bekatorou, A.; Bosnea, L. Whey Valorisation: A Complete and Novel Technology Development for Dairy Industry Starter Culture Production. *Bioresour. Technol.* **2009**, *100*(15), 3734–3739.
41. Almeida, K. E.; Tamime, A. Y.; Oliveira, M. N. Influence of Total Solids Contents of Milk Whey on the Acidifying Profile and Viability of Various Lactic Acid Bacteria. *LWT Food Sci. Technol.* **2009**, *42*(2), 672–678.
42. Linden, G.; Lorient, D. *Bioquímica Agroindustrial: Revalorización Alimentaria De La Producción Agrícola*; Editorial Acribia: Zaragoza, España, 1996; p 454.
43. L. Baro, J. Jiménez, A. Martínez, J. Bouza. Péptidos y Proteínas de la Leche Con Propiedades Funcionales. *J. Ars. Pharmaceutica.* **2001**, *42*(3–4), 135–145.
44. Hinrichs, R.; Gotz, J.; Noll, M.; Wolfschoon, A.; Eibel, H.; Weisser, H. Characterization of Different Treated Whey Protein Concentrates by Means of Low-resolution Nuclear Magnetic Resonance. *Int. Dairy J.* **2004**, *14*(9), 817–827.
45. Andrade, L. Efecto Del Flujo De Alimentación Sobre La Ultrafiltración Del Suero Pasteurizado De Queso. Engineer Thesis, Escuela Agrícola Panamericana Zamorano Honduras, 1999, pp 1–24.
46. Johnson, B. Los Concentrados De Proteína De Suero Y Sus Aplicaciones En Productos Bajos En Grasa: Alfa Editores Técnicos México 19. 2004, pp 1–3.
47. Teniza García, O. Estudio del Suero de Queso de Leche de Vaca y Propuesta Para el Reusó Del Mismo. M.S. Thesis, Instituto Politécnico Nacional, Centro de Investigación en Biotecnología Aplicada Unidad Tlaxcala México, 2008.
48. Marwaha, S. S.; Kennedy, J. F.; Sehgal, V. K. Simulation of Process Conditions of Continuous Etanol Fermentation of Whey Permeate Using Alginate Entrapped *Kluyveromyces marxianus* NCYC 179 Cells in a Packed-bed Reactor System. *Process Biochem.* **1988**, *23*(2), 17–22.
49. Castillo, F. J. Lactose Metabolism Yeasts. In *Yeast: Biotechnology and Biocatalysis;* Verachtert, H., De Mot, R., Eds.; Marcel Dekker Inc.: New York, 1990; pp 297–320.
50. Marth, E. H. Fermentation Products from Whey. In *Byproducts of Milk;* Webb, B. H., Whittier E. O., Eds.; Avi Pub. Co: Westport, NJ, 1970; pp 43–82.
51. García, M.; Quintero, R.; López, A. *Biotecnología Alimentaria.* Lamusa, S. A., Ed.; Editorial Limusa: México, 2004; pp 186–207.
52. Gosta B. *Manual de Industrias lácteas;* Ediciones Mundi-Prensa: Madrid, 2003.
53. Suárez, E.; Lobo, A.; Alvarez, S.; Riera, F. A.; Álvarez, R. Demineralization of Whey and Milk Ultrafiltration Permeate by Means of Nanofiltration. *Desalination.* **2009**, *241*(1–3), 272–280.
54. Brandelli, A.; Daroit, D. J.; Corrêa, A. P. F. Whey as a Source of Peptides with Remarkable Biological Activities. *Food Res. Int.* **2015**, *73,* 149–161.
55. Zadow, J. Protein concentrates and fractions. In *Encyclopedia of Food Science and Technology;* Francis, F., Ed.; Wiley: New York, 2003; pp 6152–6156.

56. Santana, M.; Rolim, E.; Carreiras, R.; Oliveira, W.; Medeiros, V.; Pinto, M. Obtaining Oligopeptides From Whey: Use of Subtilisin and Pancreatin. *Am. J. Food Technol.* **2008,** *3*(5), 315–324.

57. Spellman, D.; O'Cuinn, G.; FitzGerald, R. Bitterness in Bacillus Proteinase Hydrolysates of Whey Proteins. *Food Chem.* **2009,** *114*(2), 440–446.

58. Paul, G. L. The Rationale for Consuming Protein Blends in Sports Nutrition. *J. Am. Coll. Nutr.* **2009,** *28*(4), 464S–472S.

59. Tang, J.; Moore, D.; Kujbida, G.; Tarnopolsky, M.; Phillips, S. Ingestion of Whey Hydrolysate, Casein, or Soy Protein Isolate: Effects on Mixed Muscle Protein Synthesis at Rest and Following Resistance Exercise in Young Men. *Appl. Physiol.* **2009,** *107,* 987–992.

60. Phillips, S.; Hartman, J.; Wilkinson, S. Dietary Protein to Support Anabolism with Resistance Exercise in Young Men. *J. Am. Coll. Nutr.* **2005,** *24*(2), 134S–139S.

61. Bayram, T.; Pekmez, M.; Arda, N.; Yalçın, A. S. Antioxidant Activity of Whey Protein Fractions Isolated by Gel Exclusion Chromatography and Protease Treatment. *Talanta.* **2008,** *75*(3), 705–709.

62. Kong, B.; Peng, X.; Xiong, Y. L.; Zhao, X. Protection of Lung Fibroblast MRC-5 Cells Against Hydrogen Peroxide-induced Oxidative Damage by 0.1–2.8 kDa Antioxidative Peptides Isolated from Whey Protein Hydrolysate. *Food Chem.* **2012,** *135*(2), 540–547.

63. Brans, G.; Schroën, C. G. P. H.; van der Sman, R. G. M.; Boom, R. M. Membrane Fractionation of Milk: State of The Art and Challenges. *J. Membr. Sci.* **2004,** *243*(1–2), 263–272.

64. Mauri, A. N.; Añon, M. C. Effect of Solution pH on Solubility and Some Structural Properties of Soybean Protein Isolate Films. *J. Sci. Food Agr.* **2006,** *86*(7), 1064–1072.

65. Badui, S. Ed.; *Química de los alimentos;* Alhambra Mexicana: México D. F, 1999, p 648.

66. Cheftel, J. C.; Cuq, J. L.; Lorient, D. Amino Acids, Peptides and Proteins. In *Food Chemistry;* Fennema, O. R., Ed.; Marcel Dekker Inc.: New York, 1985; p 246.

67. Petruccelli, S. Modificaciones Estructurales de Aislados Proteicos de Soja Producidas por Tratamientos Reductores y Térmicos y su Relación con Propiedades Funcionales. Ph.D. Thesis, Universidad Nacional De La Plata, 1993.

68. Krochta, J. M.; Baldwin, E. A.; Nisperos-Carriedo, M. *Edible Coatings and Films to Improve Food Quality*; Technomic Publishing Co.: Lancaster, PA, 1994.

69. Tomasula, P. M.; Qi, P.; Dangaran, K. L. Structure and Function of Protein-based Edible Films and Coatings. In *Edible Films and Coatings for Food Applications;* Embuscado, M. E., Huber, K. C., Eds.; Springer Science Media: New York, 2009; pp 25–26.

70. Osés, J.; Fabregat Vázquez, M.; Pedroza Islas, R.; Tomas, S. A.; Cruz Orea, A.; Maté, J. I. Development and Characterization of Composite Edible Films Based on Whey Protein Isolate and Mesquite Gum. *J. Food Eng.* **2009,** *92,* 56–62.

71. Swaisgood, H. E. Characteristics of Milk. In *Food chemistry;* Fennema, O., Ed.; Marcel Dekker: New York, 1996; pp 841–878.

72. Fox, P. F.; McSweeney, P. L. H. *Dairy Chemistry and Biochemistry;* Springer Science and Business Media: Berlin, Germany, **1998;** 478.

73. Capitani, C. D.; Pacheco, M. T. B.; Gumerato, H. F.; Vitali, A.; Schmidt, F. L. Recuperação De Proteínas Do Soro De Leite Por Meio De Coacervação Com Polissacarídeo (Milk Whey Protein Recuperation by Coacervation with Polysaccharide). *Braz. J. Agric. Res.* **2005,** *40,* 1123–1128.

74. Walstra, P.; Geurts, T. J.; Noomen, A.; Jellema, A. *Dairy Technology: Principles of Milk Properties and Processes;* Lmarcel Dekker: New York, 1999; pp 727.

75. McHugh, T.; Krochta, J. Water Vapor Permeability Properties of Edible Whey Protein-lipid Emulsion Films. *J. Am. Oil Chem. Soc.* **1994,** *71,* 307–312.
76. Krochta, J. M.; de Milder Johnson, C. Edible Films and Biodegradable Polymer Films: Challenges and Opportunities. *Food Technol.* **1997,** *51*(2), 61–74.
77. Gallieta, G. Formación y Caracterización de Películas Comestibles en Base a Suero de Leche. M.S. Thesis, Universidad de la República Oriental del Uruguay, Montevideo. Uruguay 2001.
78. Monahan, F.; McClements, J. D. J.; Kinsella, J. E. Polymerization of Whey Protein in Whey Protein-stabilized Emulsion. *J. Agr. Food Chem.* **1993,** *41,* 1826.
79. Banker, G. S. Film Coating Theory and Practice. *J. Pharm. Sci.* **1996,** *55,* 81–89.
80. Gallieta, G.; Vanaya, F.; Ferrari, N.; Diano, W. Barrier Properties of Whey Protein Isolate Films to Carbón Dioxide and Ethylene at Cariouswater Activities. *Colonna, P. and Non Food Applications.* Les Colloques N.91 INRA Editions; Montepellier: France, 1998, pp 327–335.
81. Singh, H.; Ye, A. Interactions and Functionality of Milk Proteins in Food Emulsions. In *Milk Proteins: From Expression to Food;* Thompson, A., Boland, M., Singh, H., Eds.; Academic Press: Boston, MA, 2009; pp 321–340.
82. Dickinson, E. Structure and Composition of Adsorbed Protein Layers and the Relationship to Emulsion Stability. *J. Chem. Soc. Farad.* **1992,** *88,* 2973–2983.
83. Dickinson, E. Stability and Rheological Implications of Electrostatic Milk Protein–Polysaccharide Interactions. *Trends. Food Sci. Tech.* **1998,** *9,* 347–354.
84. Krochta, J. M. Food Emulsion and Foams. In *Theory and Practice;* Wan, P. J., Cavallo, J. L., Saleeb, F. Z., McCarthy, M. J., Eds.; American Institute of Chemical Engineers: New York, 1993; pp 57–61.
85. Fairley, P.; Krochta, J. M.; German, J. B. Interfacial Interactions in Edible Films from Whey Protein Isolate. *Food Hydrocoll.* **1997,** *11*(3), 245–252.
86. Schmitt, C.; Sanchez, C.; Desobry-Banon, S.; Hardy, J.; Structure and Technofunctiona Properties of Protein–Polysaccharide Complexes: A Review. *Crit. Rev. Food Sci.* **1998,** *38*(8), 689–753.
87. Norton, I. T.; Frith, W. J. Microstructure Design in Mixed Biopolymer Composites. *Food Hydrocoll.* **2001,** *15,* 543–553.
88. Kruif, C. G.; Tuinier, R. Polysaccharide–Protein Interactions. *Food Hydrocoll.* **2001,** *15,* 555–563.
89. Nayebzadeh, K.; Chen, J.; Mousavi, M. S. M. Interaction of WPI and Xanthan in and Rheological Properties of Protein Gels and O/W Emulsions. *Int. J. Food Eng.* **2007,** *3*(4), 1–17.
90. Mleko, S.; Li-Chan, E. C. Y.; Pikus. S. Interactions of k-carrageenan with Whey Proteins in Gels Formed at Different pH. *Food Res. Int.* **1997,** *30*(6), 427–433.
91. Turgeon, S. L.; Beaulieu, M. Improvment and Modification of Whey Protein Gel Texture Using Polysaccharides. *Food Hydrocoll.* **2001,** *15,* 583–591.
92. Doublier, J. L.; Garnier, C.; Renard, D.; Sanchez, C. Protein–Polysaccharide Interactions. *Curr. Opin. Colloid Interface Sci.* **2000,** *5,* 202–214.
93. Perissutti, G. E.; Bresolin, T. M. B.; Ganter, J. L. M. S. Interaction between the Galacto-mannan from *Mimosa scabrella* and Milk Proteins. *Food Hydrocoll.* **2002,** *16,* 403–417.
94. Pérez-Gago, M. B.; Nadaud, P.; Krochta, J. M. Water Vapor Permeability Solubility, and Tensile Properties of Heat Denatured Versus Native Whey Protein Films. *J. Food Sci.* **1999,** *64,* 1034–1037.

95. Damodaran, S. Amino Acids, Peptides and Proteins. In *Fennema's Food Chemistry;* Damodaran, S., Parkin, K. L., Fennema, O. R., Eds.; CRC Press, Taylor and Francis Group: Boca Raton, FL, **2008**, 217–330.

96. Wihodo, M.; Moraru, C. I. Physical and Chemical Methods Used to Enhance the Structure and Mechanical Properties of Protein Films: A Review. *J. Food Eng.* **2013**, *114*, 292–302.

97. Chae, S. I.; Heo, T. R. Production and Properties of Edible Film Using Whey Protein. *Biotechnol. Bioprocess Eng.* **1997**, *2*, 122–125.

98. Galani, D.; Apenten, R. K. O. Heat Induced Denaturation and Aggregation of Beta-Lactoglobulin: Kinetics of Formation of Hydrophobic and Disulphide Linked Aggregates. *Int. J. Food, Sci. Tech.* **1999**, *34*, 467–476.

99. Kinsella J. E. Milk Proteins: Physicochemical and Functional Properties. *Crt. Rev. Food Sci. Nutr.* **1984**, *21*, 197–262.

100. Dowhan, W.; Mileykovskaya, E.; Bogdanov, M. Diversity and Versatility of Lipid Protein Interactions Revealed by Molecular Genetic Approaches. *Biochim. Biophys. Acta.* **2004**, *1666*(1–2), 19–39.

101. Palsdottir, H.; Hunte, C. Lipids in Membrane Protein Structures. *Biochim. Biophys. Acta.* **2004**, *1666*, 2–18.

102. Alfaro Arismendi, M. J. Desarrollo de Metodologías de Encapsulación Utilizando Aislado Proteico de Suero Lácteo Modificado con Azúcares Para la Protección de Ingredientes Activos en Alimentos. Ph.D. Thesis, Universidad Central de Venezuela, Facultad de Ciencias en Ciencia y Tecnología de Alimentos, 2012.

103. Muñoz, J.; del C. Alfaro, M.; Zapata I. Avances en la Formulación de Emulsions. *Grasas Aceites.* **2007**, *58*(1), 64–73.

104. Weinbreck, F.; de Vries, R.; Schroogen, P.; de Kruif, C. G. Complex Coacervation of Whey Proteins and Gum Arabic. *Biomacromol.* **2003**, *4*(2), 293–303.

105. Perez, A. A.; Carrara, C. R.; Carrera, C.; Rodriguez Patino, J. M. Interactions between Milk Whey Protein and Polysaccharide in Soluton. *Food Chem.* **2009**, *116*, 104–113.

106. Dickinson, E. Hydrocolloids at Interfaces and the Influence on the Properties of Dispersed Systems. *Food Hydrocoll.* **2003**, *17*(1), 25–40.

107. Baeza, R.; Carrera Sánchez, C.; Pilosof, A. M. R.; Rodríguez Patino, J. M. Interactions of Polysaccharides with β-Lactoglobulin Adsorbed Films at the Air–Water Interface. *Food Hydrocoll.* **2005**, *19*, 239–248.

108. Funtenberger, S.; Dumay, E.; Cheftel, J. C. Pressure Induced Aggregation of Beta-Lactoglobulin in pH 7.0 Buffers. *Food Sci. Technol.* **1995**, *28*(4), 410–418.

109. Murray, B. S. Rheological Properties of Protein Films. *Curr. Opin. Colloid Interface Sci.* **2011**, *16*, 27–35.

110. Kato, A.; Sasaki, Y.; Furuta, R.; Kobayashi K. Functional Protein/Polysaccharide Conjugate Prepared by Controlled Dry Heating of Ovalbumin/Dextran Mixtures. *Agric. Biol. Chem.* **1990**, *54*, 107–112.

111. Bouyer, E.; Mekhloufi, G.; Rosilio, V.; Grossiord, J. L.; Agnely, F. Proteins, Polysaccharides, and Their Complexes Used as Stabilizers for Emulsions: Alternatives to Synthetic Surfactants in the Pharmaceutical Field. *Int. J. Pharm.* **2012**, *436*, 359–378.

112. Wüstnek, R.; Moser, B.; Muschiolik, G. Interfacial Dilational Behaviour of Adsorbed Beta-Lactoglobulin Layers at the Different Fluid Interfaces. *Colloids Surf. B.* **1999**, *15*, 263–273.

CHAPTER 8

GUAR GUM AS A PROMISING HYDROCOLLOID: PROPERTIES AND INDUSTRY OVERVIEW

CECILIA CASTRO-LÓPEZ[1], JUAN C. CONTRERAS-ESQUIVEL[1], GUILLERMO C. G. MARTINEZ-AVILA[2], ROMEO ROJAS[2], DANIEL BOONE-VILLA[3], CRISTOBEL N. AGUILAR[1], and JANET M. VENTURA-SOBREVILLA[1,*]

[1]*Autonomous University of Coahuila. School of Chemistry, Department of Food Science and Technology, 25280 Saltillo, Coahuila, México*
[2]*Autonomous University of Nuevo Leon School of Agronomy, Laboratory of Chemistry and Biochemistry, 66050 General Escobedo, Nuevo León, México*
[3]*Autonomous University of Coahuila. School of Medicine, 26090 Piedras Negras, Coahuila, México*
*Corresponding author. E-mail: janethventura@uadec.edu.mx

CONTENTS

ABSTRACT

Guar gum (GG) is a galactomannan, obtained from *Cyamopsis tetragonolobus* (Leguminosae) which has been cultivated in India and Pakistan for centuries. Chemically, GG is a hydrocolloid polysaccharide composed by mannose and galactose in molecular ratio of 2:1. It has wide applications in pharmaceutical formulations, cosmetics, foods, paper, mining, and many other industries, and is used as a natural thickener, emulsifier, stabilizer, bonding, and gelling agent, soil stabilizer, natural fiber, flocculant, and fracturing agent. GG is white to yellowish white powder, nearly odorless with bland taste that is manufactured by mechanical extraction of endosperm from the guar seed. These seeds are separated from the plant and dried. Refined guar splits are first obtained by roasting, dehusking, and polishing. These splits are then pulverized and tailor made in various mesh sizes for usage in the different industries. It is practically insoluble in organic solvents. In cold or hot water, it disperses and swells almost immediately to form some highly viscous thixotropic solutions. Through references reported in the literature about the GG, the aim of this chapter was to review the occurrence of this gum, its production, physicochemical properties, identification, and industry applications.

8.1   INTRODUCTION

The term hydrocolloid is derived from the Greek "hydro" (water) and "kola" (glue). Hydrocolloids can be defined as molecules of high molecular weight that usually have colloidal properties, capable of producing gels when are combined with the appropriate solvent. The presence of a large number of hydroxyl groups in their structure significantly increases their affinity for water molecules, making them hydrophilic compounds.[37] This term is applied to a variety of substances with gummy characteristics. However, it is more common to use the term to refer to polysaccharides or their derivatives, obtained from plants or microbiological processing and which may or may not be chemically modified to change or improve their technological capabilities.[76,77,79]

The heterogeneous group consists of polysaccharides and proteins and the researchers have mainly studied them due to their range of use in industry and the functionalities that impart to whatever system or product into which they are incorporated.[4,58,80] The commercially important hydrocolloids and their origins are given in Figure 8.1. In recent years, hydrocolloids from some seeds such as quince gum (*Cydonia oblonga*), vinal gum (*Prosopis ruscifolia*), and espina corona gum (*Gleditsia amorphoides*),[6,55,60] especially guar gum (GG, *Cyamopsis tetragonolobus*) have evoked tremendous interest due to its several industry applications.[59,63] GG have a wide range of functional properties, these include thickening, gelling, emulsifying, stabilization, and controlling the crystal growth, among others, but the functionality in these capacities can be attributed to the chemical characteristics of the monomers and the polymer's molecular structure (including the chain length and branching pattern where pertinent). These characteristics also determine how the GG yield is affected by factors, such as temperature, pH, presence of certain ions, etc.[45,46,53] Consequently, there is a large body of knowledge of how to exploit GG properties in order to improve future applications. In this chapter, information on GG, origin, manufacturing, structure, properties, and applications is described in subsequent sections to understand better its importance and potential.

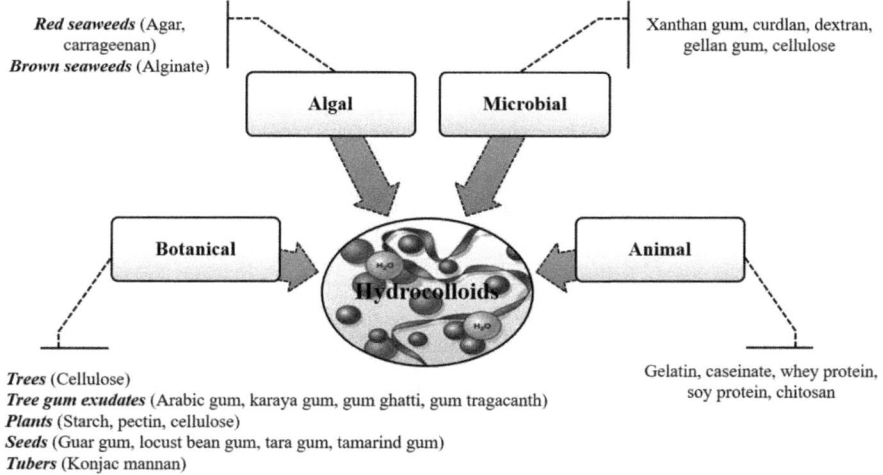

FIGURE 8.1 Sources of important hydrocolloids.

8.2 GUAR GUM

8.2.1 CULTIVATION

The guar plant, *C. tetragonolobus* (L.) Taub. (family *Leguminosae*), commonly known as guaran or cluster bean, is an important leguminous annual crop. It grows upright, reaching a height of 2–3 m. It has a main single stem with either basal branching or fine branching along the stem. It is a robust, bushy, semiupright type of plant. Guar has well developed tap root system. Stems and branches are angular, grooved, forked hairs, sometimes glaucous. Guar has branched and unbranched growth habit. It has pointed saw-toothed, alternate, trifoliate leaves with small purple and white flowers borne along the axis of spikelet. It bears hairy pods in clusters of 4–12 cm length, each pod with 7–8 seeds. Seed is hard, flinty, flattened, ovoid, and about 5 mm long, also are white, grey, or black in color (Fig. 8.2). Guar plant grows in specific climatic condition, which ensure a soil temperature around 25°C for proper germination, long photoperiod, with humid air during its growth period and finally short photoperiod with cool dry air at flowering and pod formation.[30] This crop prefers a well-drained sandy loam soil. It can tolerate saline and moderately alkaline soils with pH ranging between 7.5 and 8.0. Heavy clay soils, poor in nodulation, and bacterial activities are not suitable. Finally, the plants are sown after the first rains in July and harvested in October–November, being a short-cycle crop that is harvested within 3, 4 months of its plantation.[43,65]

FIGURE 8.2   Guar (a) pods, (b) seeds, (c) splits, and (d) powder.

Guar plant is native to the Indian subcontinent; however, this crop is also grown in other parts of the world, such as, Pakistan, Australia, Brazil, EE, UU, Malawi, and South Africa.[82] India is the world leader with 80% production of guar with its cultivation in semiarid, Northwestern parts of country encompassing States of Rajasthan, Gujarat, Haryana, and Punjab. Pakistan with 15% of world production is next to India. Remaining 5% guar is produced in rest of the world with a total production of 15,000 Tons per annum. Efforts have been made to promote cultivation of guar in Australia by the Department of Agriculture and Rural Industrial Development Agency where the research found that guar could be grown in the northern region relatively successfully. Similarly, it is reported that countries such as China and Thailand are also trying to grow guar. Therefore, in the future, economy developed by cultivation of guar not only remain monopolized by India and Pakistan.[3,40]

8.2.2 EXTRACTION

In general, the guar seeds consist of three parts: The hull (14–17%), germ (43–47%), and endosperm (35–42%). GG, the principal marketable processed product of the plant, comes from the endosperm.[61] Several methods have been used for the manufacture of different grades of GG, but due to its complex nature, the thermo-mechanical process is generally used for the manufacture of edible grade and industrial grade GG.[28,48] This process is normally undertaken by using unit operations of roasting, differential attrition, sieving, and polishing (Fig. 8.3).[28] It is very important to select guar split in this process. The split is screened to clean it and then soaked to prehydrate it in a double cone mixer. The prehydrating stage is very important because it determines the rate of hydration of the final product. Soaked splits, which have reasonably high moisture content, are passed through a flaker. Flaked guar splits are ground and then dried. Obtained powder is screened through rotary screens to deliver required particle size. Oversized particles are either recycled to main ultrafine or reground in a separate regrind plant, according to the viscosity requirement. This stage helps to reduce load at the grind step. Soaked splits are difficult to grind. Direct grinding of those generates undesirable over heating in the grinder, which reduces hydration of the final product. Through the heating, grinding, and polishing process, husk is separated from the endosperm halves and refined GG splits are obtained. After grinding process, refined guar splits are then treated and converted into powder. The split manufacturing process yields husk and germ called "guar meal," widely sold in the international market as cattle feed. Manufacturers define different quality grades of GG by its particle size, the viscosity generated with a given concentration, and the rate at which that viscosity is developed.[12,27,51,52,66]

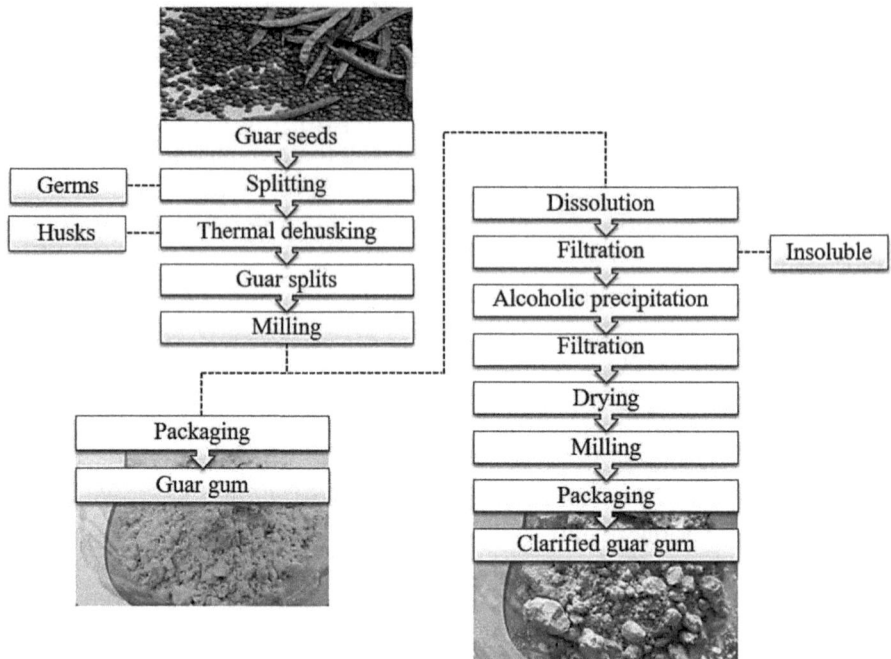

FIGURE 8.3   Guar gum flow chart.

8.2.3   DERIVATIVES

Modification of GG is done by two ways either substitution process that involves replacement of free hydroxyl group of the carbohydrate backbone by different cationic, anionic, amphoteric, and nonionic groups, or hydrolyzation process that means enzymatic, acidic, and basic hydrolysis of guar to yield low molecular weight gum. These modifications enhance properties and applications of guar in a broad spectrum of industries.[34,57,73]

Many derivatives of GG have been prepared and reported in the literature, some of these are carboxymethyl GG,[50] hydroxymethyl GG,[31] hydroxypropyl GG,[32] O-carboxymethyl-O-hydroxypropyl guar gum (CMHPG),[68] O-2-hydroxy-3-(trimethylammonia propyl) guar gum (HTPG), O-carboyxymethyl-O-2-hydroxy-3-(trimethylammonia propyl) guar gum (CMHTPG),[33] acryloyloxy GG,[67] methacryloyl GG,[81] sulfated GG,[36] and GG esters.[19] The derivatives of GG and some applications are presented in Table 8.1.

TABLE 8.1   Currently Manufactured GG Derivatives.

Guar gum derivatives				
Type of derivative	Substituent group	Ionic charge	Application	Reference
Carboxymethyl GG (CMG)	—CH$_2$—COO⁻Na⁺	Anionic	Nanoparticles for drug delivery	18
Hydroxypropyl GG (HPG)	—CH$_2$—CH(OH)CH$_3$	Nonionic	Treatment of wastewater	47
			Lubricant drops for eye treatment	54]
Carboxymethyl-hydroxypropyl GG (CMHPG)	—CH$_2$—COO⁻Na⁺ —CH$_2$—CH(OH)CH$_3$	Anionic	Fracturing fluid in mining industry and oil recovery	41

8.3   CHEMISTRY OF GUAR GUM

GG (CAS number 9000–30–0) is a galactomannan that contains 34.6% D-galactopyranosyl units and 64.4% D-mannopyranosyl units. The chemistry of the galactomannan unit confirmed that GG molecule is a linear carbohydrate polymer with a molecular weight range of 50,000–8,000,000Da.[28]

It has a straight chain of D-mannose units linked together by β (1–4) glycoside bond and D-galactose units are joined at each alternate position by an (1–6) glycosidic linkage[23] (Fig. 8.4). Therefore, GG forms a rodlike polymeric structure with a mannose backbone linked to galactose side chains, which are randomly placed on backbone with an average ratio of galactose to mannose of 1:2. The polymeric structure of GG contains numerous hydroxyl groups, which are treated for manufacturing different derivatives important in several industries. Their properties mainly depend upon their chemical features such as chain length, abundance of cis–OH group, steric hindrance, degree of polymerization, and additional substitution.[12,15,49,74]

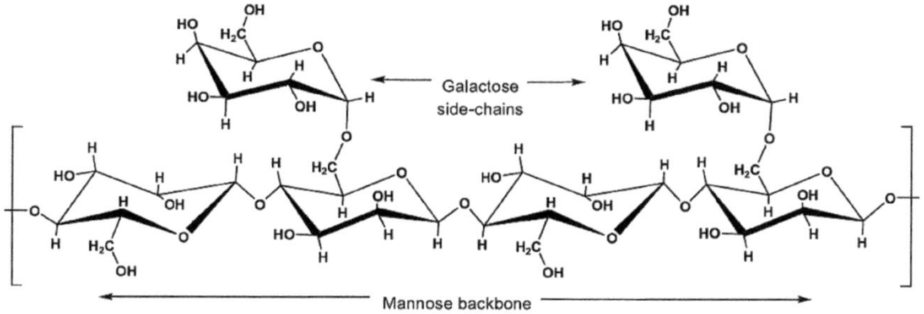

FIGURE 8.4   Chemical structure of guar gum.

8.4 PROPERTIES OF GUAR GUM

GG is a white to yellowish powder and nearly odorless. The most important property of GG is its ability to hydrate rapidly in water to attain uniform and very high viscosity at relatively low concentrations. Another advantage associated to GG is that it is soluble in hot and cold water and provides full viscosity. Apart from being the most cost-effective stabilizer and emulsifier, it provides texture improvement, and water-bonding, enhances mouth feel, and controls crystal formation. There is a wide range of physical and organoleptic properties of GG so the principal characteristics of this gum are summarized in Table 8.2 and are widely discussed in subsequent sections.

TABLE 8.2   Characteristics of Guar Gum.

	Guar gum
Origin	Extract of endosperm of seed (Leguminosae)
Chemical composition	• Medium-galactose galactomannan
	• Mannose + galactose (ratio M:G = 1.6:1)
Nutritional value (in 100 g)	292 kJ (70 kcal); slow resorption
Fiber content	Approx. 80% (contains 10% protein)
Toxicology	No health concerns, no ADI value defined
Solubility at low temperature (H_2O)	Highly soluble
Appearance of an aqueous solution	Opaque, gray, and cloudy
Viscosity of solution in water	High in cold water, lower in hot water
Impact of heat on viscosity in water (pH 7)	Viscosity decrease
Shear stability	Pseudoplastic > 0.5% concentration, shear thinning
Thickening effect	High
pH stability	High (pH 2–10)
Film formation	Low
Emulsion stabilization	Medium
Gelation	No (only with borate ions)
Crystallization control	High
Synergistic effects with other hydrocolloids	+ Starch/xanthan/CMC: viscosity increase;
	+ Gelling polysaccharides (e.g., agar): increased gel strength and elasticity
Negative interactions	Viscosity reduction with polyols
Dosage level in foods	Low–medium (0.05–2%, mostly 0.2–0.5%)

8.4.1 RHEOLOGY

GG is the most efficient aqueous thickener known. Solutions of GG and its derivatives are non-Newtonian, classified as pseudoplastic. They become fluid reversibly, when heat is applied, but irreversibly degrade when prolonged high temperature is applied at prolonged times. Some of the hydroxyalkyl-ated derivatives, recently developed, resist this degradation to a much greater degree. Solutions resist this well-shear degradation, compared with other water-soluble polymers, but degrade with time under high shear.[11]

8.4.2 VISCOSITY

The most important characteristic of GG is its ability to be dispersed in water and hydrate or swell rapidly and almost completely in cold water to form viscous colloidal dispersions. The viscosity attained is dependent on time, temperature, concentration, pH, rate of agitation, degree of purification, and practical size of the powdered gum used (Fig. 8.5). Temperature keeps a proportional relationship with viscosity of GG dispersion: The lower the temperature, lower the rate at which viscosity increases and lower the final viscosity. Heating the gum at temperatures above 60°C tends to provide a high initial viscosity but leads to an inferior stability (in terms of time-dependant changes in viscosity). The most convenient temperature depends

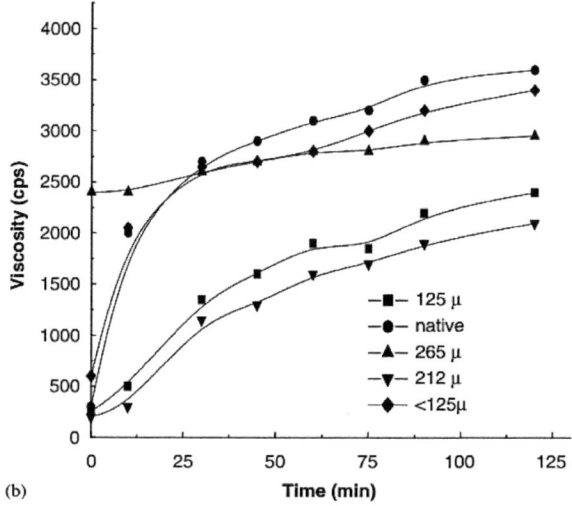

(b)

FIGURE 8.5 Changes in viscosity of guar gum as a function of time.

on the source. For example, the optimal conditions to disperse GG involve heating at 25–40°C for 2 h.[71]

The recommended use level of GG in aqueous systems is generally much less than 1%, since at higher concentrations the viscosity becomes excessive for most applications. For a typical solution, if the concentration (e.g., 1–2%) bends a tenfold increase in viscosity is obtained (4100–44,000 cps, respectively).

High viscosity products with a concentration at 3% are thick solutions that seem gels. There are guar derivatives with low viscosities for special applications, for example, when a high solids content is favored when desired electrically charged molecules thickening power and controlled behavior, or at least pseudoplasticity desired, or more Newtonian flow.[14,42,48]

8.4.3 HYDRATION RATE

Rate of hydration of GG varies. A number of factors are known to influence the hydration or dissolution process including the molecular weight and concentration of the galactomannan in the guar powder and also the environmental conditions, such as temperature and pH and the presence of co solutes, such as sucrose and salts. The major determinant of hydration kinetics is particle size, which reflects the change in surface area exposed to water. The rate and degree of hydration of GG are critical variables in influencing its biological activity.[42] A hydration time of about 2 h is required in practical applications in order to reach maximum viscosity. For some applications in which, there is a need for a quick initial viscosity, very fine mesh GG are available. However, a considerable period of time is still required for maximum hydration and viscosity to be achieved.[48]

8.4.4 HYDROGEN BONDING ACTIVITY

The hydrogen bonding activity is generally attributed to the presence and behavior of the hydroxyl group in GG molecule. The straight chain structure of GG, along with the regularity of the single membered galactose branches, produces a molecule that exhibits unusual effects on hydrated colloidal systems due to the formation of hydrogen bond. GG shows hydrogen bonding with cellulosic material and hydrated minerals. With a slight addition of GG, there are marked alterations in electro kinetic properties of any system.[20]

8.4.5   REFRACTIVE INDEX

Studies on the trends of some specific physical properties such as refractive index of GG aqueous systems at different total gum concentration, polymer ratio, and temperatures are not yet reported. The relationships among refractive index and gum concentration can be employed to determine the real gum concentration quickly and with acceptable accuracy. Recently, Moreira et al.[39] studied the values of GG aqueous dispersions at different concentrations (0.05%, 0.10%, 0.20%, 0.40%, 0.60%, and 0.80% w/w) and temperatures (15°C, 20°C, 25°C, 30°C, 35°C, and 40°C). They found that in all cases, the values decrease with increasing temperature at each gum concentration with refractive indices for solutions ranging between 1.3310 and 1.3352.

8.4.6   EFFECT OF PH

GG is stable in solution over a wide pH range. Guar solutions have an almost constant viscosity over pH range about 1.0–10.5. This stability is believed due to nonionic, uncharged nature of the molecule. While the pH does not affect the final viscosity, maximum hydration takes place at pH 8.0–9.0. Slowest hydration is present at high (above 10.0) and low (below 4.0) pH values. The preferred method of preparing guar solutions is dissolving at fastest hydration rate pH and then adjusts the pH to desired value (Fig. 8.6). Maximum viscosities achieved at both acid and alkaline pHs are the same despite the difference in hydration rates .

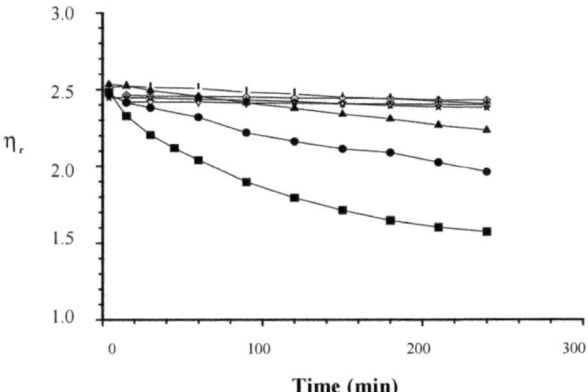

FIGURE 8.6   The relative viscosity (η_r) changes of guar gum solutions (0.07% w/w) at different pH levels, temperature = 50°C: (■) pH 1.5; (●) 2.0; (▲) 2.2; (▢) 2.5; (△) 3.0; (◊) 3.5; and (□) 4.0.

8.4.7   REACTIONS WITH SALTS

Since salt and sugar are probably the two most widely used ingredients in food other than water, their effect on GG has been extensively investigated.[8] The behavior of GG in brine is essentially the same as in water. The hydration rate is not affected, although the final viscosity is somewhat increased by the presence of sodium chloride. This property has made it a very valuable component of oil-well drilling muds, where the capacity to maintain high viscosity in the presence of brine encountered during drilling operation is absolutely essential.[21]

8.4.8   REACTIONS WITH SUGARS

In the presence of sugar, GG has a competition with it for the available water, triggering a delaying action on the hydration of the gum as sugar concentration grows.

Additionally, viscosity of GG and sugar preparations increases directly to the sugar concentration.[8] Viscosities of guar solutions containing sugar continue to increase for several days and this delay of hydration rate may be due to a reduction in the mobility of the water, proportional to the sugar concentration. In such systems, the full value of GG as a thickening and stabilizing agent is developed after about a week of storage. Sugar is effective in protecting GG against hydrolysis and loss of viscosity when heated or autoclaved. The presence of 5–10% sugar in the liquid offers maximum protection with maximum viscosity. Sugar also offers protection from the hydrolyzing effect at low pH values (down to pH 3.0) under cooking conditions.[8]

8.4.9   GEL FORMATION

Borate ion acts as a cross-linking agent with hydrated GG to form cohesive, structured gels. Formation and strength of these gels depend on the pH, temperature, and concentrations of reactants (Fig. 8.7). The optimum pH range for gel formation is 7.5–10.5. The solution-gel transformation is reversible; the gel can be liquefied by decreasing the pH below 7.0 or by heating .[21] Borated gels can also be liquefied by the addition of glycerol or mannitol, capable of reaction with the borate ion. Borate ion will inhibit the hydration of GG if it is present at the time the powdered gum is added to water. The minimum concentrations necessary to inhibit hydration are dependent on pH. For example, with 1.0% of GG, 0.25–0.5% (based on guar weight) of borax (sodium tetraborate) is needed at pH 10.0–10.5, while at pH 7.5–8.0, 1.5–2.0% of borax is required. The complexing reaction is reversible and lowering the pH below 7.0 permits the gum to hydrate normally. This technique is often used to provide better mixing and easier dispersion.[75]

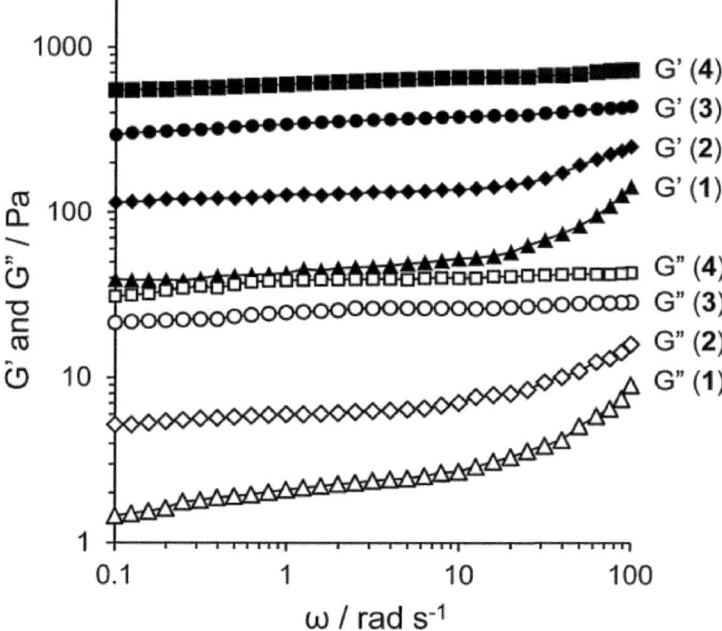

FIGURE 8.7   Viscoelastic behavior of guar gum gels (samples 1–4, 100 mg) swollen with PBS buffer, pH 7.4 (2 mL) at 298K.

8.4.10 SYNERGISTIC EFFECT

Synergisms of GG with other materials, including others gums such as xanthan gum, agar, carrageenan, or starches, are well studied.[13] The degree of the synergism is believed to be related to mannose/galactose (M/G) ratio and the galactosyl distribution on the mannan backbone which basically is determined by source and method employed for the extraction of this kind of gum.[1,10,17,35] examined the synergistic effect between sage seed gum (SSG) and GG solutions using various steady and dynamic rheological parameters.

They found that with increase in SSG fraction, the extent of viscosity reduction in the range of 0.01–316 s^{-1} increased from 58.68 for GG to 832.73 times for SSG which was not the same at different ranges of shear rate. Steady and dynamic shear tests suggested interaction with longer timescale in SSG chains in comparison with that in GG. The synergist effect of all viscoelastic parameters from frequency sweep test was observed for 3–1 SSG-GG blend, concluding that this mixture can be attractive commercially because they offer the potential to create new textures and to manipulate the rheology of products in the industry.[44]

8.5 TOXICITY

The only available data for GG that relate to absorption, distribution, metabolism, excretion, and toxicity were found in dietary studies.[78] In this way, an acute oral toxicity study on partially hydrolyzed guar gum (PHGG) was performed using groups of 16 (8 males, 8 females per group) 4-week-old Jcl:ICR mice and Jcl:SD rats. PHGG was administered by gavage at a concentration of 30% in distilled water (dose = 6000 mg kg^{-1} body weight; dose volume = 20 mL kg^{-1}) to one group per species. The control group was dosed orally with distilled water. Dosing was followed by a 14-day observation period, after which all animals were killed and examined macroscopically. Soft feces were reported for male and female mice, but no abnormal signs were reported for rats. There were no test substance-related effects on body weight (rats and mice), food consumption (rats), or necropsy findings (mice and rats). None of the animals died, and the LD50 was > 6000 mg kg^{-1} in both species.[29] On the other hand, in a 28-day oral feeding study, two groups of 10 rats (5 males, 5 females per group) were fed with PHGG in the diet (500 and 2500 mg/kg doses, respectively) daily. Body weights and food consumption were measured, and gross and microscopic pathology were evaluated. No adverse effects were observed at either administered dose.[72] The available information reveals that there are no adverse short-term toxicological consequences in animals after consuming GG in amounts exceeding those currently consumed in the normal diet. However, it should be noted that no long-term animal feeding studies of GG have been reported. It may be advisable in due course to conduct adequate feeding studies in several species, including pregnant animals, at dosage levels that approximate, and exceed the current estimated maximum daily human intakes.[24]

8.6   INDUSTRIAL APPLICATIONS OVERVIEW

8.6.1   FOOD INDUSTRY

GG stands as one of the cheapest hydrocolloids in food industry and about 40% of the total is presently used as food additive. The importance of this gum in food application is due to its various unique functional properties such as water retention capacity, reduction in evaporation rate, alteration in freezing rate, modification in ice crystal formation, regulation of rheological properties, and involvement in chemical transformation.[61] United States Food and Drug Administration (FDA) regulate the use of gums and classify them as either food additives or generally recognized as safe substances (GRAS number 2537).

The use of GG at a concentration not exceeding 2% is allowed in food application.[24] Tomato ketchup serum loss and flow values decreases when this gum is added, which makes it a novel thickener for this product.[26] It has also been incorporated in pizzas, biscuits, and pastries.[25]

GG prevents staleness and crumb formation in baked foods, provides unparalleled moisture preservation to the dough, retards fat penetration, increases the dough volume, provides greater resiliency, and improves texture and shelf life. In wheat bread dough, addition of this gum results in significant increase in loaf volume on baking.[9] The depolymerized GG is used in the preparation of low-calorie food. It enhances the creaming stability and control rheology of emulsion prepared by egg yolk.[22] In beverages, it is often used as a thickening or viscosity control agent in levels of 0.25–0.75%. GG is useful due to its resistance to break down at low pH conditions. In addition, since GG is soluble in cold water, it is easy to use in most beverage processing plants.[38] In dairy products, it thickens milk, yoghurt, and liquid cheese products and helps to maintain the homogeneity and texture of ice cream and sherbets.[5]

8.6.2   PHARMACEUTICAL INDUSTRY

GG or its derivatives are used in pharmaceutical industries as gelling, viscosifying, thickening, and suspending agent, process aid, and for stabilization, emulsification, preservation, water retention/water phase control, binding, clouding, pour control for suspensions, antacid formulations, tablet binding, disintegration agent, controlled drug delivery systems, slimming aids, nutritional foods, etc. Its hydrophilic properties and the ability to form gel make it useful in gastric ulcer treatment.

Studies revealed that a diet supplemented with this gum decreased the appetite, hunger, and desire for eating.[7] In tablets, the gum is used as binder and increases the mechanical strength of tablets during pressing, and in jellies and ointments, it is used as thickener. The depolymerized GG has been good bulking agent for dietetic food and as a food fiber; it has been used in sugar and lipid metabolic control particularly in diabetic and heart patients.[62]

GG is widely used in capsules as dietary fiber that decreases hypercholesterolemia, hyperglycemia, and obesity.[16] Sufficient gum intake as dietary fiber helps in bowel regularization, total, and LDL-cholesterol reduction, diabetes control, enhancement of mineral absorption, and prevention of digestive problems such as constipation and enhancement of bowel movement.[83] Some pharmaceutical companies are using it for making bandage paste and in dentistry formulations. The formulations such as inhalable, injectable, beads, microparticles, nanoparticles, solid monolithic matrix films, and implants also facilitate the use of the gum.[69] The high swelling characteristics of this gum sometimes hinder its use as a drug delivery carrier but it can be improved by derivatization, grafting, and network formation and can be satisfactorily used for targeted drug delivery by forming coating matrix systems, hydrogels, and nano/microparticles.[56]

8.6.3   METALLURGICAL AND MINING INDUSTRY

GG is used in froth flotation of potash as an auxiliary reagent, depressing the gangue minerals, which might be clay, talc, or shale. GG is also used as a flocculant or settling agent to concentrate ores, or tailings in the mining industry.[64]

Also, it is approved by many public organizations for use in potable water treatment as a coagulant aid in conjunction with such coagulants as alum (potassium aluminum sulfate), iron (III) sulfate, and lime (calcium oxide). In industrial waters, GG flocculates clays, carbonates, hydroxides, and silica when used alone or in conjunction with inorganic coagulants.[47]

8.6.4   PAPER INDUSTRY

GG is used as a size for paper and textiles. The major use of this gum in papermaking is in the wet end of the process. The gum is added to the pulp suspension just before the sheet is formed. GG replaces or supplements the natural hemicelluloses in paper bonding. Advantages gained by addition of guar to pulp include improved sheet formation with a more regular distribution of pulp fibers (less fiber bundles); increased mullen bursting strength; increased fold strength; increased tensile strength; increased pick; easier pulp hydration; improved finish; decreased porosity; increased flat crush of corrugating medium; increased machine speed with maintenance of test results; and increased retention of fines.[2]

8.6.5   COSMETIC INDUSTRY

The unique cosmetic properties of GG include cold solubility, viscosity enhancing, solvent resistance film forming, protective colloid, wide pH range resistance, stability, nontoxic nature, safe, etc. So, it is a choice of thickener, suspending agent, binder, and emulsifier agent in various hair/skin care cosmetic products such as creams, shampoos, premium quality soaps, lotions, conditioners, and moisturizer.[64]

In the manufacturing of toothpaste, this gum binds the aqueous phase of the paste and is used in sizeable scale to impart flowing nature so that the paste can be extruded from the collapsible tubes with the application of a little force. In shaving cream preparation, it does the same work such as imports slip during shaving and improves skin aftershave besides providing stabilization of the system.[12]

In emulsion systems, such as cream and lotions, GG prevents phase separation, sudden release of moisture, increase emulsion stability, prevent water loss, and is used as protective colloid. It stabilizes the emulsion during freeze-thaw cycle, where the water phase condenses out of the system. In lotion, it provides additional spread ability and an agreeable feel.[70] Cationic GG is used to thicken various cosmetics and toiletries products, especially to impart thickening, conditioning, foam stability, softening, and lubricity. In aerosol dispensing aqueous liquid preparation as spray or mist, it reduces fog migration. The inert and compatible nature of guar with the detergents makes it suitable for use in shampoo and cleansing preparation. Hair colorants contain guar as thickener. It is also available in self-emulsifying grades and can be used to prepare dry facemask preparation.[3]

8.7   CONCLUSIONS

Guar is an agricultural produce spreading over into Indo-Pakistan subcontinent many generations ago. GG is a high molecular weight natural hydrocolloid composed of galactose and mannan units that is obtained from the endosperm of the guar plant (*C. tetragonolobus*). Usually, it is produced as a whitish powder similar to flour and functions as thickener, stabilizer, and emulsifier. Due to its low cost, nontoxicity, biodegradability, biocompatibility, high viscosity, and high water solubility, GG has a peculiar interest for researchers due to its availability in terms of volumes and opportunities in many industrial applications, such as the food, textile, printing, explosives, pharmaceutical, and even hydraulic fracturing fields. This review states that GG has multiple physical and chemical properties that make it a strong candidate for its application in food science and technology research.

ACKNOWLEDGMENTS

The authors would like to thanks to the Mexican Council for Science and Technology (CONACyT) for providing financial support in the INFR 2015: 254178 project. This work is part of the activities in the Master degree in food science and technology of the Faculty of Chemistry Sciences.

KEYWORDS

- *Cyamopsis tetragonolobus*
- guar gum
- properties
- industry
- potential

REFERENCES

1. Ali-Razavi, S. M.; Alghooneh, A.; Behrouzian, F.; Cui, S. W. Investigation of the Interaction Between SSG and Guar Gum: Steady and Dynamic Shear Rheology. *Food Hydrocoll.* **2016,** *60,* 67–76.
2. Anderson, K. R.; Larson, B.; Thoresson, H. O. EKAAB POT Int. Applied WO 8500, 100 (CL D2H3I20 SE Appl. 84/306207), 1986, p 35 (Cf. Guar Res Ann 5:44).
3. APEDA. APEDA Annual Export Report. (Online) 2011, Available at: http://agriex-change.apeda.gov.in/index/product_description_32head.aspx?gcode=0502 (accessed Feb 28, 2016).

4. BeMiller, J. Gums and Related Polysaccharides. In *Glycoscience;* Fraser Reid, B., Tatsuta, K., Thiem, J., Eds.; Springer: Japan, 2008.

5. Brennan, C. S.; Tudorica, C. M. Carbohydrate Based Fat Replacers in the Modification of the Rheological, Textural and Sensory Quality of Yoghurt: Comparative Study of the Utilization of Barley Beta-Glucan, Guar Gum and Inulin. *Int. J. Food Sci. Technol.* **2008,** *43,* 824–833.

6. Busch, V. M.; Kolender, A. A.; Santagapita, P. R.; Buera, M. P. Vinal Gum, a Galactomannan from *Prosopis ruscifolia* Seeds: Physicochemical Characterization. *Food Hydrocoll.* **2015,** *51,* 495–502.

7. Butt, M. S.; Shahzadi, N.; Sharif, M. K.; Nasir, M. Guar Gum: A Miracle Therapy for Hypercholesterolemia, Hyperglycemia and Obesity. *Crit. Rev. Food Sci. Nutr.* **2007,** *47,* 389–396.

8. Carlson, W. A.; Zeigenfuss, E. M. The Effect of Sugar on Guar Gum as a Thickening Agent. *J. Food Technol.* **1965,** *19,* 954–958.

9. Cawley, R. W. The Role of Wheat Flour Pentosans in Baking. II. Effect of Added Flour Pentosans and Other Gums on Gluten Starch Loaves. *J. Sci. Food Agric.* **1964,** *15,* 834–838.

10. Cerqueira, M. A.; Pinheiro, A. C.; Souza, B. W. S.; Lima, A. M. P.; Teixeira, J. A.; Moreira, R. A. Extraction, Purification and Characterization of Galactomannans from Non-Traditional Sources. *Carbohydr. Polym.* **2009,** *75,* 408–414.

11. Chenlo, F.; Moreira, R.; Silva, C. Rheological Behavior of Aqueous Systems of Tragacanth and Guar Gums with Storage Time. *J. Food Eng.* **2010,** *96,* 107–113.

12. Chudzikowski, R. J. Guar Gum and its Applications. *J. Soc. Cosmet. Chem.* **1971,** *22,* 43–60.

13. Cui, S. W.; Eskin, M. A. N.; Wu, Y.; Ding, S. Synergisms between Yellow Mustard Mucilage and Galactomannans and Applications in Food Products–A Mini Review. *Adv. Colloid Interface Sci.* **2006,** *128–130,* 249–256.

14. Cunha, P. L.; Castro, R. R.; Rocha, F. A.; de Paula, R. C.; Feitosa, J. P. Low Viscosity Hydrogel of Guar Gum: Preparation and Physicochemical Characterization. *Int. J. Biol. Macromolec.* **2005,** *37* (1–2), 99–104.

15. Daas, P. J. H.; Schols, H. A.; De-Jongh, H. H. J. On the Galactosyl Distribution of Commercial Galactomannans. *Carbohydr. Res.* **2000,** *329,* 609–619.

16. Dall'alba, V.; Silva, F. M.; Antonio, J. P. Improvement of the Metabolic Syndrome Profile by Soluble Fiber-Guar Gum-in Patients with Type 2 Diabetes a Randomised Clinical Trial. *Br. J. Nutr.* **2013,** *110,* 1601–1610.

17. Dea, I. C. M.; Morris, E. R.; Rees, D. A.; Welsh, E. J.; Barnes, H. A.; Price, J. Associations of Like and Unlike Polysaccharides: Mechanism and Specificity in Galactomannans, Interacting Bacterial Polysaccharides, and Related Systems. *Carbohydr. Res.* **1977,** *57,* 249–272.

18. Dodi, G.; Pala, A.; Barbu, E.; Peptanariu, D.; Hritcu, D.; Popa, M. I.; Tamba, B. I. Carboxymethyl Guar Gum Nanoparticles for Drug Delivery Applications: Preparation and Preliminary *In-Vitro* Investigations. *Mater. Sci. Eng.* **2016,** *63,* 628–636.

19. Dong, C.; Tian, B. Studies on Preparation and Emulsifying Properties of Guar Galactomannan Ester of Palmitic Acid. *J. Appl. Polym. Sci.* **1999,** *72* (5), 639–645.

20. Doyle, J. P.; Giannouli, P.; Martin, E. J.; Brooks, M.; Morris, E. R. Effect of Sugars, Galactose Content and Chain Length on Freeze-Thaw Gelation of Galactomannans. *Carbohydr. Polym.* **2006,** *64,* 391–401.

21. El-awad, G. A Study of Guar Seed and Guar Gum Properties. (*Cyamopsis tetragonol-obus*). (Online) 1998, Available at: http://www.iaea.org/inis/collection/NCLCollection-Store/_Public/31/037/31037745.pdf (accessed Feb 10, 2016).

22. Ercelebi, E. A.; Ibanoglu, E. Stability and Rheological Properties of Egg Yolk Granule Stabilized Emulsions with Pectin and Guar Gum. *Int. J. Food Prop.* **2010,** *13,* 618–630.

23. FAO. (Online) 2016, Available at: http://www.fao.org/ag/agn/jecfa-additives/specs/monograph3/additive-218.pdf (accessed Feb 28, 2016).

24. FDA. GRAS Substances (SCOGS) Database. (Online) 2016, Available at: http://www.fda.gov/Food/IngredientsPackagingLabeling/GRAS/SCOGS/ucm2006852.htm (accessed Feb 10, 2016).

25. Griffith, A. J.; Kennedy, J. F. Biotechnology of Polysaccharide. In *Chemistry Carbo-hydrate;* Kennedy, J. I., Ed.; Oxford Science Publication, Oxford University Press: Oxford, UK, 1988; pp 597–635.

26. Gujral, H. S.; Sharma, A.; Singh, N. Effects of Hydrocolloids, Storage Temperature and Duration on the Consistency of Tomato Ketchup. *Int. J. Food Prop.* **2002,** *5,* 179–191.

27. Gunjal, B. B.; Kadam, S. S. *CRC Hand Book of World Food Legumes;* 1991; Vol. 1, pp 289–299.

28. Kawamura, Y. Prepared for the 69th Joint FAO/WHO Expert Committee on Food Additives. Guar Gum Chemical and Technical Assessment: Rome, 2008.

29. Koujitani, T.; Oishi, H.; Kubo, Y.; Maeda, T.; Sekiya, K.; Yasuba, M.; Matsuoka, N.; Nishimura, K. Absence of Detectable Toxicity in Rats Fed Partially Hydrolyzed Guar Gum (K-13) for 13 Weeks. *Int. J. Toxicol.* **1997,** *16* (6), 611–623.

30. Kumar, V. Perspective Production Technologies of Arid Legumes. In *Perspective Research Activities of Arid Legumes in India*. Kumar, D., Henary, A., Eds.; Arid Legumes Society: India, 2009; pp 119–155.

31. Lapasin, R.; Pricl, S.; Tracanelli, P. Rheology of Hydroxyethyl Guar Gum Derivatives. *Carbohydr. Polym.* **1991,** *14* (4), 411–427.

32. Lapasin, R.; Pricl, S.; Lorenzi, L. D.; Torriano, G. Flow Properties of Hydroxypropyl Guar Gum and its Long-Chain Hydrophobic Derivatives. *Carbohydr. Polym.* **1995,** *28* (3), 195–202.

33. Li-Ming, Z.; Jian-Fang, Z.; Peter, S. H. A Comparative Study on Viscosity Behavior of Water-Soluble Chemically Modified Guar Gum Derivatives with Different Functional Lateral Groups. *J. Sci. Food Agric.* **2005,** *85* (15), 2638–2644.

34. Mccleary, B. V.; Neukom, H. Effect of Enzymic Modification on the Solution and Inter-action Properties of Galactomannans. *Prog. Food Nutr. Sci.* **1982,** *6,* 109–118.

35. McCleary, B. V.; Clark, A. H.; Dea, I. C. M.; Rees, D. A. The Fine Structures of Carob and Guar Galactomannans. *Carbohydr. Res.* **1985,** *139,* 237–260.

36. Mestechkina, N. M.; Egorov, A. V.; Shcherbukhin, V. D. Synthesis of Galactomannan Sulfates. *J. Appl. Biochem. Microbiol.* **2010,** *42* (3), 326–327.

37. Milani, J.; Gisoo, M. Hydrocolloids in Food Industry, Food Industrial Processes. Methods and Equipment. In Tech. ISBN: 978-953-307-905-9. (Online) 2012, Avail-able from: http://www.intechopen.com/books/food-industrial-processes-methods-and-equipment/hydrocolloids-in-foodindustry (accessed Feb 10, 2016).

38. Miyazawa, T. Hydrocolloid Structures, Which Allow More Water Interactions through Hydrogen Bonding. *Carbohydr. Res.* **2006,** *341,* 870–877.

39. Moreira, R.; Chenlo, F.; Silva, C.; Torres, M. D.; Díaz-Varela, D.; Hilliou, L.; Argence, H. Surface Tension and Refractive Index of Guar and Tragacanth Gums Aqueous

Dispersions at Different Polymer Concentrations, Polymer Ratios and Temperatures. *Food Hydrocolloids.* **2012,** *28,* 284–290.

40. Mudgil, D.; Barak, S. Guar Gum: Processing, Properties and Food Applications-a Review. *J. Food Sci. Technol.* **2014,** *51* (3), 409–418.

41. Mukherjee, I.; Sarkar, D.; Moulik, S. P. Interaction of Gums (Guar, Carboxymethyl-Hydroxypropyl Guar, Diutan, and Xanthan) with Surfactants (DTAB, CTAB and TX-100) in Aqueous Medium. *Langmuir.* **2010,** *26,* 17906–17912.

42. Nandhini-Venugopal, K.; Abhilash, M. Study of Hydration Kinetics and Rheological Behaviour of Guar Gum. *Int. J. Pharm. Sci. Res.* **2010,** *1* (1), 28–39.

43. Nemade, S. N.; Sawarkar, S. B. Recovery and Synthesis of Guar Gum and Its Derivatives. *Int. J. Adv. Res. Chem. Sci.* **2015,** *2,* 33–40.

44. Nor-Hayati, I.; Ching, C. W.; Rozaini, M. Z. H. Flow Properties of o/w Emulsions as Affected by Xanthan Gum, Guar Gum and Carboxymethyl Cellulose Interactions Studied by a Mixture Regression Modeling. *Food Hydrocolloid.* **2016,** *53,* 199–208.

45. Nussinovitch, A. *Hydrocolloid Applications: Gum Technology in the Food and Other Industries;* Chapman and Hall: London, 1997.

46. Nussinovitch, A. *Water Soluble Polymer Applications in Foods;* Blackwell Science: Oxford, 2003.

47. Pala, S.; Ghoraia, S.; Dasha, M. K. Flocculation Properties of Polyacrylamide Grafted Carboxymethyl Guar Gum (CMG-g-PAM) Synthesized by Conventional and Microwave Assisted Method. *J. Hazard. Mater.* **2011,** *192,* 1580–1588.

48. Parija, S.; Misra, M.; Mohanty, A. K. Studies of Natural Gum Adhesive Extracts: An Overview. *Polym. Rev.* **2001,** *41,* 175–197.

49. Pasha, M.; Swamy, N. G. N. Derivatization of Guar to Sodium Carboxymethyl Hydroxyl Propyl Derivative, Characterization and Evaluation. *Pak. J. Pharm. Sci.* **2008,** *21*(1), 40–44.

50. Patel, J. J.; Karve, M.; Patel, N. K. A Novel Approach to Synthesize Carboxymethyl Guar Gum via Friedel Craft Acylation Method. *Macromol. Indian* J. **2014,** *10* (1), 18–22.

51. Patel, M. B.; McGinnis, J. The Effect of Autoclaving and Enzyme Supplementation of Guar Meal on the Performance of Chicks and Laying Hens. *Poult. Sci.* **1985,** *64,* 1148–1156.

52. Pathak, R. Clusterbean: Physiology, Genetics and Cultivation; DOI 10.1007/978-981–287–907–3_3.

53. Peleg, M. On Fundamental Issues in Texture Evaluation and Texturization. *Food Hydrocolloid.* **2006,** *20,* 405–414.

54. Petricek, I.; Berta, A.; Higazy, M. T.; Németh, J.; Prost, M. E. Hydroxypropyl-Guar Gellable Lubricant Eye Drops for Dry Eye Treatment. *Exp. Opin. Pharmacother.* **2008,** *9* (8), 1431–1436.

55. Perduca, M. J.; Spotti, M. J.; Santiago, L. G.; Judis, M. A.; Rubiolo, A. C.; Carrara, C. R. Rheological Characterization of the Hydrocolloid From *Gleditsia amorphoides* Seeds. *LWT-Food Sci. Technol.* **2013,** *51,* 143–147.

56. Prabhaharan, M. Prospective of Guar Gum and Its Derivatives as Controlled Drug Delivery System. *Int. J. Biol. Macromol.* **2011,** *49* (2), 117–124.

57. Prabhanjan, H.; Gharia, M. M.; Srivastava, H. C. Guar Gum Derivatives. Part II: Preparation and Properties. *Carbohydr. Polym.* **1990,** *12,* 1–7.

58. Rana, V.; Rai, P.; Tiwary, A. K.; Singh, R. S.; Kennedy, J. F.; Knells, C. J. Modified Gums: Approaches and Applications in Drug Delivery. *Carbohydr. Polym.* **2011,** *83,* 1031–1047.

59. Reddy, K.; Krishna-Mohan, G.; Satla, S.; Gaikwad, S. Natural Polysaccharides: Versatile Excipients for Controlled Drug Delivery Systems. *Asian J. Pharm. Sci.* **2011,** *6* (6), 275–286.

60. Ritzoulis, C.; Marini, E.; Aslanidou, A.; Georgiadis, N.; Karayannakidis, P. D.; Koukiotis, C.; Filotheou, A.; Lousinian, S.; Tzimpilis, E. Hydrocolloids from Quince Seed: Extraction, Characterization, and Study of Their Emulsifying/Stabilizing Capacity. *Food Hydrocolloid.* **2014,** *42,* 178–186.

61. Rodge, A. B.; Sonkamble, S. M.; Salve, R. V. Effect of Hydrocolloid (Guar Gum) Incorporation on the Quality Characteristics of Bread. *J. Food Process Technol.* **2012,** *3* (2), 1–7.

62. Saeed, S.; Mosa-Al-Reza, H.; Fatemeh, A. N. Antihyperglycemic and Antihyperlipidemic Effects of Guar Gum on Streptozotocin-Induced Diabetes in Male Rats. *Pharmacogn. Mag.* **2012,** *8,* 65–72.

63. Scherz, H. *Hydrocolloids: Stabilizers, Thickening and Gelling Agents in Food Products. Food Chemistry and Food Quality. Food Chemical Society GDCh,* Behr's Verlag GmbH: Hamburg, Germany, 1996; Vol. 2.

64. Sharma, B. R.; Chechani, V.; Dhuldhoya, N. C.; Merchant, U. C. Guar Gum. (Online) 2007. Available at: http://www.lucidcolloids.com/pdf/814_guar-gum.pdf (accessed Feb 10, 2016).

65. Sharma, P. *Guar Industry Vision 2020: Single Vision Strategies*; CCS National Institute of Agricultural Marketing: Jaipur, 2010.

66. Sharma, P.; Gummagolmath, K. C. Reforming Guar Industry in India: Issues and Strategies. *Agric. Econ. Res. Rev.* **2012,** *2* 5(1), 37–48.

67. Shenoy, M. A.; D'Melo, D. J. Synthesis and Characterization of Acryloyloxy Guar Gum. *J. Appl. Polym. Sci.* **2010,** *117* (1), 148–154.

68. Shi, H. Y.; Li, M. Z. New Grafted Polysaccharide Based on O-Carboxymethyl-O-Hydroxypropyl Guar Gum and Nisopropylacrylamide: Synthesis and Phase Transition Behavior in Aqueous Media. *Carbohydr. polym.* **2007,** *67* (3), 337–342.

69. Soumya, R. S.; Ghosh, S.; Abraham, E. I. Preparation and Characterization of Guar Gum Nanoparticles. *Int. J. Biol. Macromol.* **2010,** *46* (2), 267–269.

70. Srichamroen, A. Influence of Temperature and Salt on the Viscosity Property of Guar Gum. *J. Commun. Dev. Res.* **2007,** *15* (2), 55–62.

71. Srivastava, M.; Kapoor, V. P. Seed Galactomannans: An Overview. *Chem. Biodivers.* **2005,** *2* (3), 295–317. DOI: 10.1002/cbdv.200590013.

72. Takahashi, H.; Yang, S.; Fujiki, M.; Kim, M.; Yamamoto, T.; Greenberg. N. A. Toxicity Studies of Partially Hydrolyzed Guar Gum. *J. Am. Coll. Toxicol.* **1994,** *13* (4), 273–278.

73. Thomas, T. A.; Dabas, B. S.; Chopra, D. D. Guar Gum has Many Uses. *Ind. Farm.* **1980,** *32,* 7–10.

74. Tripathy, S.; Das, M. K. Guar Gum: Present Status and Applications. *J. Pharm. Sci. Innov.* **2013,** *2* (4), 24–28.

75. Whistler, R. L.; Hymowitz, T. *Guar Agronomy, Production. Industrial Use and Nutrition;* R. L. Whistler, R. L.; Hymowitz, Eds.; Purdue University Press: West Lafayette, IN, 1979; pp 1–96.

76. Whistler, R. L.; Daniel, J. R. Carbohydrates. In *Food Chemistry;* 2nd ed.; Fennema, O. R., Ed.; Marcel Dekker: New York, 1985.

77. Whistler, R. L. Factors Influencing Gum Costs and Applications. In *Industrial Gums;* 2nd ed.; Whistler R. L., BeMiller J. N., Eds.; Academic Press: San Diego, 1973. pp 5–18.

78. WHO Toxicological Evaluation of Certain Food Additives and Contaminants: WHO Food Additives Series 21. Cambridge University Press: Cambridge, UK (Online) Feb 10, 2003, Available at: http://www.inchem.org/documents/jecfa/jecmono/v21je01.htm (accessed Feb 10, 2016).

79. Williams, P. A.; Phillips, G. O. Handbook of Hydrocolloids. In *Introduction to Food Hydrocolloid;* Phillips, G, O., Williams, P. A., Eds.; CRC Press: New York, NY, 2000; pp 1–19.

80. Wüstenberg, T. *Cellulose and Cellulose Derivatives in the Food Industry: Fundamentals and Applications;* 1st ed.; Wiley-VCH Verlag GmbH & Co. KGaA: Weinheim, Germany, 2015; pp 1–2.

81. Xiao, W.; Dong, L. In *Novel Excellent Property Film Prepared from Methacryloyl Chloride-Graft-Guar gum matrixes*, Xian Ning, April, 16–18, 2011; 1442–1445, 2011.

82. Yadav, H.; Shalendra, N. *An Analysis of Guar Crop in India.* United States Department of Agriculture (USDA): New Delhi, India. GAIN Report Number, IN4035, 2014.

83. Yoon, S. J.; Chu, D. C.; Juneja, L. R. Chemical and Physical Properties. Safety and Application of Partially Hydrolyzed Guar Gum as Dietary Fiber. *J. Clin. Biochem. Nutr.* **2008,** *42,* 1–7.

CHAPTER 9

BIOFUNCTIONAL PEPTIDES: BIOLOGICAL ACTIVITIES, PRODUCTION, AND APPLICATIONS

GLORIA ALICIA MARTÍNEZ-MEDINA[1], ARELY PRADO-BARRAGÁN[2], JOSÉ L. MARTÍNEZ[1], HÉCTOR A. RUIZ[1], ROSA MA. RODRÍGUEZ[1], JUAN C. CONTRERAS[1], and CRISTÓBAL N. AGUILAR[1,*]

[1]*Food Research Department, Chemistry School, Coahuila Autonomous University, Saltillo Unit 25280, Coahuila, México*

[2]*Biotechnology Department, Biological and Health Sciences Division, Metropolitan Autonomous University, Iztapalapa Unit, Ciudad de México 09340, Mexico*

Corresponding author. E-mail: cristobal.aguilar@uadec.edu.mx

CONTENTS

ABSTRACT

Biofunctional peptides are molecules with biological properties, which represent an attractive, innovative and original alternative, in the pharmaceutical, and food industries, for use as nutraceutical additives with high added value. In the last decade, the most important challenges are standardization and implementation of production methodologies for peptide generation at industrial scale and further the purification of this compound. The aim of this review to have a close look on the production, biological properties characterization, and possible application described so far on biofunctional peptides.

9.1 INTRODUCTION

The relationship between food and health has been recognized since Hippocrates time, [26] secondary from this fact, arises concepts such as "nutraceutical" or "functional food." In agreement with Institute of Food Technologist (IFT), functional food it is a component that provides, besides the basic nutrition a health benefit. These components can afford essential elements in quantities that exceed the amount for induvial maintenance, growth, and standard development; also can supply other biological active components with desirable health effects.[27] Whereas that a nutraceutical, in accordance with Canadian Pharmacopeia is a compound purified or isolated from food matrix, generally sold in pharmaceutical form and has been demonstrated present positive physiological effects, preventive, or protective function against a chronic disease.[16]

Proteins represent an integral food component that provides essential amino acids that brings energy for healthy body's growth and maintenance, additionally, various proteins possess specific biological activities, making them potential functional food ingredients.[11] Recent researches on functional food has special interest on bioactive compounds including functional peptides,[32] this terminus includes short sequences from 2 to 40 amino acids units that are inactive in precursor protein, however, at be released can exert a broad range of biological activities.[10]

Functional peptide can be liberated by hydrolysis or fermentative process starting from different protein sources, from animal origin among which are, bovine blood, gelatin, meat, egg, some fishes such as sardine and salmon, or vegetable protein such as wheat, maize, soy, rice, mushroom, pumpkin, and sorghum.[41] Amino acid sequences in peptides confer different biological properties, besides some peptides can present multifunctionality.[20] Functional peptides can exhibit antioxidant, antihypertensive, immunomodulatory even antimicrobial, antifungal, or anticoagulant activities.[53]

After oral administration, peptides can act locally in gastrointestinal system or can be transported, overtake, and impact in peripheral tissues through circulatory system, and exert directly their physiological properties in cardiovascular, digestive, immunologic, or nervous system,[43] and therefore, can have therapeutic role and act as alternative to other pharmacological molecules in body systems; offer numerous advantages over conventional therapeutics due to their bioactivities, biospecificity, broad spectrum, low toxicity, structural diversity, and low accumulation levels in body tissues.[2]

9.2 BIOFUNCTIONALITIES

Functional peptides derived from food protein have been studied and show biological activities on digestive, nervous, and immune system, manifesting a positive effect in health.[11] Within biological activities most reported listed antimicrobial, antihypertensive, immunomodulatory, and antioxidant activities.

9.2.1 ANTIMICROBIAL ACTIVITY

Antimicrobial peptides relevance comes from the generated concerns for antibiotic excessive use, motivates new molecules searching with antimicrobial activity, and effectiveness but which involves less collateral effects. Antimicrobial peptides are characterized for being positive charged molecules and presence of hydrophobic amino acid residues that can act against a microorganism broad spectrum, included Gram positive and Gram negative bacteria, viruses and fungi; this because an electrostatic attraction exist between peptides and microorganism cytoplasmic membrane, which is negatively charged, and subsequently peptides oligomerize and form transmembrane pores, inducing a cellular content leak, even though can proceed through other mechanism such as interruption of essential process for microorganism as DNA, protein an cell wall synthesis.[7]

9.2.2 ANTIHYPERTENSIVE ACTIVITY

Cardiovascular disease (CVD's) are a set of conditions where heart and blood vessel can be affected; between we can found coronary heart disease, stroke, and heart failures further constitutes worldwide leading death cause.[55] Chronic disease such as hypertension is considered the major risk in some CDV's develop.[46]

Within mechanism that regulating blood pressure in organism can be found Angiotensin I Converting Enzyme (ACE I) classified such as carboxidipeptidase (EC3.4.15.1) that plays a substantial role, being that their action generates a potent vasoconstrictor peptide.[20] Hypertension treatment mainly has been used a set of synthetic drugs inhibiting ACE action and generally produce secondary effects such as cough, cutaneous reaction, or taste perturbation;[60] actually an alternative is find molecules with ACE inhibition capacity from food origin sources,[15] ACE inhibitor peptides, act binding to the enzyme active site, coupling to an inhibitor site that promotes a change on conformational structure in protein, even joining to enzyme–substrate complex, avoiding enzyme functionality.[30] These molecules submit advantage such as supply sources that are cheap and have the capacity to be incorporated in functional foods.

9.2.3 IMMUNOMODULATORY ACTIVITY

Peptides with immunomodulatory capacity exists, for instance in the case of some peptides derived from casein and whey milk proteins that have capacity to improve and increase immune cells functions, promoting their proliferation, as well as antibodies, and cytokines production.[36] Immunomodulatory activities fomented by peptides depends on the structure and amino acid type and charge present in this molecules.[3]

9.2.4 ANTIOXIDANT ACTIVITY

Free radical normally is generated in the organism during respiration, also can be produced by external stimulus such as environmental pollution, tobacco components, or radiation, and they can act against infections, nevertheless this type of compounds excess may result on damage in biological molecules that integrate tissues such as protein, lipids even DNA, and they translated in to developing disease such as atherosclerosis, arthritis, diabetes, or cancer,[57] also they can be added to food products in order to retard no desirable reactions such as lipo-peroxidation that promote color, flavor, and aroma changes, in substitution to synthetic antioxidant which are attributed toxicity and DNA damage.[11]

Antioxidant activity of this molecules its influenced for diverse structural factors such as molecular size, amino acid composition, and sequence,[59] they can work through different mechanism, but most commons are electron donation, lipid radical neutralization and promote oxidation ion or metal chelation.[56]

9.2.5 OTHER FUNCTIONALITIES

Functional peptide also can exert broad diversity activities, such as opiate activity, where that molecules present affinity for opiate receptors and act such as hexogen modulators in hormone release, intestinal motility, and emotional behavior.[6]

In the other hand antihypercholesterolemic peptides, in particular obtained from soy hydrolysis, have capacity to inhibit cholesterol absorption, for their solubility repression.[43]

9.3 BIOFUNCTIONAL PEPTIDES PRODUCTION

Due to the high therapeutic relevance attributed to emergent molecules such as functional peptides, has been developed a set of methodologies for their production, among which may be mentioned, chemical synthesis, microbial fermentation, or enzymatic hydrolysis even combinations of these techniques, also sought alternatives as pretreatments or implementation of new technologies for production as Table 9.1 demonstrates.

Critical parameters understanding for peptide with physiological activity generation its relevant; within these parameters can found: protein source and their characteristics, amino acid composition, process conditions, essentially as temperature, pH, specificity, reaction time, and in the vegetal origin peptide, protein variation, may be compromised by environmental factors such as temperature, moisture, and fertility of soils in which their growth,[22] nonetheless, adequate control of this points can generate multifunctional peptides or with a biological specific activity.[56]

9.3.1 MICROBIAL FERMENTATION

Fermentation is one of the most used process for biofunctional peptide generation and Lactic Acid Bacteria (LAB) constitute most commonly used strains such as: *Lactobacillus helveticus*, *Lactobacillus delbrueckii ssp. bulgaricus*, *Lactobacillus lactis ssp. diacetylactis*, *Lactococcus lactis ssp. cremoris,* or *Streptococcus salivarius ssp. Thermophilus*,[11] but also has been employed organisms such as *Kluyveromyces marxianus* yeast,[25] fungus *Fusarium tricinctum*[62] and also has been reported synthetic gene development for this kind of functional products,[69] using microorganisms such as *Streptococcus thermophiles* [50] or *Escherichia coli*.[33]

Functional peptide can be founded naturally on dairy products due to are rich on precursor protein for this physiological active compounds type[11] that derived from LAB proteolysis, conducted through; complex proteolytic system that consists in three major components: (1) Cell wall binding protease, that promote initial proteolysis, turning protein in oligopeptides, (2) Specific transporters, that carry out peptides to cytoplasm, and (3) Intracellular peptidases that transform low-weight oligopeptides into free amino acid.[13] Likewise, has been presented a report set where peptide generation has been studied in traditional fermented food such as Kefir[19] or Kapi (Thai traditional fermented shrimp pastes).[35]

9.3.2   ENZYMATIC HYDROLYSIS

Protein enzymatic hydrolysis it's one of the most used methodologies for biofunctional peptide generation,[36] this process can be optimized, through certain physicochemical parameters control, such as pH or temperature, providing ideal conditions for protease action.[34] Proteinases wide variety such as chymotrypsin, alcalase, pepsin, thermolysin, or even enzymes from bacteria, or fungal sources, and also combination are employed in biofunctional peptides,[40] generally from animal or vegetable protein substrates, being milk proteins the most used;[61] but also bovine blood, meat, gelatin, egg, wheat, soy, rice, mushroom, pumpkin, and sorghum,[41] further, has been obtained peptides from marine substrates such as algae, fishes, mollusks, crustaceans, or by-products such as viscera, substandard muscles, or skin.[44]

This peptide manufacture method have advantages in the generation of peptide defined profiles and also, highly concentrated alkalis or acid residues are not generated, that limits their use in products intended to human consumption, how takes place in chemical hydrolysis; as well as the fact that improve the generation of L-form amino acids that constitutes molecules that promote biological activities, making them viable for their application in functional food formulation or nutraceuticals.[5]

Also exist methodologies that improves enzyme immobilization during the production process, in which highlights advantages such as easy enzyme recuperation and high purity products.[48,66]

9.3.3 EMERGENT TECHNOLOGIES

Recent advances in functional peptides generation field, are focus in search alternative protein sources and new production methodologies, such as ultrasound (US) and microwave (MW), supercritical-fluids hydrolysis (SPF), and high hydrostatic pressure (HHP) procedures.

9.3.3.1 ULTRASOUND

US, its defined such as acoustic wave, with above 20 kHz frequency, overpassing de human ear limits and that needs an external medium to propagate; this waves comes from a vibrational body, and when they propagated in surrounding medium, this start to oscillate, where by oscillation too the particles transmit energy to each other.[38]

High intensity US technologies application in food matrix such as proteins, implies the generation of physicochemical changes on this molecules,[45] due to cavitational, mechanic, and thermal effects generated, microjets formation, microturbulence, high velocity interparticles collisions, and micropore perturbation that induces the generation of elevated pressures and temperatures in medium, over 5000K or 500 atm, also free radical generation, and result in a improve and increase in quality and velocity of extraction[32] that can be exploited in food and pharmaceutical industry for functional peptide production, where US its implemented such as a pretreatment[30,71] or their simultaneous application during enzymatic hydrolysis[68], due to the fact of this technology, have capacity to realize structural and conformational modifications in protein, affecting hydrogen bonds, or hydrophilic interactions and therefore, tertiary and quaternary structures, also, the physical and mechanical, caused by US can cause protein unfolding which can promote exposure of proteolysis susceptible sites and causing a major degreed of hydrolysis. But, between the adverse effects are conformation damage in the enzyme and, therefore, in the hydrolytic activity.[47]

9.3.3.2   MICROWAVE

In the other hand, technologies such as MW, that basically, allows that molecules involves in reaction, fluctuate under magnetic field, where the energy is converted in heat and therefore, its possible carry out chemical reactions,[9,49] which are accelerated, with increased yields and high pure products,[28] has been reported be useful, such as pretreatment or in assisted enzymatic, or chemical hydrolysis for peptide production with functional potential from food proteins.

9.3.3.3   SUBCRITICAL WATER

Employing fluids such as water under subcritical condition, that means use of water over the boiling point but under their critical point, keeping at liquid form with pressure,[51] this conditions modify natural characteristics of water regarding hydrogen bonds, ionic products, and dielectric constant, supplying a medium that facilities some reactions such as hydrolysis, this technology gained importance in extraction of products with functional properties such as peptides, because is an eco-friendly alternative.[8]

9.3.3.4   HIGH HYDROSTATIC PRESSURE

Nonthermal treatments such as HHP, where materials are subjected to pressures from 100 to 1000 MPa in a vessel that promotes no covalent bonds rupture that improves set of structural modifications;[63] this is an alternative technology for food industry in interest compounds extraction form protein, using them such as pretreatment, because this process have an effect where the protein structure is altered and exposing to enzyme action.[64]

9.4   RECOVERY PROCESS

Purification and separation process in functional peptide production allowing successful recovery, represent crucial in process, the purification method selection, depends largely of methodologies to generate, generally has been used methodologies widely applied on protein purification such as solvent selective precipitation, ultrafiltration techniques, and chromatographic methods,[1] also properties such as charges, variation in molecular weight, and affinity during purification methodologies are the most significant barriers, for this reason scientific propose methodologies that use complex

TABLE 9.1 Recent Works in Peptide Production.

Author	Method	Microorganism/enzyme	Substrate	Peptide functionality
42	Smf	*Aeribacillus pallidus* SAT4	Synthetic medium	Antimicrobial
24	Smf	*Debaryomyces hansenii*	YPG medium supplemented with casein	ECA-I inhibitor
67	SSF	*Bacillus subtilis*	Walnut protein meal	Antioxidant
39	Smf	*Bacillus subtilis* A14h	Mineral medium supplemented with tomato pomace	Antioxidant and antibacterial
12	EH	Pepsin, trypsin, and alcalase	Roe egg	Antioxidant
58	EH	Protease from hepatopancreas of Pacific white shrimp	Sea bass skin	Antioxidant
4	EH	Pepsin	Caseins and whey from goat milk	Antioxidant
17	EH	Subtilisin, trypsin, and combination	Goat milk ultrafiltrated proteins	Antioxidant
21	EH	Subtilisin, trypsin, and combination	Casein and whey from goat milk	ECA-I Inhibitor
29	EH	Papain and flavourzyme	Fish scales	Iron binding
54	EH	Alcalase and trypsin	Amaranth protein isolate	Antithrombotic
14	EH	Trypsin, papain, neutrase, flavourzyme, and alcalase	Egg white	Antioxidant
52	EH assisted by MW	Bromelain	Gingko protein	Antioxidant
65	MW, US Pretreatment & HE	Pepsin and trypsin	Milk protein	Antioxidant
8	SWH	-	Squid muscle	Antioxidant
70	HHP pretreatment & EH	Alcalase, Savinase, Corolase 7089, and Protamex	Squid	Anti-inflammatory and antioxidant
23	EH assisted by HHP	Alcalase, Savinase, Corolase 7089, and Protamex	Lentil protein	ECA-I inhibitor and antioxidant

Smf = Submerged fermentation, **SSF** = Solid State Fermentation, **EH** = Enzymatic Hydrolysis, **HHP** = High Hydrostatic Pressure, **SWH** = Subcritical Water Hydrolysis, **MW** = Microwave, **US** = Ultrasound instrumentation to solve this challenge set,[53] likewise has been applied combination on this type of methodologies, Lafarga et al. (2016)[37,] and Jin et al. (2016)[31] employed successively ultrafiltration and chromatographic methodologies, por peptides with antihypertensive and antioxidant activity, respectively; while other authors separate peptides with apparent anticancer properties through technologies named electrodialysis with a ultrafiltration membrane,[18] Usually this technologies implementation, are successful at laboratory level but when are treated to scale up, have repercussions in enormous increase in bioprocess costs, being this stage demanding on new procedures and technologies that simplify, and reduces cost on this phase.

9.5 CONCLUSIONS

Bioprocess implicated in functional peptide generation, implicate a set of complex steps, in which scientific research perform an effort to optimize the process and generate new production techniques, and also that *in vitro* functional activities expressed by compounds, remains after consumption, coupled to the intention of reduce their bitter taste; but especially the most outstanding challenge, as in other bioprocess, the raised cost implied in purification techniques of functional peptides intended to human consumption, whose propose its present a positive effect in health.

KEYWORDS

- **biofunctional peptides**
- **food proteins**
- **biological modulator**
- **antioxidants**
- **nutraceuticals**

REFERENCES

1. Agyei, D., et al. Bioprocess Challenges to the Isolation and Purification of Bioactive Peptides. *Food Bioprod. Process.* **2016,** *98,* 244–256. Available at: http://dx.doi.org/10.1016/j.fbp.2016.02.003. [Accessed April 5, 2016].

2. Agyei, D.; Danquah, M. K. Industrial-Scale Manufacturing of Pharmaceutical-Grade Bioactive Peptides. *Biotechnol. Adv.* **2011,** *29* (3), 272–277. Available at: http://linkinghub.elsevier.com/retrieve/pii/S0734975011000024. [Accessed May 18, 2016].

3. Agyei, D.; Danquah, M. K. Rethinking Food-derived Bioactive Peptides for Antimicrobial and Immunomodulatory Activities. *Trends Food Sci. Technol.* **2012,** *23* (2), 62–69. Available at: http://linkinghub.elsevier.com/retrieve/pii/S0924224411001610. [Accessed January 15, 2016].

4. Ahmed, A. S., et al. Identification of Potent Antioxidant Bioactive Peptides from Goat Milk Proteins. *Food Res. Int.* **2015,** *74,* 80–88. Available at: http://www.sciencedirect.com/science/article/pii/S0963996915001957. [Accessed March 20, 2016].

5. Aluko, R. Bioactive Peptides. *Functional Food and Nutraceuticals;* Springer-Verlag New York, Inc.: New York, 2012; pp 37–61. Available at: http://link.springer.com/10.1007/978-1-4614-3480-1_3 [Accessed January 18, 2016].

6. Alvarado Carrasco, C.; Guerra, M. Lactosuero Como Fuente de Péptidos Bioactivos. *An. venez. nutr.* **2010,** *23* (1), 45–50. Available at: http://www.scielo.org.ve/scielo.php?script=sci_arttext&amp\npid=S0798-07522010000100007. [Accessed January 6, 2016].

7. Aoki, W.; Kuroda, K.; Ueda, M. Next Generation of Antimicrobial Peptides as Molecular Targeted Medicines. *J. Biosci. Bioeng.* **2012,** *114* (4), 365–370. Available at: http://dx.doi.org/10.1016/j.jbiosc.2012.05.001. [Accessed February 12, 2016].

8. Asaduzzaman, A. K. M.; Chun, B. S. Recovery of Functional Materials with Thermally Stable Antioxidative Properties in Squid Muscle Hydrolyzates by Subcritical Water. *J. Food Sci. Technol.* **2013,** *52* (2), 793–802.

9. Budarin, V. L., et al. The Potential of Microwave Technology for the Recovery, Synthesis and Manufacturing of Chemicals from Bio-wastes. *Catal. Today* **2015,** *239,* 80–89.

10. Carrasco-Castilla, J., et al. Use of Proteomics and Peptidomics Methods in Food Bioactive Peptide Science and Engineering. *Food Eng. Rev.* **2012,** *4* (4), 224–243. Available at: http://link.springer.com/10.1007/s12393-012-9058-8. [Accessed January 25, 2016].

11. de Castro, R. J. S.; Sato, H. H. Biologically Active Peptides: Processes for Their Generation, Purification and Identification and Applications as Natural Additives in the Food and Pharmaceutical Industries. *Food Res. Int.* **2015,** *74,* 185–198. Available at: http://www.sciencedirect.com/science/article/pii/S0963996915300028?via%3Dihub [Accessed February 12, 2016].

12. Chalamaiah, M. et al. Antioxidant Activity and Functional Properties of Enzymatic Protein Hydrolysates from Common Carp (*Cyprinus carpio*) Roe (Egg). *J. Food Sci. Technol.* **2015,** *52,* 8300–8307. Available at: http://www.scopus.com/inward/record.url?eid=2-s2.0-84921364741&partnerID=tZOtx3y1. [Accessed March 15, 2016].

13. Chaves-López, C., et al. Impact of Microbial Cultures on Proteolysis and Release of Bioactive Peptides in Fermented Milk. *Food Microbiol.* **2014,** *42,* 117–121. Available at: http://dx.doi.org/10.1016/j.fm.2014.03.005. [Accessed Febryary 16, 2016].

14. Chen, C., et al. Purification and Identification of Antioxidant Peptides from Egg White Protein Hydrolysate. *Amino Acids* **2012,** *43* (1), 457–466. [Accessed May 13, 2016].

15. Chen, J., et al. Comparison of Aanalytical Methods to Assay Inhibitors of Angiotensin I-Converting Enzyme. *Food Chem.* **2013,** *141* (4), 3329–3334. Available at: http://www.sciencedirect.com/science/article/pii/S030881461300825X?via%3Dihub [Accessed January 12, 2016].

16. Cicero, A. F. G.; Parini, A.; Rosticci, M. Nutraceuticals and Cholesterol-Lowering Action. *IJC Metab Endocr.* **2015**, *6*, 1–4. Available at: http://www.sciencedirect.com/science/article/pii/S2214762414000516?via%3Dihub [Accessed Febryary 12, 2016].

17. De Gobba et al. Antioxidant peptides from goat milk protein fractions hydrolysed by two commercial proteases. *Int. Dairy J.* **2014**. *39* (1). 28–40. Available at: http://dx.doi.org/10.1016/j.idairyj.2014.03.015.

18. Doyen, A., et al. Demonstration of *In Vitro* Anticancer Properties of Peptide Fractions from a Snow Crab By-products Hydrolysate after Separation by Electrodialysis with Ultrafiltration Membranes. *Sep. Purif. Technol.* **2011,** *78* (3), 321–329. Available at: http://dx.doi.org/10.1016/j.seppur.2011.01.037. [Accessed March 25, 2016].

19. Ebner, J., et al. Peptide Profiling of Bovine Kefir Reveals 236 Unique Peptides Released from Caseins During Its Production by Starter Culture or Kefir Grains. *J. Proteomics.* **2015,** *117,* 41–57. Available at: http://www.sciencedirect.com/science/article/pii/S1874391915000135. [Accessed April 15, 2016].

20. Erdmann, K.; Cheung, B. W. Y.; Schröder, H. The Possible Roles of Food-Derived Bioactive Peptides in Reducing the Risk of Cardiovascular Disease. *J. Nutr. Biochem.* **2008,** *19* (10), 643–654. Available at: http://www.sciencedirect.com/science/article/pii/S0955286307002756?via%3Dihub. [Accessed February 18, 2016].

21. Espejo-Carpio, F. J., et al. Angiotensin I-Converting Enzyme Inhibitory Activity of Enzymatic Hydrolysates of Goat Milk Protein Fractions. *Int. Dairy J.* **2013,** *32* (2), 175–183. Available at: http://linkinghub.elsevier.com/retrieve/pii/S0958694613001143. [Accessed Februaru 12, 2016].

22. Fernandez Figares, I., et al. Amino-Acid Composition and Protein and Carbohydrate Accumulation in the Grain of Triticale Grown under Terminal Water Stress Simulated by a Senescing Agent. *J. Cereal Sci.* **2000,** *32* (3), 249–258. Available at: http://www.sciencedirect.com/science/article/pii/S0733521000903291. [Accessed April 28, 2016].

23. Garcia-Mora, P., et al. High-Pressure Improves Enzymatic Proteolysis and the Release of Peptides with Angiotensin I Converting Enzyme Inhibitory and Antioxidant Activities from Lentil Proteins. *Food Chem.* **2015,** *171,* 224–232. Available at: http://dx.doi.org/10.1016/j.foodchem.2014.08.116. [Accessed February 25, 2016].

24. García-Tejedor, A., et al. Dairy Debaryomyces Hansenii Strains Produce the Antihypertensive Ccasein-derived Peptides LHLPLP and HLPLP. *LWT–Food Sci.Technol.* **2015,** *61* (2), 550–556. [Accessed April 20, 2016].

25. Hamme, V., et al. Crude Goat Whey Fermentation by *Kluyveromyces Marxianus* and *Lactobacillus Rhamnosus*: Contribution to Proteolysis and ACE Inhibitory Activity. *J. Dairy Res.* **2009,** *76* (2), 152–157. Available at: http://www.ncbi.nlm.nih.gov/pubmed/19121243. [Accessed March 14, 2016].

26. Hasler, C. M. Functional Foods: Benefits, Concerns and Challenges—A Position Paper from the American Council on Science and Health. *J. Nutr.* **2002,** *132* (12), 3772–3781. [Accessed January 12, 2016].

27. Hasler, C. M.; Brown, A. C. Position of the American Dietetic Association: Functional Foods. *J. Am. Diet Assoc.* **2009,** *109* (4), 735–746. [Accessed January 12, 2016].

28. Hoz, A. de la; Diaz-Ortiz, A.; Moreno, A. Microwaves in Organic Synthesis. Thermal and Non-thermal Microwave Effects. *Chem. Soc. Rev.* **2005,** *34,* 164–178. Available at: http://pubs.rsc.org/en/content/articlehtml/2005/cs/b411438h. [Accessed March 18, 2016].

29. Huang C.Y. et al. Evaluation of iron-binding acticity of collagen peptides prepared from the scales of four cultivates fishes in Taiwan. *J. Food Sci. Tecnolo.* **2013**. 23, 671–678. Available at: http://www.sciencedirect.com/science/article/pii/S0023643814004460

30. Jao, C. L.; Huang, S. L.; Hsu, K. C. Angiotensin I-Converting Enzyme Inhibitory Peptides: Inhibition Mode, Bioavailability, and Antihypertensive Effects. *BioMedicine.* **2012,** *2* (4), 130–136. Available at: http://www.sciencedirect.com/science/article/pii/S2211802012000526. [Accessed February 23, 2016].

31. Jia, J., et al. The Use of Ultrasound for Enzymatic Preparation of ACE-Inhibitory Peptides from Wheat Germ Protein. *Food Chem.* **2010,** *119* (1), 336–342. Available at: http://dx.doi.org/10.1016/j.foodchem.2009.06.036. [Accessed March 20, 2016].

32. Jin D.X., et al. Preparation of antioxidative corn protein hydrolysates, purification and evaluation of three novel corn antioxidant peptides. *Food Chem.* **2016,** *204,* 427–436 Available at: http://www.sciencedirect.com/science/article/pii/S0308814616302886

33. Kadam, S. U., et al. Ultrasound Applications for the Extraction, Identification and Delivery of Food Proteins and Bioactive Peptides. *Trends Food Sci. Technol.* **2015,** *46* (1), 60–67. Available at: http://www.sciencedirect.com/science/article/pii/S092422441500179X?via%3Dihub [Accessed April 25, 2016].

34. Kim, H. K., et al. Expression of the Cationic Antimicrobial Peptide Lactoferricin Fused with the Anionic Peptide in *Escherichia coli. Appl. Microbiol. Biotechnol.* **2006,** *72* (2), 330–338.

35. Kim, S. K.; Wijesekara, I. Development and Biological Activities of Marine-derived Bioactive Peptides: A Review. *J. Funct. Foods.* **2010,** *2* (1), 1–9. [Accessed February 25, 2016].

36. Kleekayai, T., et al. Extraction of Antioxidant and ACE Inhibitory Peptides from Thai Traditional Fermented Shrimp Pastes. *Food Chem.* **2015,** *176,* 441–447. Available at: http://www.sciencedirect.com/science/article/pii/S0308814614019281. [Accessed January 16, 2016].

37. Korhonen, H.; Pihlanto, A. Bioactive Peptides: Production and Functionality. *Int. Dairy J.* **2006,** *16* (9), 945–960. Available at: http://www.sciencedirect.com/science/article/pii/S0958694605002426?via%3Dihub. [Accessed January 6, 2016].

38. Lafarga et al. Identification of bioactive peptides from a papain hydrolysate of bovine serum albumin and assessment of an antihypertensive effect in spontaneously rats. *Food Res. Int.* **2016,** *81,* 91–99.

39. Mason, T. J.; Lorimer, J. P. *Applied Sonochemistry;* Wiley-VCH Verlag GmbH & Co. KGaA: Weinheim, FRG, **2002.** Available at: http://doi.wiley.com/10.1002/352760054X [Accessed April 18, 2016].

40. Moayedi, A.; Hashemi, M.; Safari, M. Valorization of Tomato Waste Proteins through Production of Antioxidant and Antibacterial Hydrolysates by Proteolytic *Bacillus subtilis*: Optimization of Fermentation Conditions. *J. Food Sci. Technol.* **2016,** *53* (1), 391–400.

41. Mohanty, D. P., et al. Milk Derived Bioactive Peptides and Their Impact on Human Health—A Review. *Saudi J. Biol. Sci.* **2015,** *23,* 577–583. Available at: http://linkinghub.elsevier.com/retrieve/pii/S1319562X15001382. [Accessed January10, 2016].

42. Möller, N. P., et al. Bioactive Peptides and Proteins from Foods: Indication for Health Effects. *Eur. J. Nutr.* **2008,** *47* (4), 171–182. Available at: http://link.springer.com/10.1007/s00394-008-0710-2. [Accessed January 12, 2016].

43. Muhammad, S. A.; Ahmed, S. Production and Characterization of a New Antibacterial Peptide Obtained from *Aeribacillus pallidus* SAT4. *Biotechnol. Rep.* **2015,** *8,* 72–80.

44. Mulero Cánovas, J., et al. Péptidos Bioactivos. *Clin. Invest. Arteriosclerosis.* **2011,** *23* (5), 219–227.

45. Ngo, D. H., et al. Biological Activities and Potential Health Benefits of Bioactive Peptides Derived from Marine Organisms. *Int. J. Biol. Macromolecules.* **2012,** *51* (4), 378–383. Available at: http://dx.doi.org/10.1016/j.ijbiomac.2012.06.001. [Accessed March 25, 2016].

46. O'Donnell, C. P., et al. Effect of Ultrasonic Processing on Food Enzymes of Industrial Importance. *Trends Food Sci. Technol.* **2010,** *21* (7), 358–367.

47. Ogedegbe, G.; Pickering, T. G. Epidemiology of Hypertension. In *Hurst's the Heart,* 13th ed.; McGraw-Hill: New York, 2011. Available at: http://mhmedical.com/content. aspx?aid=7823628. [Accessed February 6, 2016].

48. Ozuna, C., et al. Innovative Applications of High-Intensity Ultrasound in the Development of Functional Food Ingredients: Production of Protein Hydrolysates and Bioactive Peptides. *Food Res. Int.* **2015,** *77,* 685–696. Available at: http://dx.doi.org/10.1016/j. foodres.2015.10.015. [Accessed March 25, 2016].

49. Pedroche, J., et al. Obtaining of *Brassica carinata* Protein Hydrolysates Enriched in Bioactive Peptides Using Immobilized Digestive Proteases. *Food Res. Int.* **2007,** *40* (7), 931–938. [Accessed March 18, 2016].

50. Reddy, P. M., et al. Evaluating the Potential Nonthermal Microwave Effects of Microwave-assisted Proteolytic Reactions. *J. Proteomics.* **2013,** *80,* 160–170. Available at: http://dx.doi.org/10.1016/j.jprot.2013.01.005. [Accessed April 25, 2016].

51. Renye, J. A.; Somkuti, G. A. Cloning of Milk-derived Bioactive Peptides in *Streptococcus thermophilus. Biotechnol. Lett.* **2008,** *30* (4), 723–730.

52. Rogalinski, T., Herrmann, S.; Brunner, G. Production of Amino Aacids from Bovine Serum Albumin by Continuous Sub-critical Water Hydrolysis. *J. Supercrit. Fluids.* **2005,** *36* (1), 49–58.

53. Ruan, G., et al. The Study on Microwave Assisted Enzymatic Digestion of Ginkgo Protein. *J. Mol. Catal B: Enzym.* **2013,** *94,* 23–28. Available at: http://dx.doi. org/10.1016sss/j.molcatb.2013.04.010. [Accessed August 25, 2016].

54. Saadi, S., et al. Recent Advances in Food Biopeptides: Production, Biological Functionalities and Therapeutic Applications. *Biotechnol. Adv.* **2014,** *33* (1), 80–116. Available at: http://www.sciencedirect.com/science/article/pii/S0734975014001906. [Accessed March 25, 2016].

55. Sabbione A.C., et al. Potential antithrombotica activity detected in amaranth proteins and its hydrolysates. *LWT-Food Sci.* **2015.** 60, 171–177. Avaliable at: http://www.sciencedirect.com/science/article/pii/S0023643814004460 [Accessed April 25, 2016].

56. Salehi-Abargouei, A., et al. Effects of Dietary Approaches to Stop Hypertension (DASH)-Style Diet on Fatal or Nonfatal Cardiovascular Diseases-Incidence: A Systematic Review and Meta-Analysis on Observational Prospective Studies. *Nutrition.* **2013,** *29* (4), 611–618. Available at: http://dx.doi.org/10.1016/j.nut.2012.12.018. [Accessed February 12, 2016].

57. Samaranayaka, A. G. P.; Li-Chan, E. C. Y. Food-derived Peptidic Antioxidants: A Review of Their Production, Assessment, and Potential Applications. *J. Funct. Foods.* **2011,** *3* (4), 229–254.

58. Sarmadi, B. H.; Ismail, A. Antioxidative Peptides from Food Proteins: A Review. *Peptides.* **2010,** *31* (10), 1949–1956. Available at: http://linkinghub.elsevier.com/ retrieve/pii/S0196978110002640. [Accessed February 10, 2016].

59. Senphan, T.; Benjakul, S. Antioxidative Activities of Hydrolysates from Seabass Skin Prepared Using Protease from Hepatopancreas of Pacific White Shrimp. *J. Funct.*

Foods. **2014,** *6,* 147–156. Available at: http://linkinghub.elsevier.com/retrieve/pii/ S1756464613002284. [Accessed April 25, 2016].

60. Sila, A.; Bougatef, A. Antioxidant Peptides from Marine By-products: Isolation, Identification and Application in Food Systems. A Review. *J. Funct. Foods.* **2016,** *21,* 10–26. Available at: http://dx.doi.org/10.1016/j.jff.2015.11.007. [Accessed March 15, 2016].

61. Spiller, H. A. Angiotensin Converting Enzyme (ACE) Inhibitors. In *Encyclopedia of Toxicology* ; Central Ohio Poison Center: Columbus, OH, 2014; pp 14–16. [Accessed February 20, 2016].

62. Srinivas, S.; Prakash, V. Bioactive Peptides from Bovine Milk α-Casein: Isolation, Characterization and Multifunctional properties. *Int. J. Peptide Res. Ther.* **2010,** *16* (1), 7–15. Available at: http://link.springer.com/10.1007/s10989-009-9196-x. [Accessed March 24, 2016].

63. Tejesvi, M. V., et al. An Antimicrobial Peptide from Endophytic *Fusarium Tricinctum* of *Rhododendron Tomentosum Harmaja*. *Fungal Divers.* **2013,** *60* (1), 153–159.

64. Tian, Y., et al. Effect of High Hydrostatic Pressure (HHP) on Slowly Digestible Properties of Rice Starches. *Food Chem.* **2014,** *152,* 225–229. Available at: http://dx.doi.org/10.1016/j.foodchem.2013.11.162. [Accessed April 25, 2016].

65. Toldrà, M., et al. Hemoglobin Hydrolysates from Porcine Blood Obtained through Enzymatic Hydrolysis Assisted by High Hydrostatic Pressure Processing. *Innov. Food Sci. Emerg. Technol.* **2011,** *12* (4), 435–442. Available at: http://dx.doi.org/10.1016/j.ifset.2011.05.002. [Accessed May 1, 2016].

66. Uluko, H., et al. Effects of Thermal, Microwave, and Ultrasound Pretreatments on Antioxidative Capacity of Enzymatic Milk Protein Concentrate Hydrolysates. *J. Funct. Foods.* **2015,** *18* (2), 1138–1146. Available at: http://dx.doi.org/10.1016/j.jff.2014.11.024. [Accessed April 28, 2016].

67. Wang, Y., et al. Preparation of Active Corn Peptides from Zein through Double Enzymes Immobilized with Calcium Alginate-chitosan Beads. *Process Biochem.* **2014,** *49* (10), 1682–1690. [Accessed April 18, 2016].

68. Wu, W., et al. Optimization of Production Conditions for Antioxidant Peptides from Walnut Protein Meal Using Solid-state Fermentation. *Food Sci. Biotechnol.* **2014,** *23* (6), 1941–1949.

69. Yang, B., et al. Amino Acid Composition, Molecular Weight Distribution and Antioxidant Activity of Protein Hydrolysates of Soy Sauce Lees. *Food Chem.* **2011,** *124* (2), 551–555. Available at: http://dx.doi.org/10.1016/j.foodchem.2010.06.069. [Accessed April 2, 2016].

70. Zambrowicz, A., et al. Manufacturing of Peptides Exhibiting Biological Activity. *Amino Acids* **2013,** *44* (2), .315–320.

71. Zhang, Y., et al. *In vitro* Anti-Inflammatory and Antioxidant Activities and Protein Quality of High Hydrostatic Pressure Treated Squids (*Todarodes pacificus*). *Food Chem.* **2016,** *203,* 258–266. Available at: http://dx.doi.org/10.1016/j.foodchem.2016.02.072.

72. Zou, Y., et al. Enzymolysis Kinetics, Thermodynamics and Model of Porcine Cerebral Protein with Single-Frequency Countercurrent and Pulsed Uultrasound-Assisted Processing. *Ultrason. Sonochem.* **2016,** *28,* 294–301. Available at: http://dx.doi.org/10.1016/j.ultsonch.2015.08.006.

CHAPTER 10

BIOPRODUCTS OBTAINED FROM THE BIOPROCESSING OF THE BANANA PEEL WASTE: AN OVERVIEW

SÓCRATES PALACIOS-PONCE[1], ANNA ILYINA[1,*],
RODOLFO RAMOS-GONZÁLEZ[2], HÉCTOR A. RUIZ[1],
JOSÉ L. MARTÍNEZ-HERNÁNDEZ[1], ELDA P. SEGURA-CENICEROS[1],
MIGUEL A. AGUILAR[3], OLGA SÁNCHEZ[4], and
CRISTÓBAL N. AGUILAR[1]

1Department of Food Research, School of Chemistry, Universidad Autónoma de Coahuila, Blvd. V. Carranza e Ing. José Cárdenas Valdés, Saltillo 25280, Coahuila, México
2CONACYT, Universidad Autónoma de Coahuila, Blvd. V. Carranza e Ing. José Cárdenas Valdés, Saltillo 25280, Coahuila, México
3Centro de Investigación y de Estudios Avanzados del Instituto Politécnico Nacional (Unidad Saltillo), Av. Industria Metalúrgica 1062, Parque Industrial Saltillo-Ramos Arizpe, Ramos Arizpe 25900, Coahuila, México
4Facultad de Ingeniería Química, Instituto Superior Politécnico José Antonio Echeverría, 127 Marianao, La Havana, Cuba
*Corresponding author. E-mail: anna_ilina@hotmail.com

CONTENTS

ABSTRACT

The constant disposal of large volumes of untreated wastes, emanating from agro-industrial activities, has resulted in environmental problems. However, as a result of their low cost, these agro-industrial wastes are exploited as a source of raw materials for the industrial sector, thereby mitigating the harmful environmental impact caused by their deposition. Banana peels constitute around 30–40% of the total weight of the fruit and contain a significant amount of proteins, carbohydrates, and fiber. Due to its organic and nutrient-rich nature, banana peels are considered as raw materials for the manufacture of products with high added value. Hence, this chapter analyzes diverse alternatives to the valorization and applications of banana peels for the production of compounds with high added value, including antioxidants and antimicrobial products, pectins, biosorbents, bioethanol and biogas production, and nanoparticles.

10.1 INTRODUCTION

Bananas and plantains serve as food and are one of the most consumed fruits in the world and it is commercially produced in about 120 countries. Banana is cultivated all over the world, especially in tropical and subtropical areas, where it is grown in a sustainable way, contributing to the economy of countries dedicated to its production.[1–3] Currently, it is the second most produced fruit in the world. The estimated production in 2012 was approximately 101 million tons.[4] India is the world's largest producer, accounting for over 25% of the world's production.[5] However, most of the produce are consumed locally (India has very low exportation rates compared with producer countries and exports leaders such as Ecuador, Philippines, and Costa Rica). The main importers and consumers are the United States of America and the European Union.[6]

Years of dedicated research on natural and selective breeding have resulted in feasible transformation of edible bananas into hundreds of varieties with improvements, among which we can highlight: reduced seed size, sterility, oversized pulp, and fruit development without fertilization.[7,8] Banana is an annual crop of warm and humid regions with approximately 1200 seedless fruit varieties.

There are different commercial uses for this fruit, but traditionally, the whole fruit is sold for direct consumption. Some more common forms related to the development of industrial scale processes are: dehydrated (slices, dices), flour, fried, mashed (jellies, clarified juices, and unfermented and fermented beverages), as well as frozen pulp (sliced, whole, and halves).[9] Banana production processes generate a significant amount (about 40%) of banana peels.[10]

Commercial banana production generates a several proportion of waste. In countries such as India, approximately 1.6 million metric tons of dried banana peels are produced yearly and do not find any commercial application. Generally, these residues are disposed in municipal waste due to lack of adequate infrastructure for handling.[2,11] Therefore, it is important to find applications for this by-product, as it constitutes a real environmental problem.[1] The environmental impact is due to the fact that the residue has a high content of nitrogen and phosphorus, as well as water, which makes it susceptible to modification by microorganisms.[12]

Recent studies have focused on using industrial by-products of banana, to meet the increasing demand for raw materials in several industries.[13–16] This chapter shows how by-products of the industrial production of banana (ripe peels) can be utilized to produce high value-added products.

10.2 BANANA PEELS

Banana peels constitute about 30–40% of the total weight of the fruit and contain a significant amount of carbohydrate, protein, fiber, and low amounts of lignin.[17,18] They are considered as raw materials for the manufacture of products with high added value.[11]

10.2.1 NUTRITIONAL COMPOSITION

Banana is a climacteric fruit, once harvested (preclimacteric) at green-stage maturation, the fruit shows several changes, some of which are: (1) related to metabolic rates such as physical and chemical changes (composition, color, texture, and odor) and (2) related to biochemical reactions (respiration, ripening, and senescence).[9] Indeed, few reports have focused on quantification of the nutritional components of banana peels. Moreover, different varieties of bananas exist, which have not been reported in the literature.

From previous reports, it is evident that the values of different parameters are similar to certain green and ripe peels. The peels contain mainly nonstructural carbohydrates (NSC) in organic matter (58–77.2%), crude fiber (2–10%), ash (7.2–15%), ether extract (5.1–10.2%), and crude protein

(6.7–8.6%).[19–22] Studies conducted on seven Cavendish *Musa* varieties culti-vated in Cuba, show that the peel accounts for 41 and 23% of the total weight, on a wet and dry weight basis, respectively. On the basis of dry weight, significant quantities of the following were obtained : Ash (13–19%), crude fiber (8–11%), crude protein (6.1–7.9%), soluble sugars (8–11.4%), and minerals such as potassium (0.09–0.17%) and calcium (0.28–0.40%).[23] The results of studies conducted in Nigeria also show that the green and ripe banana peels contain predominantly NSC in organic matter (56.8–59.4%), ash (13–16.5%), crude fiber (6.4–13%), ether extract (6–11.6%), and crude protein (7.7–8.1%).[24–26] Among the most current information are the results of a study of six varieties of bananas and plantains: dessert banana (*Musa* AAA), plantain (*Musa* AAB), cooking banana (*Musa* ABB), and hybrid banana (*Musa* AAAB) at different stages of ripeness. It shows that dry weight values vary according to the variety: For bananas (7.7–12.9%), and for plantains (12.6–18.7%). Cooking bananas had the highest content (22.4%), while hybrids had values similar to those reported for bananas. The latter were shown to be high in total dietary fiber (40–50%) with insoluble dietary fiber constituting the dominant fraction, and the values reported in dry weight to proteins (8–11%), lipids and fatty acids (2.2–10.9%), and ash (6.4–12.8%). Its chemical composition shows the presence of significant amounts of essential amino acids such as leucine, valine, phenylalanine, just as fatty acids expressed as the polyunsaturated linoleic acid (omega-6) and α-linoleic acid (omega-3).[10] Except for lysine, the detected values corre-spond to the highest levels that set the standards of the food and agriculture organization (FAO). The most abundant mineral (as a chemical element) is potassium, followed by phosphorus, calcium, and magnesium. The sugar content of banana is directly related to the degree of fruit ripeness: a higher maturation stage will increase the soluble sugar contents (glucose, fructose), while starch and hemicellulose contents decrease, perhaps caused by the action of endogenous enzymes.[10] It has been found that banana peels are a potential source of dietary fibers and pectin.[17]

10.2.2   ALTERNATIVE USES

Considering the importance of reducing the environmental pollution caused by this type of agro-industrial waste (peels), it is necessary to study the potential use of this feedstock, in producing economic value-added products. Figure 10.1 highlights the use of banana peels for the production of enzymes, pectins, ethanol, methane, compounds with antioxidant and antimicrobial activity, organic fertilizers, biosorbents[6,9] as well as in the synthesis of palladium and silver nanoparticles (NPs).[27,28]

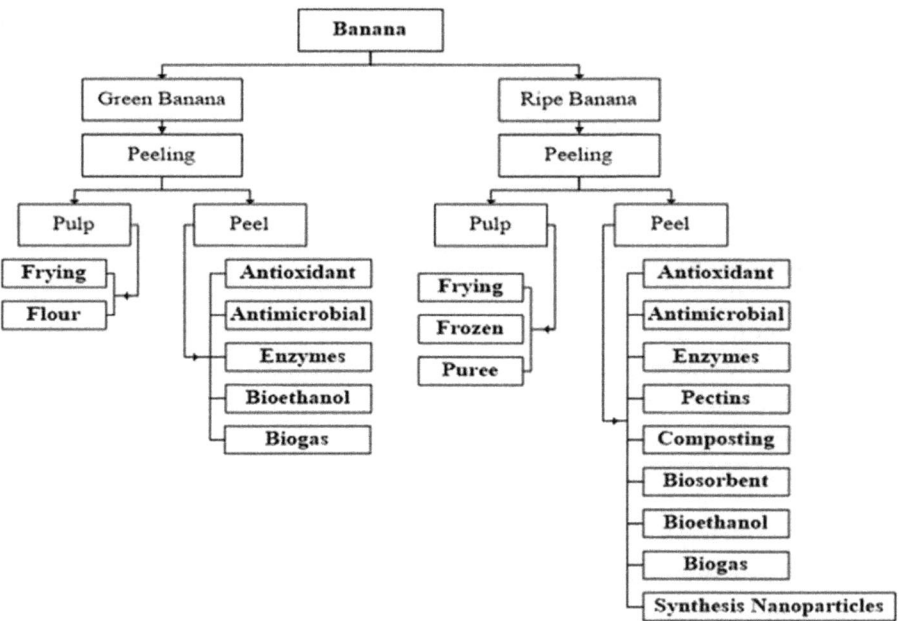

FIGURE 10.1   Potentials uses for pulp and peel banana generated in the agro-industrial sector as raw materials for biotechnological processes.

10.2.2.1 ANTIOXIDANTS SOURCE

Banana peels are rich in phytochemicals and antioxidants. Thus, its extracts exhibit high activities against lipid peroxidation. For banana peels, of variety *Musa acuminata* AAA, total free phenolics had values in the range of 0.9–3.0 mg/g fresh peel.[29]

Antioxidants such as gallocatechin, catechin, and epicatechin were found in Cavendish banana peels. Gallocatechin was determined at a concentration of 160 mg/100 g DW (dry weight), much higher than 29.6 mg/100 g DW in the pulp, expressing high antioxidant power.[30] Dopamine was found in high amounts in commercial Cavendish banana peels, with values ranging from 80 to 560 and 2.5 to 10 mg/100 g in peel and pulp, respectively.[31] Ripe banana peels also contain other compounds such as cyanidin and delphinidin.[32] Likewise, the presence of carotenoids such as β-carotene, α-carotene, and xanthophylls (violaxanthin, auroxanthin, neoxanthin, isolutein, β-cryptoxanthin, and α-cryptoxanthin) was demonstrated in a range of 3–4 µg/glutein equivalent. It was reported that most of the carotenoids are in esterified form, mainly myristate and laurate, with caprate or palmitate in lesser amounts. Also, banana peel extracts were found to contain sterols and triterpenes such as β-sitosterol, stigmasterol, campesterol, cycloeucalenol, cycloartenol, and 24-methylene cycloartanol. Between the sterols and triterpenes, 24-methylene cyclo-artanol palmitate and an unidentified triterpene ketone were the main components.[33]

The extracts obtained from green and ripe banana peels (*Musa,* AAA, *cv.* Cavendish) using ethyl acetate and water as solvent, showed significant antioxidant activity. Extracts from green peels showed greater activity when compared with yellow peels. On application of the ß-carotene bleaching method, linoleic acid was inhibited by a polar fraction of 70% acetone extracts. Using the ferric thiocyanate method, at a concentration of 0.5 mg/mL, antioxidant activity was recorded and by the DPPH di ((phenyl)-(2, 4, 6-trinitrophenyl) iminoazanium) radical method, free radical scavenging activity was obtained mainly from the aqueous–acetone extract, followed by the acetone extracts. The antioxidant activity of aqueous extracts was comparable to synthetic antioxidants, such as BHA (butylated hydroxyanisole) and BHT (butylated hydroxytoluene).[34]

By standard chemical methods (DPPH, beta-carotene bleaching) and in vitro biochemical method (induced lipid peroxidation in liver tissue and erythrocytes) for extracts of banana peels, *Musa paradisiaca* demonstrated a considerable potential for radical scavenging at concentrations of 25, 50, 100 µg/mL and 0.25, 0.5, 1.0, 2 µg/mL, respectively. The polyphenol content was determined as 89 ± 2.12 (mg gallic acid equivalent/100 g DW-extract), flavonoids 47.85 ± 3.75 (mg quercetin equivalent/100 g DW-extract) and ascorbic acid 756.45 ± 10.89 (mg /100 g DW-excerpt).[35]

For banana peel extracts, *M. acuminata Colla* AAA ("Gran Naine" and "Gruesa") was reported to have a high capacity to scavenge DPPH and 2, 2′-azino-bis (3-ethylbenzthiazoline)-6-sulfonic acid (ABTS), free radicals, and are also good lipid peroxidation inhibitors. The antioxidant activities of extracts obtained from the different cultivar crops were similar.[12] The solvents isopropyl alcohol, hexane or dichloromethane, were used to obtain extracts from *Musa sapientum,* yellow banana peels. The extract obtained with isopropyl alcohol (a medium polar compound) exhibited the best antioxidant activity.[36]

A study was performed with nine varieties of banana peels *M. sapientum* species: *Musa balbisiana* that is, Monthan (*Musa spp.*—Bluggoe—AAB), Karpooravalli (*Musa spp.*—Karpooravalli—ABB), Nendran (*Musa spp.*—French Plantain—AAB), Kadali (*Musa spp.*—Ney Poovan—AB), *M. acuminata* such as—Pach—ainadan (*Musa spp.*—Pachanadan—AABS), Poovan (*Musa spp.*—Mysore—AAB), Rasthali (*Musa spp.*—Rasthali—AAB), Robusta—Cavendish sub group (*Musa spp.*—Robusta—AAB), and Sevvazhai (*Musa spp.*—Redbanana—AAA), revealed that the ethanolic extract obtained for each one showed a significant antioxidant activity when evaluated in vitro with free radical scavenging assays such as DPPH, ABTS, and lipid peroxidation inhibition assays.[37]

In another study conducted with three varieties of banana, namely "Pachabale" (*M. paradisiaca* cv. Dwarf Cavendish, AAA), "Yelakkibale" (*M. paradisiaca* cv. Ney poovan, AB) and "Nendranbale" (*M. paradisiaca* cv. Nendran, AAB), antioxidant activity was also reported. An assessment of the three extracts obtained with methanol, ethanol, and aqueous medium was conducted. The polyphenol content in the methanol extract of these three banana varieties was in the range of 520–850 mg equivalents tannic acid/100 g of sample while the ethanol and aqueous extracts were in the range of 200–700 mg equivalents of tannic acid/100 g of sample. Similarly, the flavonoid content of the methanolic extracts was between 385 and 1035 mg equivalents of tannic acid/100 g sample while the aqueous medium extracts was between 350 and 714 mg equivalents of tannic acid/100 g sample. These values are high compared with ethanol extracts of 222–818 mg equivalents of tannic acid/100 g sample. The aqueous medium extracts of all three varieties showed the highest antioxidant capacity, followed by the methanolic and ethanolic extracts.[38]

10.2.2.2   EXTRACTS WITH ANTIMICROBIAL EFFECT

Extracts obtained from green and ripe banana peels (*Musa*, AAA, cv. Cavendish) using ethyl acetate, proved to have significant antimicrobial activity. It was shown that extracts from green peels have higher antimicrobial activity compared with ripe peels, in the case of five species of selected bacteria: Gram positive (*Bacillus cereus* IFO 13597, *Bacillus subtilis* IFO3009, *Staphylococcus aureus* IFO 3761), Gram negatives (*Salmonella enteritidis* IFO 3313, and *Escherichia coli* IFO 13168). The isolated compounds β-sitosterol, D-malic acid, palmitic acid and 12-hydroxystearic acid evaluated by the agar disc diffusion method (DD) and the minimum inhibitory concentration (MIC), exhibited activity against all tested bacteria. The MIC values obtained for β-sitosterol, malic acid, and succinic acid, used as a control, varied between 140 and 750 ppm, respectively. D-malic acid had the highest antimicrobial activity against all bands of bacteria species, while 12-hydroxystearic acid only recorded antimicrobial activity by the agar DD method.[34]

The extracts obtained from yellow banana peels (*M. sapientum*) using isopropyl alcohol, hexane, dichloromethane, and acetone also showed antimicrobial activity. The extract obtained with isopropyl alcohol (a solvent of medium polarity) showed greater antimicrobial activity, which was evaluated at different dilutions (1:100, 1:1000; 1:10,000) against *Escherichia coli*, *Staphylococcus sp.*, and *Klebsiella sp.*[36]

10.2.2.3   PECTINS EXTRACTION

With acid extraction, a higher content of galacturonic acid, as well as a higher degree of methylation and acetylation was achieved in banana peels compared with plantain peels. The molecular weights of extracted pectins ranged from 132.6 to 573.8 kDa. This parameter did not change significantly, based on fruit variety. Thus, maturation states did not affect the composition of polysaccharides (in terms of galacturonic acid, rhamnose, arabinose, xylose, mannose, galactose, and glucose) in a consistent manner.[17]

A study was conducted on the influence of parameters such as pH (1.5 and 2), time (1 and 4 h), and temperature (80°C and 90°C), on the extraction of high quality pectin from banana peels (Musa, genotype AAA, "Gran Naine"). At stage five of ripeness (more yellow than green), the results show that pH is the most important parameter due to its influence on the yield and chemical composition of pectin. Low values negatively affect the galacturonic acid content (a less pure molecule), but increases the yield of the extraction process. This can be attributed to certain impurities or degraded pectins, depending on the experimental conditions. Additionally, long periods and temperatures of extraction cause a significant decrease in the degree of methylation. Galactose, rhamnose, and arabinose were the major neutral sugars detected in the pectins. The molecular weights of extracted biopolymers ranged from 87 to 248 kDa, and was mainly influenced by pH and extraction time. A pH value of 2, time of 1 h and temperature of 90°C, are considered suitable for extraction of pectins from banana peels.[15]

Qiu et al.[39] used response surface methodology to estimate the optimal parameters for pectin extraction from banana peels (species not reported). They made cuts of size 4 × 4 mm, enzymatically pretreated with α-amylase and neutrase enzymes. The optimal conditions for the extraction were a pH of 1.5, extraction temperature of 85°C, extraction time of 2 h and a precipitation temperature of 70°C.

10.2.2.4   ANIMAL FEEDING USE

Banana peels are considered as a promising raw material for the production of animal feed because of the high content of carbohydrates, proteins, lipids, fiber, essential amino acids, and minerals, such as sodium, calcium, potassium, iron, and manganese (Tables 10.1–10.3).[40] The dietary fiber content is approximately 30–50% DW, both for banana and plantain peels (Table 10.1).[10] These values are similar or higher than those reported for other fruit peels. Banana peels (*Musa*, AAA genotype, Grand Nain) (*Musa*, ABB genotype, French Clair) in different ripening stages: 1 (green), 5 (more yellow than green), and 7 (yellow with brown stains) showed that the concentrations of neutral detergent fiber and acid detergent fiber are high in both varieties for the ripening stages described. The lignin content is different for both varieties during the ripening stages. It increases during ripening in banana peels of the AAA genotype (14.3 to 5.4%) but decreases in banana peels of the ABB genotype (from 12.1 to 7%). Cellulose levels increased from 7.6 to 9.6% and from 6.5 to 7.8% from ripening stages 1–5 and decreased from 9.6 to 7.5% and from 7.8 to 6.4% from ripening stages 5–7 to Grand Nain Clair and French, respectively. The hemicellulose content was higher in banana peels (6.4–8.4%) and in plantain peels (0.6–2.0%). The hemicellulose content increases with ripeness in bananas but not in plantains, as a result of changes in enzymatic activity. High contents of cellulose compared to hemicellulose for ripening states and varieties described are related to values obtained by Bardiya et al.[41] for ripened bananas variety not reported.[17] These features make this product a good material to be considered for livestock and poultry feeding.[42]

Banana peels may be used as culture media for fungal biomass growth, enriching the content of protein, and fatty acids of the solid mixture. Throughout solid-state fermentation (SSF) with *Aspergillus niger*, *Aspergillus flavus* and *Pennicilium sp.*, the protein content increased by 34%. Increases in levels of sugars up to 142% can be achieved by *Aspergillus flavus* fermentation.[43,44] The microbial fermentation process in banana peels, the low quality, and the low cost of this raw material, constitute an important step in the obtention of high nutritional value food. The protein and sugar contents may be compared with soy flour which is a common ingredient in most feed stuff.[45] This and other applications of banana peels in

TABLE 10.1 Nutritional Composition for Banana Peels of Different Varieties and Maturation Stages on a Dry Basis.

Stage/peel	Variety	Dry matter	Ash	Fiber	Fat	NSC	Protein	Soluble sugars	P	Ca	Reference
Green peel	–	88.5*	13	6.5	10.2	63	7.4	–	–	–	19
	–	–	–	10.6	–	–	6.7	–	–	–	20
	–	8.6	15.8	10.2	7.3	59	8.2	–	–	–	21
Ripe peel	–	–	14.8	13.1	5.1	58	8.6	–	–	–	22
–	Mysore	10.8	7.2	2.3	5.2	77	8.1	–	–	–	21
–	Parecido al Rey	–	15.7	9.1	–	–	6.9	11.4	0.11	0.28	23
	Tetraploide	–	14.6	10.2	–	–	7.3	11.4	0.17	0.49	
	Cavendish Gigante	–	17.4	11.9	–	–	6.9	10.7	0.15	0.39	
	Cavendish Enano	–	19.2	11.2	–	–	6.6	11.4	0.15	0.4	
	Robusta	–	15.8	10	–	–	7.2	9.7	0.15	0.42	
	UC-RS	–	14.2	10.1	–	–	6.6	9.7	0.09	0.29	
	Valery	–	13.8	8.3	–	–	6.1	8	0.09	0.34	
Green peel	–	–	16.5	13	6	57	7.7	–	–	–	24
	–	–	15.3	6.4	–	–	8.1	–	–	–	26
Ripe peel	–	14.1	13.4	7.7	11.6	59	7.9	–	–	–	25

*Sun drying.
All values are quantified on a dry basis.
"–" no reported data in cited article.

TABLE 10.2 Chemical Composition for Banana Peels of Different Varieties and Maturation Stages by Variety.[10]

Varieties Species group Subspecies/subgroup Type fruit	French Clair AAB Plantain Cooking		Grande Naine AAA Cavendish Dessert		CRBP039 AAAB Plantain hybrid Cooking	
*Stage	1	7	1	7	1	7
×Dry matter	12.6 ± 0.8	14.8 ± 0.2	8.7 ± 0.1	10.2 ± 0.2	9.6 ± 0.1	11.2 ± 0.5
×Crude protein	8.3 ± 0.1	9.1 ± 0.2	3.3 ± 0.1	8.1 ± 0.2	9.8 ± 0.4	11.0 ± 0.2
×Crude fat	4.6 ± 0.1	5.9 ± 0.1	3.8 ± 0.1	5.7 ± 0.1	5.3 ± 0.2	7.8 ± 0.2
×Ash	8.8 ± 0.2	8.2 ± 0.3	9.6 ± 0.2	12.8 ± 0.3	7.5 ± 0.2	8.2 ± 0.3
×Dietary fiber	32.9 ± 0.8	46.9 ± 0.9	43.2 ± 0.5	49.7 ± 0.3	39.7 ± 0.7	36.9 ± 0.4
×Starch	35.4 ± 0.0	3.2 ± 0.0	11.1 ± 0.1	3.3 ± 0.1	37.6 ± 0.6	0.3 ± 0.03
×*Sugars*						
Glucose	0.8 ± 0.1	15.6 ± 0.2	1.0 ± 0.3	14.0 ± 0.2	0.3 ± 0.1	15.2 ± 0.1
Fructose	0.7 ± 0.1	26.6 ± 0.2	1.2 ± 0.1	18.4 ± 0.2	0.1 ± 0.1	22.5 ± 0.2
Sucrose	0.2 ± 0.1	ND	1.5 ± 0.3	ND	0.5 ± 0.1	ND
×*Amino acids*						
Essential amino acid						
Leucine	0.40 ± 0.01	0.40 ± 0.00	0.30 ± 0.00	0.38 ± 0.00	0.44 ± 0.00	0.45 ± 0.00
Valine	0.29 ± 0.00	0.33 ± 0.00	0.24 ± 0.01	0.32 ± 0.02	0.40 ± 0.1	0.45 ± 0.01
Threonine	0.30 ± 0.00	0.31 ± 0.00	0.20 ± 0.02	0.28 ± 0.00	0.29 ± 0.04	0.34 ± 0.03
Phenylalanine	0.30 ± 0.01	0.31 ± 0.01	0.19 ± 0.00	0.25 ± 0.00	0.34 ± 0.00	0.34 ± 0.01
Nonessential amino acid						
Glutamic	1.30 ± 0.01	0.99 ± 0.02	0.70 ± 0.02	1.10 ± 0.03	1.48 ± 0.00	0.14 ± 0.00
Aspartic acid	0.92 ± 0.00	0.89 ± 0.00	0.54 ± 0.02	0.69 ± 0.01	1.20 ± 0.03	0.80 ± 0.03

TABLE 10.2 *(Continued)*

Varieties Species group Subspecies/subgroup Type fruit	French Clair AAB Plantain Cooking		Grande Naine AAA Cavendish Dessert		CRBP039 AAAB Plantain hybrid Cooking	
*Stage	1	7	1	7	1	7
Glycine	0.40 ± 0.00	0.45 ± 0.01	0.27 ± 0.01	0.37 ± 0.01	0.38 ± 0.02	0.41 ± 0.00
Alanine	0.39 ± 0.01	0.40 ± 0.01	0.27 ± 0.01	0.38 ± 0.02	0.41 ± 0.01	0.40 ± 0.01
*Fatty acid						
Saturated fatty acid						
Palmitic	38.1 ± 1.4	37.2 ± 3.4	38.2 ± 2.2	41.5 ± 3.8	38.2 ± 0.4	38.2 ± 0.5
Stearic	3.7 ± 0.0	3.5 ± 0.8	5.3 ± 1.2	3.4 ± 0.8	2.2 ± 0.8	5.2 ± 1.0
Polyunsaturated fatty acid						
Linoleic	23.9 ± 0.6	22.2 ± 3.2	22.7 ± 4.9	23.9 ± 4.0	23.0 ± 0.0	26.3 ± 0.6
α-linolenic	20.2 ± 1.1	29.0 ± 1.7	21.1 ± 1.9	21.2 ± 1.9	27.6 ± 0.3	25.5 ± 0.1
**Minerals						
Potassium (K)	45,891 ± 08	55,234 ± 68	57,246 ± 93	63,521 ± 57	47,150 ± 820	54,820 ± 67
Phosphorus (P)	2020 ± 70	3364 ± 16	2099 ± 38	2081 ± 47	2600 ± 42	2530 ± 438
Calcium (Ca)	1742 ± 0	1654 ± 0	5954 ± 81	3769 ± 80	3950 ± 707	1700 ± 282
Magnesium (Mg)	828 ± 0	1101 ± 73	1047 ± 0	1378 ± 8	1070 ± 0	1020 ± 14

*Stage: stages of ripeness, 1 (green); 7 (yellow/a few brown spots)
ND: not detected
All values are quantified in a dry basis

× Values reported in percentage (%)
• • Values reported in mg/kg (ppm)

TABLE 10.3 Peels Characterization for Five Varieties of Banana.

References Varieties	73	40		38	
Species group	Musa sapientum	Musa sapientum	Musa paradisiaca Dwarf Cavendish	Musa paradisiaca Ney poovan	Musa paradisiaca Nendran
	—	—	AAA	AB	AAB
Stage	Fruit	Ripe fruit	—	—	—
Moisture	78.48*	—	88.9 ± 0.00**	88.2 ± 0.15**	82.6 ± 0.04**
Dry matter	14.08*	6.70 ± 2.22**	1.71 ± 0.01**	1.45 ± 0.01**	1.47 ± 0.00**
Crude fiber	7.68*	31.70 ± 0.25**	—	—	—
Proteins	7.87*	0.90 ± 0.25**	6.77 ± 0.00**	7.76 ± 0.08**	4.60 ± 0.08**
Crude fat	11.6*	1.70 ± 0.10**	—	—	—
Ash	13.44*	8.50 ± 1.52**	12.90 ± 0.01**	12.96 ± 0.05**	8.98 ± 0.01**
Carbohydrate	59.51*	59.0 ± 1.36**	26.40 ± 0.00**	9.8 ± 0.00**	41.9 ± 0.00**
Minerals					
Potassium (K)	44*	78.10 ± 6.58+	—	—	—
Calcium (Ca)	7*	19.2 ± 0.00+	166.54 ± 0.93+	244.68 ± 0.88+	204.8 ± 0.00+
Sodium (Na)	34*	24.3 ± 0.12+	—	—	—
Iron (Fe)	0.93*	0.61 ± 0.22+	10.00 ± 0.00+	3.33 ± 0.00+	4.0 ± 0.00+
Manganese (Mn)	—	76.20 ± 0.00+	—	—	—
Bromine (Br)	—	0.04 ± 0.00+	—	—	—
Rubidium (Rb)	—	0.21 ± 0.05+	—	—	—
Strontium (Sr)	—	0.03 ± 0.01+	—	—	—
Zirconium (Zr)	—	0.02 ± 0.00+	—	—	—

TABLE 10.3 (Continued)

References Varieties	73	40	38		
Species group	Musa sapientum	Musa sapientum	Musa paradisiaca Dwarf Cavendish	Musa paradisiaca Ney poovan	Musa paradisiaca Nendran
	–	–	AAA	AB	AAB
xStage	Fruit	Ripe fruit	–	–	–
Niobium (Nb)	–	$0.02 \pm 0.00^+$	–	–	–
Phosphorus (P)	40•	–	$145.66 \pm 0.57^+$	$212.00 \pm 0.00^+$	$140.00 \pm 0.00^+$
Sulphur	12•	–	–	–	–
Magnesium (Mg)	26•	–	–	–	–
Ascorbic acid	18•	–	–	–	–

*Values reported in % dry matter. •Minerals: mg/100 g peels.
**Values reported in % +Minerals: mg/g peels.

xStage: stages of ripeness.

biotechnological processes are listed in Table 10.4. The use of whole ripened bananas (peel and pulp) was tested narrowly through a combined diet with other ingredients to feed pigs, such that a combined intake of peel and pulp was achieved.[46]

10.2.2.5   BIOSORBENT

In different studies, banana peels have been shown biosorption of heavy metals and are regarded as an excellent source of biomass for the chelation of metals. The application of these residues can remove the aforementioned contaminants from subsurface water, sewage, and industrial effluents, as well as reduce organic pollution.[47–49] However, the adsorptive capacity depends on factors, such as the pH of the solution, the adsorbent dose, the metal concentration, the exposure time, and the agitation speed.[50]

Banana peels dust (*Musa paradisiaca*), with particle size of 1 and 2 mm, has been applied to heavy metals biosorption, such as cadmium, lead and copper. Tests revealed a high metal retention capacity (12% or more) with material of an average size of 1 mm. Banana peels dust was treated with 0.5 N NaOH for 20 min and the reported biosorptive capacities are 67.2 mg/g for cadmium, 65.5 mg/g for lead, and 36.6 mg/g for copper, after employing a particle size of 1 mm. It was found that a treatment with lemon and orange peels made in the same study, under the same conditions, resulted in better biosorptive capacity compared with the banana peel for copper and lead removal, but not for cadmium.[47]

Another study reported the use of pulverized banana peels with particle size between 35 and 45 μm, for heavy metals removal (Cu (II) and Pb (II)) from raw river water. The uptake kinetics of heavy metals reached equilibrium after 10 min and a pH greater than three favors the extraction of metal ions. The maximum absorptive capacity was 20.97 mg/g for Cu (II) and 41.44 mg/g for Pb (II).[48] Likewise, the removal of Cd (II) and Pb (II) from sewage using dried, cut, ground, and sieved banana peels 60 mesh (250 μm) (unreported species), has been reported. Absorptive capacity values of 5.71 mg/g for Pb (II) and 2.18 mg/g for Cd (II), were obtained with an optimal absorbent material dose of 40 and 30 g/L, for lead and cadmium removal, respectively.[49]

Removal of heavy metal, Cr (VI), from industrial wastewater has been reported with cut-dried-pulverized banana peels, sieved with 120 mesh (125 μm), achieving a quick and efficient absorption rate (95%) in 10 min, with the application of 10 mg/l of absorbent, in a medium with optimum

TABLE 10.4 Potential Applications of Banana Peels in Biotechnological Processes.

Uses	Species: variety	Products	Fermentation process	Microorganism/inoculum	Application	Reference
Animal feed	–	Feeding stuffs	Solid state fermentation	*Aspergillus niger, Aspergillus flavus, Penicilium sp.*	Animal feed	43,44
Enzyme	N/D	a-amylase	Submerged fermentation Solid state fermentation	*Helminthosporium oxysporium, Aspergillus niger, Aspergillus fumigatus, Aspergillus flavus, and Penicillium frequestans.*	Food, brewing, textile, detergent and pharmaceutical industries.	60
	N/D	Laccase	Solid state fermentation	*Trametespubescens*	Paper Industry	58
	N/D	Xylanase	Submerged fermentation	*Trichoderma harzianum* 1073	N/D	59
	N/D	Laccase	Solid state fermentation	*Aspergillus fumigatus* VKJ2.4.5	Paper Industry	57
	N/D	Cellulase	Solid state fermentation	*Trichoderma viride* GIM 3.0010	Textile, paper, pulp and food industry	55
	N/D	Lipase	Submerged fermentation	N/D	N/D	56
Bioethanol/ biogas	–	Ethanol	Submerged fermentation	N/D yeast	Ethanol production	18
	–	Ethanol	Simultaneous saccharification and fermentation	*Saccharomyces cerevisiae Pachvsolen tannophilus*	Ethanol production	64
	Musa acuminata	Ethanol	Simultaneous saccharification and fermentation	*Saccharomyces cerevisiae*	Ethanol production	11
	–	Ethanol	Submerged fermentation	*Saccharomyces cerevisiae*	Ethanol production	65

TABLE 10.4 (Continued)

Uses	Species: variety	Products	Fermentation process	Microorganism/inoculum	Application	Reference
	–	Ethanol	Submerged fermentation	N/D yeast	Ethanol production	61
	Musa acuminata	Glucose and reducing sugars	Enzymatic Pretreatment	–	Ethanol production, vitamins, organic acids.	62
	Musa cavendishi	Ethanol	Submerged fermentation	Saccharomyces cerevisiae	Ethanol production	66
	–	Methane gas	Anaerobic digestion	Cattle dung	Biogas production	41
	Robusta, Rasthali, Virupakshi, Red banana, Poovan, Naadan, Nendran and Karpuravalli	Methane gas	Anaerobic digestion	Mesophilic daily fee	Biogas production	69
	–	Methane gas	Anaerobic digestion	Piggey dung	Biogas production	70
		Methane gas	Anaerobic digestion	Leachate banana waste	Biogas production	13

pH of 2. The absorptive capacity quantified for Cr (VI) was 131.56 mg/g.[51] Using banana peels with particle size of 125 μm, an efficient absorption or removal of Cd (II) and Cr (VI) from industrial wastewater was achieved, at an optimum pH of 8. The quantified, absorption capacity for Cd (II) was 35.52 mg/g, under these conditions.[52] The removal of dyes from aqueous solutions has also been studied. Thus, with the use of dried banana peels, cut and sieved through 5 mm mesh, dye absorbing capacities were reported in the following decreasing order: methyl orange (MO) > methylene blue (MB) > Rhodamine B (RB) > Congo red (CR) > Crystal Violet (MV) > amido Black 10B (AB), 100 mg experiencing conditions color/l of solution and application rate of 1 g of absorbent/l of solution. Maximum absorption was obtained under pH conditions ranging from 6 to 7.[53] There are other works by the same author, in which banana peels used for heavy metals removal had absorption capacities of 7.97, 5.80, 6.88, and 2.55 mg/g for Pb (II), Zn (II), Ni (II), and Co (II), respectively.[50]

10.2.2.6   FUNGAL ENZYMES PRODUCTION

Banana peels have also been identified as a good substrate for the production of cellulolytic enzymes, using filamentous mold in solid fermentation systems, leading to the production of extracellular enzymes.[6,54]

The use of ripe banana peels for the production of cellulases by *Trichoderma viride* 3.0010 GIM, through the process of solid fermentation, has also been reported. Under conditions of 65% initial moisture content, with an inoculum size equal to 1.5×10^9 spores/flask, at 30°C and a time of 144 h, it was possible to quantify the maximum activity detected with paper (FPA) as 5.56 U/g of dry substrate, carboxy methyl cellulase, 10.31 U/g of dry substrate (CMCase), and β-glucosidase 3.01 U/g of dry substrate. Additional carbon or nitrogen sources exert an inhibitory effect on enzymes production. Thus, banana peels provide sufficient nutrients for microorganism growth and cellulase synthesis.[55]

Using banana peels alone or combined with potato skins (food wastes from restaurants and fruit markets), it has been reported that lipases can be obtained, during a lipolytic organism growth (name not stated). The results demonstrate that the application of combined substrates (banana peels and potato skins), enriched with 40% of a medium composed of peanut flour extract for 48 h leads to maximum lipase activity of 6.2 IU/mL. In a system that contained only banana peels, production decreased to 1.2 IU/mL.[56]

Production of laccase and manganese peroxidase enzymes has been studied in tests in a trail column bioreactor, evaluating various substrates including banana peels, sugarcane bagasse, wheat bran, poplar leaves, wheat straw, and rice bran. The maximum yield levels of production with banana peels was obtained, using *Aspergillus fumigatus* strain VKJ2.4.5, by solid fermentation. The maximum laccase and manganese peroxidase levels (5792 ± 40.95 U/L; 1334.66 ± 167.32 U/L), were obtained in conditions of 80% humidity, 6 days of incubation time, 6% of inoculum level, and an aeration level of 2.5 L/min. It is remarkable that the ripening state of banana peels was not mentioned in this work.[57] A study of laccase production by *Trametes pusbescens* was performed using chopped banana peels (*Musa cavendish*) (particle size equal 7.5 × 7.5 mm), under SSF, quantifying the maximum enzyme activity achievement to be 1600 U/L in 22 days.[58]

Xylanase production by *Trichoderma harzianum* 1073 D3 was reported in other works and the yields recorded in the presence of banana peels show activities (between 15 and 20 U/mg protein).[59] The production of α-amylase under submerged and SSF using banana peels from domestic sources, applied in medium for strains of amylolytic mold isolation *(Helminthosporium oxysporium, Aspergillus fumigatus, and Penicillium frequestans)* has been reported. High enzyme activity expressed as enzyme units (EU), was achieved in submerged fermentation with *H. oxysporium* strains (1.98 ± 0.02 EU), *A. fumigatus* (1.02 ± 0.04 EU) and *A. niger* (1.50 ± 0.01 EU) and in solid medium, results were obtained with *P. frequestans* (3.47 ± 0.00 EU), *A. niger* (0.46 ± 0.00 EU), and *A. flavus* (0.19 ± 0.00 EU). However, the values were lower than those obtained with yucca and potato peels quantified in the same study.[60]

10.2.2.7 *BIOETHANOL AND BIOGAS PRODUCTION*

Banana peel is considered as a residual lignocellulosic biomass requiring hydrolytic pretreatment for conversion of starch, cellulose and hemicellulose to simple sugars, such as glucose and xylose, which can later be fermented to ethanol.[3,61]

Cellulolytic and pectinolytic enzymes have been used to treat banana peels (*M. acuminata*) at ripening stage 5 (more yellow than green), for the hydrolysis of cellulose, hemicellulose, and pectin. Optimization and validation of concentrations of cellulases, β-glucosidase, and pectinase enzymes as well as hydrolysis time for the production of glucose and reducing sugars, was achieved. The optimized concentrations of enzymes used were 8 FPU/g cellulose, β-glucosidase (15 IU/g cellulose) and pectinase (66 IU/g pectin), radii of (1:2:8), with 15 h in a laboratory scale fermenter, reaching the highest performance levels of glucose (28.2 g/L) and reducing sugars (48 g/L) in 9 h, saving 40% of hydrolysis time of the substrate. Therefore, the availability of sugars from lignocellulosic materials such as banana peels, is important to obtain high value-added products, such as ethanol through fermentation processes.[62,63]

The use of hydrothermal treatment on dried and ground banana peels (*M. acuminata*), at ripening state five (more yellow than green), supplemented simultaneously by saccharification and fermentation (SSF) in a batch type fermenter, has been evaluated for ethanol production. The optimized parameters were: cellulase enzyme concentration (9 FPU/g-cellulose), pectinase enzyme concentration (72 IU/g-pectin) (supported on the basis of cellulose and pectin concentration of pretreated material), temperature (37°C) and incubation time (15 h). The concentration of ethanol produced was 28.2 g/L, and ethanol productivity was 2.3 g/L/h. This study shows that two processes such as hydrothermal pretreatment and simultaneous saccharification and fermentation, can be performed in the same equipment, resulting in the economic improvement of the process by reducing unit operations, saving energy and operating costs.[11]

Energy analysis of the anhydrous ethanol production process has also been reported, by acid hydrolysis of amylaceous material of whole bananas (fruit and pulp) and enzymatic hydrolysis of the lignocellulosic wastes (peels and stem), which revealed similar positive results in the evaluated production routes. The highlighted rates of mass yield and net energy value obtained in the amylaceous material were for pulp (388.7 L ethanol/ton in a dry biomass, 9.86 MJ/L of ethanol) and for fruit (346.5 L of ethanol/ton in a dry biomass, 9.94 MJ/L of ethanol), respectively, compared to those obtained for lignocellulosic material, peels (86.1 L ethanol/ton dry biomass and 5.24 MJ/L of ethanol), and for stem (123.5 L ethanol/ton in a dry biomass and 8.79 MJ/L of ethanol). These results show that these processes can be considered energetically viable and exposed to optimization.[61]

The combined use of agro-industrial wastes such as kinnow and banana peels, in a 4:6 ratio in a simultaneous fermentation and saccharification process has been reported. Cellulases, cocultures of *Saccharomyces cerevisiae* G (6% v/v) and *Pachysolen tannophilus* MTCC 1077 (4% v/v) at 30°C and after 48 h of incubation, with continuous stirring for the first 24 h, has also been reported. Fermentation process efficiency was achieved with ethanol production of 26.84 g/L and ethanol yield of 0.426 g/g and

83.52%.[64] Similarly, the use of ripe banana peels (obtained from market, without specification of species) and beet peels (obtained from juice centers) as substrates for the production of ethanol with *Saccharomyces cerevisiae* strains has been reported. Also, ethanol production levels were determined as 2.15 and 1.90%, during a period of four days for banana and beet peels, respectively.[65]

Hammond et al.[18] also reported the use of bananas (pulp and peel) at different stages of ripening (green, normal, and overripe) without specifying species, for ethanol production. Ethanol volumes obtained at 15.56°C in a normal state of ripening, using an enzymatic complex (α-amylase and gluco-amylase) in pulp, had a higher yield (0.116 L/kg in a wet sample) compared with that obtained with peel (0.019 L/kg in a wet sample).

In another study, which focused on the production of bioethanol, the potential use of pulp and peels of ripe bananas (*M. cavendish*) in a natural state was evaluated. Also, the use of waste, previously hydrolyzed by acids and enzymes as raw material was demonstrated. The maximum yields and productivities of ethanol were 0.47 ± 0.03 g/g of total sugars and 3.0 ± 0.7 g/L/h, respectively, in pulp and 0.34 ± 0.11 g/g of total sugars and 1.32 ± 0.03 g/L/h, respectively, in peels.[66]

Biogas generation under anaerobic digestion conditions using banana and plantain peels has also been reported. An exogenous microbial inoculum was not applied; the process was performed in the presence of the natural microflora of the material. The processing time period was between 30 and 100 days.[13,67]

Bardiya et al.[41] reported the use of chopped banana peels between 5 and 10 mm in size and pulverized peel in an anaerobic digestion process conducted in a digester, with a capacity of 2 L at 37°C and 40 days of hydraulic retention time (HRT). The highest rate of gas production in chopped peels (1210 mL/day) and pulverized (1160 mL/day) were observed in a period of 25 days, compared with those observed for 40 days (875 mL/day) in chopped peels (925 mL/day) in pulverized peels, representing 38 and 25% of gas production in less time, respectively. However, for the period of days evaluated (25 and 40 days), yields varied in chopped peels from 188 to 219 L/kg and in pulverized peels from 181 to 231 L/kg. It was observed that powdered peels produced more total gas per day compared with chopped peels, despite having the same hydraulic retention time (HRT).

In a study conducted on eight varieties of fresh and ripe banana peels (Robusta, Rasthali, Virupakshi, Red Banana, Poovan, Naadan, Nendran, and Karpuravalli) obtained from a domestic kitchen, the biochemical methane potential (BPM) was evaluated.[68] In test, 0.5 g of powder sample was added to 135 mL flasks with 75 mL of nutrient and an inoculum solution. The BPM profiles obtained from the peels showed variability in the methane yield. Methane production rates were higher compared with other fruit wastes. Single-phase biogas production curves were obtained, itemizing the 90% yield of methane between 40 and 50 days of fermentation.[69]

It has been reported that banana peels generate higher volumes of biogas compared with plantains peels; thus, production yield can be increased by the combined use of these two residues in a 1:1 ratio. In the presence of a mixture of peels (banana–plaintain) with 0.5 kg of breeding pigs waste in a 10 L capacity anaerobic digester, a total of 13,365 L of biogas was obtained in 35 days. It is a process in which a latency period of 3 h was observed before gas production. This study also reported that from the employed digesters, nine species of aerobic and anaerobic bacteria were isolated, and seven yeast species were also isolated.[70]

Laboratory studies were conducted to evaluate the methane performance and digestion radii of green bananas and peduncle under conditions expected to exist in a large-scale plant. A 200 L digester with a working volume of 160 L was applied. In tests with a load condition of 0.6 kg volatile solids/m^3 d, 398 ± 20 L CH$_4$/kg of volatile solids in 70 days were obtained; and with a load of 1.6 kg volatile solids/m^3 d, 210 L CH$_4$/kg of volatile solids in 23 days of operation were obtained, this decrease in performance was due to an increase in the charging rate. With the performance of the lowest load tested, if 1 ton/day is used, this could generate up to 7.5 kW of electricity. The residue obtained at the end of the digestion contained more than 4000 mg/L of potassium, 200 mg/L of nitrogen and 75 mg/L of phosphorus levels. These values exceed the acceptable limits for general agricultural irrigation, but could be used as a biofertilizer.[13]

10.2.2.8 NP SYNTHESIS

The use of banana peels is also related with the field of synthesis of NPs. Specifically, the extracts obtained are considered as nontoxic and eco-friendly materials, and are applicable for silver, palladium, cadmium sulfide (CdS), and hydroxyapatite (HAP) NPs synthesis.

It has been reported that extracts from the *M. paradisiaca* banana peels are used in the palladium and silver NPs synthesis. Synthesizing NPs of these elements is performed by altered reaction conditions such as extract concentration, pH, incubation temperature, palladium chloride concentration, and silver nitrate concentration. Evaluated conditions for synthesis of these NPs were pH = 3, 10 mg banana peel, 2 mL of solution of palladium silver nitrate (1 mm), at different incubation temperatures of 40°C, 60°C, 80°C, and 100°C, respectively, with the aim of estimating the particle size. It has been reported that a temperature of 80°C was able to produce a particle size of 50 nm, in the palladium NPs synthesis. In addition, it has been reported that silver NPs exhibit antifungal and antimicrobial properties. The associated functional groups in the synthetic process of NPs are carboxyl, amine and hydroxyl.[27,28] Otherwise, cadmium sulfide NPs have been synthesized by Zhou et al.[71] They used banana peels extract as a nontoxic and eco-friendly capping agent. The CdS NPs were obtained from cadmium nitrate and sodium sulfide. The CdS NPs prepared by this technology revealed that the average size of the NPs was around 1.48 nm. In other hand, Gopi et al.[72] reported a novel nontoxic "green" route for the synthesis of HAP NPs mediated by banana peel pectin. The banana peel pectin extracted, played a fundamental role on the control of the crystallinity and crystallite size of HAP NPs.

10.3 CONCLUSIONS

Research related to the utilization of agro-industrial wastes, such as banana and plantain peels, emphasize a range of processes and high-value products obtainable from these raw materials. Another important fact is the reduction of pollution caused by the deposition of this residue in the environment without pretreatment. It is of remarkable importance to continue improving the processes that utilize these wastes, especially when combined with emerging technologies, to achieve the most productive and environmentally friendly techniques.

ACKNOWLEDGMENTS

The authors would like to thank the Mexican Council of Science and Technology (CONACYT) for its financial support to carry out this investigation project, grant No. 213844 (PDCPN2013-01 CONACYT-Mexico). Also acknowledges CONACYT for the undergraduate and graduate scholarships to carry out their studies and for the financial support under the program "Cátedras CONACYT-2015" (Project No. 729).

KEYWORDS

- banana peels
- bioprocessing
- agricultural residue
- lignocellulosic material
- banana fiber
- peel extracts

REFERENCES

1. Zhang, P.; Whistler, R. L.; BeMiller, J. N.; Hamaker, B. R. Banana Starch: Production, Physicochemical Properties, and Digestibility—A Review. *Carbohydr. Polym.* **2005,** *59,* 443–458. doi:http://dx.doi.org/10.1016/j.carbpol.2004.10.014

2. Bello, R. H.; Linzmeyer, P.; Franco, C. M. B.; Souza, O.; Sellin, N.; Medeiros, S. H. W.; Marangoni, C. Pervaporation of Ethanol Produced from Banana Waste. *Waste Manage.* **2014,** *34*(8), 1501–1509. doi:http://dx.doi.org/10.1016/j.wasman.2014.04.013

3. Gabhane, J.; Prince William, S. P. M.; Gadhe, A.; Rath, R.; Vaidya, A. N.; Wate, S. Pretreatment of Banana Agricultural Waste for Bio-ethanol Production: Individual and Interactive Effects of Acid and Alkali Pretreatments with Autoclaving, Microwave Heating and Ultrasonication. *Waste Manage.* **2014,** *34*(2), 49–503. doi:http://dx.doi.org/10.1016/j.wasman.2013.10.013

4. FAO (Food and Agriculture Organization of the United Nations) (2012), "Production/Yield quantities of Bananas in World + (Total)" *Statistic* FAOSTAT. Available at: http://www.fao.org/faostat/es/#data/QC/visualize [Accessed 20 Aug. 2014].

5. Global fruit production in 2012, v. (2012). *Fruit: world production by type 2012 | Statistic.* [online] Statista. Available at: http://www.statista.com/statistics/264001/worldwide-production-of-fruit-by-variety/ [Accessed 20 Aug. 2014].

6. Padam, B.; Tin, H.; Chye, F.; Abdullah, M. Banana By-products: An Under-utilized Renewable Food Biomass with Great Potential. *J. Food Sci. Technol.* **2014,** *51,* 3527–3545. doi:10.1007/s13197-012-0861-2

7. Arvanitoyannis, I. S.; Mavromatis, A. Banana Cultivars, Cultivation Practices, and Physicochemical Properties. *Crit. Rev. Food Sci. Nutr.* **2008,** *49,* 113–135. doi:10.1080/10408390701764344

8. Ploetz, R. C.; Kepler, A. K.; Daniells, J.; Nelson, S. C. In *Banana and Plantain—An Overview with Emphasis on Pacific Island Cultivars, ver. 1, Species Profiles for Pacific Island Agroforestry;* Elevitch, C. R., Ed.; Permanent Agriculture Resources (PAR): Hawai, 2007.

9. Mohapatra, D.; Mishra, S.; Singh, C.; Jayas, D. Post-harvest Processing of Banana: Opportunities and Challenges. *Food Bioprocess Tech.* **2011,** *4,* 327–339. doi:10.1007/s11947-010-0377-6.

10. Happi Emaga, T.; Andrianaivo, R. H.; Wathelet, B.; Tchango, J. T.; Paquot, M. Effects of the Stage of Maturation and Varieties on the Chemical Composition of Banana and Plantain Peels. *Food Chem.* **2007**, *103,* 590–600. doi:http://dx.doi.org/10.1016/j.foodchem.2006.09.006

11. Oberoi, H. S.; Vadlani, P. V.; Saida, L.; Bansal, S.; Hughes, J. D. Ethanol Production from Banana Peels Using Statistically Optimized Simultaneous Saccharification and Fermentation Process. *Waste Manage.* **2011**, *31,* 1576–1584. doi:http://dx.doi.org/10.1016/j.wasman.2011.02.007

12. González-Montelongo, R.; Gloria Lobo, M.; González, M. Antioxidant Activity in Banana Peel Extracts: Testing Extraction Conditions and Related Bioactive Compounds. *Food Chem.* **2010**, *119,* 1030–1039. doi:http://dx.doi.org/10.1016/j.foodchem.2009.08.012

13. Clarke, W. P.; Radnidge, P.; Lai, T. E.; Jensen, P. D.; Hardin, M. T. Digestion of Waste Bananas to Generate Energy in Australia. *Waste Manage.* **2008**, *28,* 527–533. doi:http://dx.doi.org/10.1016/j.wasman.2007.01.012

14. Doran, I.; Sen, B.; Kaya, Z. The Effects of Compost Prepared from Waste Material of Banana on the Growth, Yield and Quality Properties of Banana Plants. *J. Env. Biol.* **2005**, *26,* 7–12.

15. Happi Emaga, T.; Ronkart, S. N.; Robert, C.; Wathelet, B.; Paquot, M. Characterisation of Pectins Extracted from Banana Peels (Musa AAA) Under Different Conditions Using an Experimental Design. *Food Chem.* **2008**, *108,* 463–471. doi:http://dx.doi.org/10.1016/j.foodchem.2007.10.078

16. Kuo, J. M.; Hwang, A.; Yeh, D. B.; Pan, M. H.; Tsai, M. L.; Pan, B. S. Lipoxygenase from Banana Leaf: Purification and Characterization of an Enzyme That Catalyzes Linoleic Acid Oxygenation at the 9-Position. *J. Agric. Food Chem.* **2006**, *54,* 3151–3156.. doi:10.1021/jf060022q

17. Happi Emaga, T.; Robert, C.; Ronkart, S. N.; Wathelet, B.; Paquot, M. Dietary Fibre Components and Pectin Chemical Features of Peels During Ripening in Banana and Plantain Varieties. *Bioresour. Technol.* **2008**, *99,* 4346–4354. doi:http://dx.doi.org/10.1016/j.biortech.2007.08.030

18. Hammond, J. B.; Egg, R.; Diggins, D.; Coble, C. G. Alcohol from Bananas. *Bioresour. Technol.* **1996**, *56,* 125–130. doi:http://dx.doi.org/10.1016/0960-8524(95)00177-8

19. Barnett, W. L. Grasses and Forage Crops in Jamaica. *J. Jamaica Agric. Soc.* **1956**, *40,* 16–26.

20. LY, J. Bananas y Plátanos Para Alimentar Cerdos: Aspectos De La Composición Química De Las Frutas y De Su Palatabilidad. En: Revista Computadorizada de Producción Porcina. 2004. Vol. 11, no. 3, p. 5-24

21. Devendra, C.; Göhl, B. I. The Chemical Composition of Caribbean Feedingstuffs. *Trop. Agric.* **1970**, *47,* 335–342.

22. De Camargo, M. R. T.; Sturion, G. L.; Bicudo, M. H. Avaliaçao Química e Biológica da Casca de Banana Madura. *Arch. Latinoam. Nutr.* **1996**, *46,* 320–324.

23. Llanes, A.; López, A.; Fonseca, P. L.; Viltres, E.; Arias, A. Composición Química de la Pulpa y la Cáscara de Siete Clones de Plátano Fruta. *Ciencia Agrícola.* **1985**, *39,* 181–184.

24. Maymone, B.; Tiberio, M. Ricerche Sulla Composizione Chimica, Sulla Digestibilitá e Sull Valore Nutritivo si Alloni Carcami Della Cultivazione dei Banano (*Musa sapientum*, L.; *Musa cavendishi*, L.). In *Annali Sperimentalli di Agraria;* Graphic Arts Laterza: Bari, 1951; Vol. 5; pp 133–156.

25. Oyenuga, V. A. *Nigeria's Foods and Feedingstuffs;* Ibadan University Press: Nigeria, 1968.

26. Onwuka, C. F. I.; Adetiloye, P. O.; Afolami, C. A. Use of Household Wastes and Crop Residues in Small Ruminant Feeding in Nigeria. *Small Ruminant. Res.* **1997,** *24,* 233–237. doi:http://dx.doi.org/10.1016/S0921-4488(96)00953-4

27. Bankar, A.; Joshi, B.; Kumar, A. R.; Zinjarde, S. Banana Peel Extract Mediated Novel Route for the Synthesis of Palladium Nanoparticles. *Mater. Lett.* **2010,** *64,* 1951–1953. doi:http://dx.doi.org/10.1016/j.matlet.2010.06.021

28. Bankar, A.; Joshi, B.; Kumar, A. R.; Zinjarde, S. Banana Peel Extract Mediated Novel Route for the Synthesis of Silver Nanoparticles. *Colloids Surf. A Physicochem. Eng. Asp.* **2010,** *368,* 58–63. doi:http://dx.doi.org/10.1016/j.colsurfa.2010.07.024

29. Nguyen, T. B. T.; Ketsa, S.; van Doorn, W. G. Relationship Between Browning and the Activities of Polyphenoloxidase and Phenylalanine Ammonia Lyase in Banana Peel During Low Temperature Storage. *Postharvest Biol.Technol.* **2003,** *30,* 187–193. doi:http://dx.doi.org/10.1016/S0925-5214(03)00103-0

30. Someya, S.; Yoshiki, Y.; Okubo, K. Antioxidant Compounds from Bananas (*Musa Cavendish*). *Food Chem.* **2002,** *79,* 351–354. doi:http://dx.doi.org/10.1016/S0308-8146 (02)00186-3

31. Kanazawa, K.; Sakakibara, H. High Content of Dopamine, A Strong Antioxidant, in Cavendish Banana. *J. Agric. Food Chem.* **2000,** *48,* 844–848. doi:10.1021/jf9909860

32. Seymour, G. B. Banana. In *Biochemistry of Fruit Ripening;* Seymour, G., Taylor, J., Tucker, G., Eds.; Chapman and Hall: London, 1993; pp 95–98.

33. Knapp, F. F.; Nicholas, H. J. The Sterols and Triterpenes of Banana peel. *Phytochemistry.* **1969,** *8,* 207–214. doi:http://dx.doi.org/10.1016/S0031-9422(00)85814-8

34. Mokbel, M. S.; Hashinaga, F. Antibacterial and Antioxidant Activities Banana (Musa, AAA cv. Cavendish) Fruits Peel. *Am. J. Biochem. Biotechnol.* **2005,** *1,* 125–131.

35. Parmar, H. S.; Kar, A. Comparative Analysis of Free Radical Scavenging Potential of Several Fruit Peel Extracts by In Vitro Methods. *Drug. Discov. Ther.* **2009,** *3,* 49–55.

36. Krithika, R.; Lavayan, P.; Swomya, L. Study of Antimicrobial and Antioxidant Properties of Yellow Banana Peel (*Musa sapientum*) to Use It as a Natural Food Preservative. *Inter. J. Medicobiol. Res.* **2012,** *1,* 338–341.

37. Baskar, R.; Shrisakthi, S.; Sathyapriya, B.; Shyampriya, R.; Nithya, R.; Poongodi, P. Antioxidant Potential of Peel Extracts of Banana Varieties (*Musa sapientum*). *Food Nutr. Sci.* **2011,** *2,* 1113–1128. doi:doi:10.4236/fns.2011.210151

38. Nagarajaiah, S. B.; Prakash, J. Chemical Composition and Antioxidant Potential of Peels from Three Varieties of Banana. *Asian J. Food Agro-Indus.* **2011,** *4,* 31–46.

39. Qiu, L. P.; Zhao, G. L.; Wu, H.; Jiang, L.; Li, X. F.; Liu, J. J. Investigation of Combined Effects of Independent Variables on Extraction of Pectin from Banana Peel Using Response Surface Methodology. *Carbohyd. Polym.* **2010,** *80,* 326–331. doi:http://dx.doi.org/10.1016/j.carbpol.2010.01.018

40. Anhwange, B. A. Chemical Composition of *Musa sapientum* (Banana) Peels. *J. Food Technol.* **2008,** *6,* 263–266.

41. Bardiya, N.; Somayaji, D.; Khanna, S. Biomethanation of Banana Peel and Pineapple Waste. *Bioresour. Technol.* **1996,** *58,* 73–76. doi:http://dx.doi.org/10.1016/S0960-8524(96)00107-1

42. Mohapatra, D.; Mishra, S.; Sutar, N. Banana and Its By-product Utilisation: An Overview. *J. Sci. Ind. Res.* **2010,** *69,* 323–329.

43. Akinyele, B. J.; Agbro, O. Increasing the Nutritional Value of Plantain Wastes by the Activities of Fungi Using the Solid State Fermentation Technique. *Res. J. Microbiol.* **2007,** *2,* 117–124.

44. Yabaya, A.; Ado, S. A. Mycelial Protein Production by Aspergillus Niger Using Banana Peels. *Sci.World. J.* **2008,** *3,* 9–12.

45. Hong, K. J.; Lee, C. H.; Kim, S. W. *Aspergillus Oryzae* GB-107 Fermentation Improves Nutritional Quality of Food Soybeans and Feed Soybean Meals. *J. Med. Food,* **2004,** *7,* 430–435. doi:10.1089/jmf.2004.7.430

46. Clavijo, H., Maner, J. H. The Use of Waste Bananas for Swine Feed. In *Animal Feeds of Tropical and Subtropical Origin*; Tropical Production Institute: London, 1975; pp. 99–106.

47. Kelly-Vargas, K.; Cerro-Lopez, M.; Reyna-Tellez, S.; Bandala, E. R.; Sanchez-Salas, J. L. Biosorption of Heavy Metals in Polluted Water, Using Different Waste Fruit Cortex. *Phys. Chem. Earth Parts A/B/C.* **2012,** *37–39,* 26–29. doi:http://dx.doi.org/10.1016/j.pce.2011.03.006

48. Castro, R. S. D.; Caetano, L.; Ferreira, G.; Padilha, P. M.; Saeki, M. J.; Zara, L. F.; Martines, M. A. U.; Castro, G. R. Banana Peel Applied to the Solid Phase Extraction of Copper and Lead from River Water: Preconcentration of Metal Ions with a Fruit Waste. *Indust. Eng. Chem. Res.* **2011,** *50,* 3446–3451. doi:10.1021/ie101499e

49. Anwar, J.; Shafique, U.; Waheed uz., Z.; Salman, M.; Dar, A.; Anwar, S. Removal of Pb(II) and Cd(II) from Water by Adsorption on Peels of Banana. *Bioresour. Technol.* **2010,** *101,* 1752–1755. doi:http://dx.doi.org/10.1016/j.biortech.2009.10.021

50. Arunakumara, K.; Walpola, B. C.; Yoon, M. Banana Peel: A Green Solution for Metal Removal from Contaminated Waters. *Korean J. Environ. Agric.* **2013,** *32,* 108–116.

51. Memon, J. R.; Memon, S. Q.; Bhanger, M. I.; El-Turki, A.; Hallam, K. R., Allen, G.C. Banana peel: A Green and Economical Sorbent for the Selective Removal of Cr(VI) from Industrial Wastewater. *Colloids Surf. B.* **2009,** *70,* 232–237. doi:http://dx.doi.org/10.1016/j.colsurfb.2008.12.032

52. Memon, J. R.; Memon, S. Q.; Bhanger, M. I.; Memon, G. Z.; El-Turki, A.; Allen, G. C. Characterization of Banana Peel by Scanning Electron Microscopy and FT-IR Spectroscopy and Its Use for Cadmium Removal. *Colloids Surf. B.* **2008,** *66,* 260–265. doi:http://dx.doi.org/10.1016/j.colsurfb.2008.07.001

53. Annadurai, G.; Juang, R. S.; Lee, D. J. Use of Cellulose-Based Wastes for Adsorption of Dyes from Aqueous Solutions. *J. Hazard. Mater.* **2002,** *92,* 263–274. doi:http://dx.doi.org/10.1016/S0304-3894(02)00017-1

54. Boberg, J.; Finlay, R. D.; Stenlid, J.; Näsholm, T., Lindahl, B. D. Glucose and Ammonium Additions Affect Needle Decomposition and Carbon Allocation by the Litter Degrading Fungus Mycena Epipterygia. *Soil Biol. Biochem.* **2008,** *40,* 995–999. doi:http://dx.doi.org/10.1016/j.soilbio.2007.11.005

55. Sun, H.; Li, J.; Zhao, P.; Peng, M. Banana peel: A Novel Substrate for Cellulase Production under Solid-State Fermentation. *Afr. J. Biotechnol.* **2011,** *10,* 17887–17890.

56. Jadhav, S.; Chougule, D.; Rampure, S. Lipase Production from Banana Peel Extract and Potato Peel Extract. *Int. J. Pure Appl. Microbio.* **2013,** *3,* 11–13.

57. Vivekanand, V.; Dwivedi, P.; Pareek, N.; Singh, R. Banana Peel: A Potential Substrate for Laccase Production by Aspergillus fumigatus VkJ2.4.5 in Solid-State Fermentation. *Appl. Biochem. Biotechnol.* **2011,** *165,* 204–220. doi:10.1007/s12010-011-9244-9

58. Osma, J. F.; Toca Herrera, J. L.; Rodríguez Couto, S. Banana Skin: A Novel Waste for Laccase Production by Trametes Pubescens Under Solid-state Conditions.

Application to Synthetic Dye Decolouration. *Dyes Pigm.* **2007,** *75,* 32–37. doi:http://dx.doi.org/10.1016/j.dyepig.2006.05.021

59. Seyis, I., Aksoz, N. Xylanase Production from Trichoderma Harzianum 1073 D3 with Alternative Carbon and Nitrogen Sources. *Food Technol. Biotechnol.* **2005,** *43,* 37–40.

60. Adeniran, A. H.; Abiose, S. H. Amylolytic Potentiality of Fungi Isolated from Some Nigerian Agricultural Wastes. *Afr. J. Biotechnol.* **2009,** *8,* 667–672.

61. Velásquez-Arredondo, H. I.; Ruiz-Colorado, A. A.; De Oliveira junior, S. Ethanol Production Process from Banana Fruit and Its Lignocellulosic Residues: Energy analysis. *Energy.* **2010,** *35,* 3081–3087. doi:http://dx.doi.org/10.1016/j.energy.2010.03.052

62. Oberoi, H. S.; Sandhu, S. K.; Vadlani, P. V. Statistical Optimization of Hydrolysis Process for Banana Peels Using Cellulolytic and Pectinolytic Enzymes. *Food Bioprod. Process.* **2012,** *90,* 257–265. doi:http://dx.doi.org/10.1016/j.fbp.2011.05.002

63. Naranjo, J. M.; Cardona, C. A.; Higuita, J. C. Use of Residual Banana for Polyhydroxybutyrate (PHB) Production: Case of Study in an Integrated Biorefinery. *Waste. Manage.* **2014,** *34*(12), 2634–2640. doi:http://dx.doi.org/10.1016/j.wasman.2014.09.007

64. Sharma, N.; Kalra, K. L.; Oberoi, H.; Bansal, S. Optimization of Fermentation Parameters for Production of Ethanol from Kinnow Waste and Banana Peels by Simultaneous Saccharification and Fermentation. *Indian J. Microbiol.* **2007,** *47,* 310–316. doi:10.1007/s12088-007-0057-z

65. Dhabekar, A.; Chandak, A. Utilization of Banana Peels and Beet Waste for Alcohol Production. *Asiatic J. Biotechnol. Res.* **2010,** *1,* 8–13.

66. Souza, O.; Schulz, M. A.; Fischer, G. A. A.; Wagner, T. M.; Sellin, N. Energia Alternativa de Biomassa: Bioetanol A Partir de Casca e Polpa de Banana. *Rev.Bras. Eng. Agríc. Ambient.* **2012,** *16,* 915–921.

67. Chanakya, H. N.; Sharma, I.; Ramachandra, T. V. Micro-Scale Anaerobic Digestion of Point Source Components of Organic Fraction of Municipal Solid Waste. *Waste. Manage.* **2009,** *29,* 1306–1312. doi:http://dx.doi.org/10.1016/j.wasman.2008.09.014

68. Owen, W. F.; Stuckey, D. C.; Healy Jr, J. B.; Young, L. Y.; McCarty, P. L. Bioassay for Monitoring Biochemical Methane Potential and Anaerobic Toxicity. *Water Res.* **1979,** *13,* 485–492. doi:http://dx.doi.org/10.1016/0043-1354(79)90043-5

69. Gunaseelan, V. N. Biochemical Methane Potential of Fruits and Vegetable Solid Waste Feedstocks. *Biomass Bioenerg.* **2004,** *26,* 389–399. doi:http://dx.doi.org/10.1016/j.biombioe.2003.08.006

70. Ilori, M. O.; Adebusoye, S. A.; Iawal, A. K.; Awotiwon, O. A. Production of Biogas from Banana and Plantain Peels. *Adv. Environ. Biol.* **2007,** *1,* 33–38.

71. Zhou, G. J.; Li, S. H.; Zhang, Y. C.; Fu, Y. Z. Biosynthesis of CdS Nanoparticles in Banana Peel Extract. *J. Nanosci. Nanotechnol.* **2014,** *14*(6), 4437–4442. doi:10.1166/jnn.2014.8259

72. Gopi, D.; Kanimozhi, K.; Bhuvaneshwari, N.; Indira, J.; Kavitha, L. Novel Banana Peel Pectin Mediated Green Route for the Synthesis of Hydroxyapatite Nanoparticles and Their Spectral Characterization. *Spectrochim. Acta Mol. Biomol. Spectrosc.* **2014,** *118,* 589–597. doi:http://dx.doi.org/10.1016/j.saa.2013.09.034

CHAPTER 11

CURRENT TRENDS IN THE BIOTECHNICAL PRODUCTION FRUCTOOLIGOSACCHARIDES

ORLANDO DE LA ROSA[1], DIANA B. MUÑIZ MÁRQUEZ[2],
JORGE E. WONG PAZ[2], RAÚL RODRÍGUEZ[1],
ROSA MA. RODRÍGUEZ[1], JUAN C. CONTRERAS[1], and
CRISTÓBAL AGUILAR[1*]

[1]*Food Research Department, School of Chemistry, University Autonomous of Coahuila, Saltillo CP25280, Coahuila, Mexico*

[2]*Engineering Department, Technological Institute of Ciudad Valles, National Technological of Mexico, Ciudad Valles 79010, San Luis Potosí, Mexico*

Corresponding author. E-mail: cristobal.aguilar@uadec.edu.mx

CONTENTS

ABSTRACT

Nutritional and therapeutic benefits of prebiotics have captured the interest of consumers and the food industry for use as food ingredients, in order to create functional foods. Fructooligosaccharides (FOS) are alternative sweeteners, consisting of 1-kestose, 1-nystose, and 1β-fructofuranosilnystose produced from sucrose by the action of fructosyltransferase (2.4.1.9) and β-fructofuranosidase (3.2.1.26) from plant bacteria, yeasts, and fungi. FOS are low caloric, non-cariogenic, and they aid in the absorption of minerals such as calcium and magnesium in the gut; reduce levels of cholesterol, triglycerides, and phospholipids; and stimulate the development of the intestinal and colon microflora. This review is focused on reviewing the functional properties of FOS, biotechnological production, and recent trends.

11.1   INTRODUCTION

Nowadays, there is a growing interest in people to improve their health through good nutrition. Nutraceutical ingredients and functional foods have attracted special attention in the development of new products due to the human health benefits observed.[1]

Prebiotics are considered as nutraceutical ingredients and they are used in the functional foods processing (falta cita). Prebiotics are indigestible ingredients by humans and have a positive influence on the body of the host by selectively stimulating the growth and/or activity of bacteria or a limited number of bacterial species in the colon because they are substrates for growth and metabolism of probiotic bacteria. It is of great interest to the general public because this provides a better balance in the intestinal ecosystem and improves host health.[2] Great efforts in developing strategies in the daily diet for modulating the composition and activity of the microbiota, using prebiotics, probiotics, and a combination of both (symbiotic) has been explored.[3–6]

There are three essential criteria for a food ingredient to be classified as a prebiotic: (1) must not be hydrolyzed or absorbed in the upper gastrointestinal tract, (2) must be a selective substrate for one or a limited number of probiotics, and (3) must be able to alter the colonies of microflora to a better and healthier composition.[2,7]

Among prebiotics, fructooligosaccharides (FOS) in addition to meet the above requirements, they attract attention due to its properties and its great economic potential for the sugar industry.[8] Having a sweetness of 0.4–0.6

times compared with sucrose, these being used in the pharmaceutical industry as artificial sweeteners.[9]

The health benefits and applications of FOS in nutrition have been well documented and these include activation of the immune system and resistance to infection, FOS are low caloric because they are rarely hydrolyzed by digestive enzymes and are not used as an energy source in the body. So they can be safe for inclusion in products for diabetics, are non-cariogenic by which they could be used in chewing gums and dental products, and playing an important role in reducing cholesterol, triglycerides, and phospholipids, as well as help improve the absorption of minerals such as calcium and magnesium in the gut.[7,10,11]

11.1.1 PREBIOTICS

A prebiotic is a food ingredient that beneficially affects the host by selectively stimulating the growth and/or activity of one or a limited number of "probiotics" bacteria and thus improves host health. Prebiotics are short-chain carbohydrates which are not digested by human digestive enzymes and selectively stimulate the activity of certain groups of beneficial bacteria for the body.[13] In the intestine, prebiotics are fermented by beneficial bacteria to produce short-chain fatty acid (SCFA). Prebiotics also have many other health benefits, such as reduce risk of suffering cancer of the large intestine and increase absorption of calcium and magnesium. Among the best-known prebiotics are carbohydrates, particularly oligosaccharides such as galactooligosaccharides (GOS), maltooligosaccharides, FOS, xylooligosaccharides, inulin, and hydrolysates.[12]

Prebiotics are found in many vegetables and fruits and, also are also considered as components of functional foods which have significant technological advances. Their addition to these foods improves the sensory characteristics such as taste and texture, which also enhances the stability of foams and emulsions.

According to the definition of a functional food, it is one, that is, part of the human diet and is shown to provide additional health benefits and reduce the risk of chronic diseases through its additional benefits.

A functional food can be classified if it meets one of the following: (1) foods with natural bioactive substances (e.g., dietary fiber), (2) food supplemented with bioactive substances (e.g., probiotics and antioxidants), and (3) food ingredients derived and introduced to conventional foods (e.g., prebiotics).[14]

Carbohydrate prebiotics are short chain non-digestible by human digestive enzymes and are called short-chain carbohydrates resistant. They are
also called non-digestible oligosaccharides (NDOs) which are soluble in
80% ethanol. A prebiotic is a non-active constituent of food that reaches the
colon and is selectively fermented. The benefit to the host is mediated by
selectively stimulating the growth and/or activity of one or a limited number
of bacteria.[15]

Prebiotics pass through the small intestine to the colon and become
accessible for probiotic bacteria without having been exploited by intestinal
bacteria. Lactulose, GOS, FOS, inulin and its hydrolysates, maltooligosaccharides, resistant starch, and prebiotic are normally used in the human diet,
including FOS are the most studied and widely marketed[14] (Table 11.1).

TABLE 11.1   Non-digestible Oligosaccharides with Bifidogenic Properties Available in the
Market[16].

Compound	Molecular Structure
Cyclodextrins	(Gu)n
Fructooligosaccharides	(Fr)n–Gu
Galactooligosaccharides	(Ga)n–Gu
Gentiooligosaccharides	(Gu)n
Glycosylsucrose	(Gu)n–Fr
Isomaltooligosaccharides	(Gu)n
Isomaltulose (or palatinose)	(Gu–Fr)n
Lactosucrose	Ga–Gu–Fr
Lactulose	Ga–Fr
Maltooligosaccharides	(Gu)n
Raffinose	Ga–Gu–Fr
Soybean oligosaccharides	(Ga)n–Gu–Fr
Xylooligosaccharides	(Xy)n

Ga, galactose; Gu, glucose; Fr, fructose; Xy, xilose.

11.2   FRUCTOOLIGOSACCHRIDES (FOS)

FOS also known as oligofructose and usually are called as oligosaccharides
derived from inulin, mainly composed of 1-kestose (GF2), 1-nystose (GF3),
and 1-β-fructofuranosylnystose (GF4), in which fructosyl units are linked
in the β-(2-1) position of a sucrose molecule.[17,18a,b] The formula GFn which

indicates the degree of polymerization by the number of fructose molecules which are present and linked to glucose (Fig. 11.1).

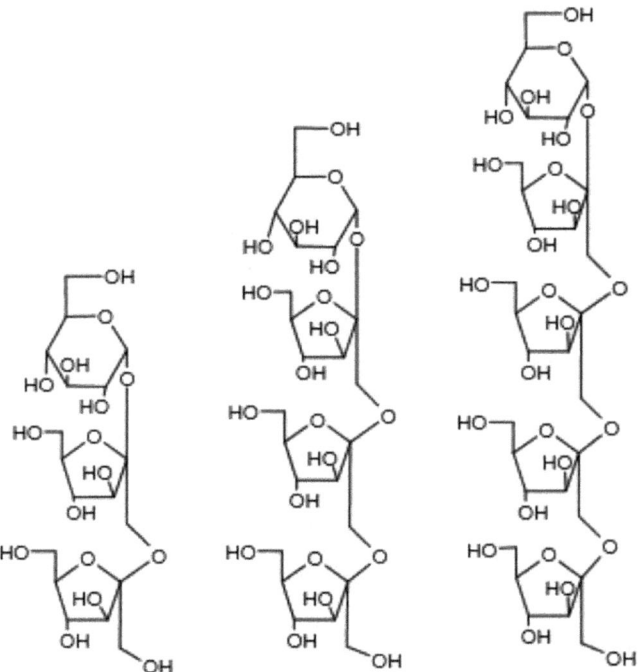

FIGURE 11.1 FOS structures 1-kestose (GF2, left), 1-nystose (GF3, middle), 1-fructofuranosyl nystose (GF4, right).

11.2.1 FUNCTIONAL PROPERTIES

Among the functional properties of the FOS, the ability to reduce the risk of chronic diseases such as colon cancer, ulcerative colitis, intestinal cancer, cardiovascular disease, and obesity,[18a,b] this is due to the large part digestibility of these prebiotics in the body, because they are hardly digested by digestive enzymes and gastric juice, that enables the most prebiotics arrive intact the intestine and colon where they are susceptible to be fermented by probiotic bacteria and beneficial intestinal microflora which produce metabolites such as SCFA and propionate between that stand out because of its anti-cancer effect.[19] Butyrate can be a source of energy for colon epithelial cells that also is thought to promote proliferation and differentiation of cells in the intestine, it has effects of inhibiting cancer cells, colonic adenomas and carcinomas and induction of apoptosis, preventing the tumor formation.[14,20–23] These effects are regulated by the expression of differentiation markers (alkaline phosphatase) and glutathione S-transferase and other response genes as well as the suppression of expression of 2-cyclooxygenase and also alters the epigenome through inhibition diacetylases of hystone. Propionate may have the anti-inflammatory ability over cancer cells in the colon.[23–25]

Among other properties, the improvement and regulation of the immune system is highlighted, this is due to the activation of proliferation and differentiation of intestinal epithelial cells and colon promoted by the formation of butyrate, development and growth of the epithelial barrier which provides protection against pathogens because it hinders their attachment to the gut, in addition to the development and regulation of gut-associated lymphoid tissue (GALT) which forms the largest area of immune tissue in the human body, comprising the innate immunity with important roles from NOD and toll-like receptors (TLR).[23,26]

Among other benefits for its prebiotic nature and as soluble fiber, FOS have the ability to have an impact on obesity, it is now emerging an overview of how the consumption of FOS can make an impact in reducing the consumption of food and energy for people obese leading to weight loss and improving health. This is because they are non-digestible by human digestive enzymes by binding type possessing $\beta(2–1)$, it is thought that these prebiotics fermentation by the beneficial microflora achieves immune regulation with anti-inflammatory effects improving intestinal permeability and metabolism. In a study conducted by Dehghan,[27] 52 women with type-2 diabetes were given a dose of 10 g oligofructose B+ Inulin per day, reduction decreased glucose levels in plasma was observed and a decrease in glycosylated hemoglobin levels. In a similar study[28] other beneficial health effects of oligosaccharides were observed, the reduction of the levels of cholesterol and triglycerides was observed after 12 weeks.[29] This is thought to be due to inhibition of lipogenic enzyme in the liver, resulting from the action of propionate produced by the fermentation of prebiotics, this reaches the liver via the portal vein and inhibits the pathways of cholesterol by inhibition HMG-CoA reductase.[30]

FOS also have a beneficial impact on mineral absorption, in studies by Ref. [14,31] with the administration of 15 g/day oligofructose or inulin 40 g/day an increase in apparent calcium absorption was observed. Also has been observed the increase in magnesium absorption caused by the ingestion of FOS.[14,32]

Prebiotics such as FOS are considered safe for inclusion in traditional diets because their presence as natural ingredients in food and plants. According to data from the US Department of Agriculture it is estimated that the average daily consumption of FOS from chicory ranges between 1 and 4 g/day.[14,33] A study showed that FOS from chicory has no toxicity to organs and that these compounds are not mutagenic, carcinogenic, or teratogenic.[34] Other results show that these fructans are well tolerated in amounts up to 20 g/day, can trigger diarrhea if doses of 30 g/day or more.[14,35]

11.2.2   FOS NATURAL SOURCES

FOS are naturally found in vegetables such as onions and garlic and, also found in tomatoes, in honey, accompanied by the inulin from chicory roots (*Cichoriumi intibus* L.) bulbs in Jerusalem artichoke (*Helianthus tuberosus*), some monocots such as rye , barley, rice, wheat, and banana,[36,37] but they are found in very small amounts and their presence depends the season. Because of this currently it has drawn attention to the production of FOS through biotechnology.[38,39]

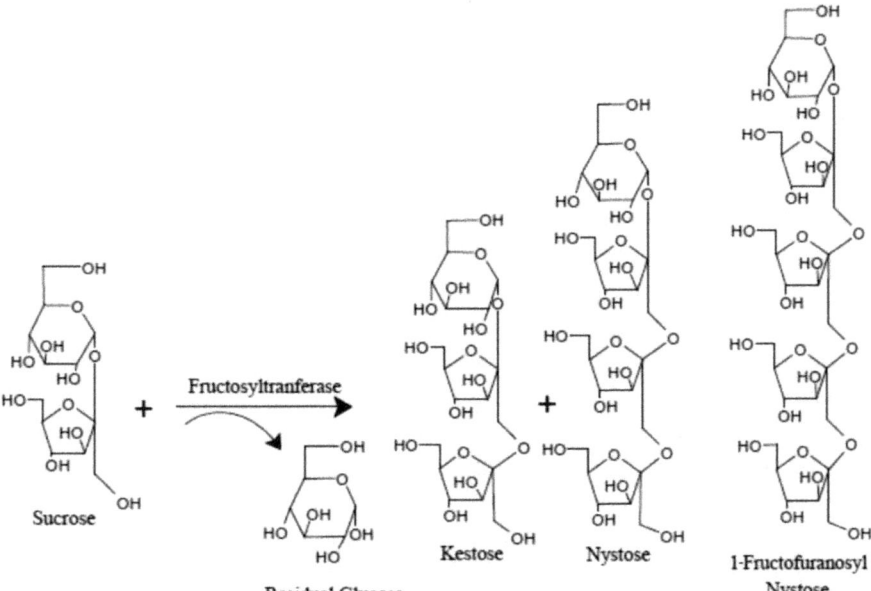

FIGURE 11.2   FOS (kestose, nystose, and 1-fructofuranosyl nystose) synthesis as from sucrose by the action of the enzyme fructosyltranferase.

FOS can be produced from sucrose by the action of enzymes with the transfructosylating enzyme fructosyltransferase (FTase) (EC 2.4.1.9) (Fig, 11.2) and the β-fructofuranosidase (FFase) (EC 3.2.1.26) derived from plants and microorganisms. It has been found that microorganisms of genera *Aureobasidium* spp., *Penicillium* spp., *Aspergillus* spp., and *Fusarium* spp. have the ability to produce these enzymes.[12,37,40–44]

11.3 FOS PRODUCTION

For the production of FOS high initial substrate concentration is required for efficient tranfructosylation.[45] Other products of this reaction are fructose and glucose which has been found to be an inhibitor of the reaction of transfructosylation when it accumulates in the media.

Transferase activity acts on sucrose breaking the β-(1,2) link and transferring fructosyl group to an acceptor molecule such as sucrose, releasing glucose. This reaction produces FOS which units are linked by a β-(2,1) bond in position of sucrose.[6] Due to the enzymes that catalyze the above reaction there is a difference in the opinion regarding nomenclature by different authors referred to both FOS producing enzyme fructofuranosidase (FFase) (EC 3.2.1.26)[42,46,47] and fructosyltransferase (FTase) (EC 2.4.1.9).[37,38,40,48–50] Both enzymes have been reported in literature as FOS producing enzymes and has also been shown that these have both activities, hydrolyzing activity (U_N), and transfructosylating activity (U_T). Their activities varies greatly (ratio, U_T / U_N) depending on the nature of these enzymes, if they are isolated from plants, bacteria, yeasts, and fungi, the genus, species, and strain. The manner in which the enzyme activity is defined varies according to the author, the activity units are defined as the amount of enzyme that transfers them 1 μmol of fructose per minute,[37,51,52] or as the amount of enzyme which liberates 1 μmol of glucose per minute,[53,54] or by 1 μmol nitrophenol liberated per minute per p-nitrophenol-α-D-glucopyranoside,[55] or as the amount of enzyme which produces 1 μmol of kestose per minute.[37,56,57] Besides these there are more different analytical methods that are used to determine the production of FOS by this enzyme so it is difficult to establish a specific definition for comparison.[37] Activity units may also vary greatly if these enzymes are intracellular or extracellular. Nguyen et al.[46] reported the production of β-fructofuranosidase FOS by intracellular and extracellular *Aspergillus niger* IMI 303 386 showing the best results transfructosylation intracellular β-fructofuranosidase.

In 1988, Hidaka et al.[45] evaluated the ratio (U_T / U_N) for producing different microorganisms showing FOS resulted in *A. niger* ATCC 20611 strain with high productivity having a transfer activity much greater than its hydrolytic activity $(U_T / U_N) = 14.2$ after one day incubation, 12.2 after three days.

In 2001, Antošová and Polakovic[50] showed different characteristics of different enzymes with transferase activity from different sources, both micro-organisms and plants had different characteristics of these oligosaccharides.

The molecular mass of fungal FTase is in anhomopolymer formed from 2 to 6 monomeric units. Many papers have defined temperature and pH optimum for activity of these enzymes between 50 and 60°C, and a pH of 4.5–6.5, respectively.[6,37] Fructosyltransferases from plants have different pH range and temperature as FTase from Jerusalem artichoke which optimal pH is between 3.5 and 5 and the optimum temperature between 20 and 25°C in the case of 1-SST and pH 5.5–7 and temperature 25–35°C for 1-FFT.[50]

Yoshikawa et al.[42] reported in a work, the production of at least five types of β-fructofuranosidase in the cell wall of *Aureobasidium pullulans* DSM2404 catalyzing the reaction of transfructosylation.

In this test the FFaseI was predominant in the period of formation of FOS while FFase levels II–V increased in the period of degradation of FOS. Ratios (U_T / U_N) of FFases IV were 14.3, 12.1, 11.7, 1.28, and 8.11, respectively, where FFaseI proved to have the best (U_T / U_N) ratio further was the only enzyme showed activity with glucose in the medium, the other enzymes being strongly inhibited by the presence of glucose.

Figure 11.3 shows the action of both activities Hidrolyzing (U_N) and tranfructosylating (U_T).

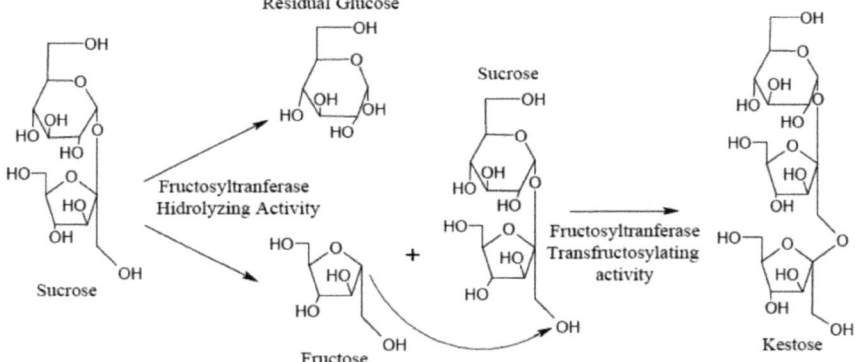

FIGURE 11.3 Kestose formation by the action of fructosyltranferase enzyme hydrolyzing activity (U_N) and transfructosylating activity (U_T), hydrolyzing a sucrose molecule and transferring a fructosyl group to another acceptor molecule of sucrose, respectively.

There are mainly two methods that can be used for the production of these enzymes with transfuctosylation activity and FOS production by fermentation, submerged fermentation (SmF) and solid-state fermentation (SSF) which will be described below.

11.3.1 FOS PRODUCTION BY SMF

The most common and studied method for production of FOS is the trans-fructosylation of sucrose in two steps, in the first step by SmF enzyme is produced, and in the second step the enzyme is reacted with the carbon source for production of FOS under controlled conditions.[12]

The variables studied to define generally the best operating conditions for the production of the enzyme are the source of carbon and nitrogen its concentration, time of cultivation, agitation, and aeration. Other important factors are the addition of various minerals, small amounts of amino acids, polymers, and surfactants.[6,37]

In 2011, Silva et al.[57] evaluated at three different levels important factors for the production of FOS, carbon and nitrogen sources, percentage of sucrose and yeast extract, respectively, inoculum percentage, pH, temperature, agitation, concentration of urea and the average concentration of various mineral salts K_2HPO_4 (NH_4) $2SO_4$, $MgSO_4$, $ZnSO_4$, and $MnSO_4$. Where the sucrose concentration proved to be a positive parameter for the formation of FOS because the enzymes catalyze the reaction of transfructosylation at high substrate concentrations. Higher productivity conversion was 54.7% and 223 g/L total FOS with an initial concentration of 400 g/L sucrose. As mineral salts MnSO4 proved to be the only mineral presented a significant effect stimulant for production of FOS. The K_2HPO_4 is described as a source of micronutrient for cell growth as well as being a buffer solution. Its optimal concentration varies between 4 and 5 g/L.[53,54,58–60]

The effect of pH on the average production of fructosyltransferase and microorganism growth has been reported. A pH of 5.5 has been found as the best initial value for the production of *Aspergillus oryzae* fructosyltransferase CFR202,[54,59] *Aspergillus japonicus* JN19,[61] and *Penicillium purpurogenum*.[62]

As for the 1-step process,[12] evaluated a one-step system for the production of FOS in which obtained under optimal production conditions 64.1 gFOS/g sucrose, being the temperature the most significant parameter in this trial.

11.3.2 FOS PRODUCTION BY SSF

The SSF presents a growing interest and high potential for small-scale units. Some advantages of this process are the simplicity of operation, high volumetric productivity, product concentration, an initial investment inexpensive, requires low power is required in addition to that there is less risk of contamination due to its high concentration of inoculum and low humidity in the reactor, and a simpler separation process.[12,17,59,63]

Among the disadvantages of the SSF are the difficult measurement of parameters such as pH and aeration, the type of sampling is destructive, and many complications for fermentations in larger scale arise due to problems of heat transfer and oxygen in the media and a homogeneous diffusion of the substrate.

Food, agriculture, and forestry industry produces large volumes of waste that can be utilized as materials for SSF. Examples include potato waste, corncob, tapioca bagasse, sugar cane bagasse, wheat grain waste, among others.[64,65] This makes the process costs are low compared to SmF and an alternative to the disposal of industrial wastes.

Various agroindustrial by-products of wheat cereal products, corn, sugarcane bagasse, and by-products of the processing of coffee and tea have been used as substrates for the production of FTase in SSF by *A. oryzae* CFR 202.[41]

In 2009, Mussatto et al.[17] studied the ability to colonize different synthetic materials (polyurethane foam, stainless steel sponge, vegetable fiber, pumice, zeolites, and glass fiber) from *A. japonicus* ATCC 20236 to produce FOS from sucrose (165 g/L).

Dietary fiber was the best support for the growth of *A. japonicus* (1.25 g/g carrier) producing 116.3 g/L FOS (56.3 g/L 1-kestose, 46.9 g/L 1-nystose, and 13.1 g/L 1-β-fructofuranosyl nystose) with 69% of yield (78% based only on the amount of sucrose consumed), reporting high activity of the enzyme β-fructofuranosidase (42.9 U/mL).

Mussatto and Teixeira[63] assessed various media for the production of FOS looking to reduce process costs studying the use of supplemented and non-supplemented media, and seeking an increase in the production yield of FOS resulting in a high production of FOS 128.7 g/L of β-FFase activity (71.3 U/mL), using as support coffee silverskin showing similar results in trials with the supplemented and unsupplemented support.

It has been sought to increase the performance and productivity of the FOS in the SSF assessing conditions that affect this process, in one study[66] conducted fermentations with coffee husk evaluating different moisture levels 60, 70 and 80% sucrose solution with 240 g/L, and a solution of spores of *A. japonicus* 2×10^5, 2×10^6, or 2×10^7 spores/g dry support, and different temperature tests 26, 30, and 34°C for 20 h , showing that the humidity did not influence the production of FOS or FFase enzyme, the temperature being between 26 and 30°C and the inoculum of 2×10^7 esp/g material which maximized the production of FOS to 208.8 g /L FOS and a productivity of 10.44 g/L * h FFasa 64.12 U/mL and with a productivity of 4 U/mL * h.

11.3.3   IMPROVED PRODUCTION YIELDS OF FOS

A considerable disadvantage for the production of large scale FOS is that the resulting mixture of the bioreactor consists of different carbohydrates: monosaccharides, unreacted disaccharides, and oligosaccharides. The incomplete conversion creates great challenges to producers of FOS because a purer product would increase their value and their utility in other food and pharmaceuticals. To generate a purer mixture several studies had tried to remove the digestible carbohydrates from FOS mix, in which they have used different bioengineering strategies for accomplishing this. In such strategies it has been applied different separation techniques using different technologies as well as additional steps of using enzymes and selective bioconversion fermentations. Production yields generally range from 55% to 60%, based on the initial sucrose concentration due to inhibition of the reaction by-products generated during the reaction.

One of the factors which directly influence the production of FOS by the reaction of transfructosylation, is the accumulation of residual glucose in the medium during the reaction which strongly inhibits production of FOS.[6,10,42,47]

Because of this, different authors have sought to remove the glucose produced during the reaction by means of systems with mixed-enzymes, in such systems the goal is to eliminate the reaction by-product which is an inhibitor, and thus the use of the substrate can be maximized. Tanriseven and Gokmen[49] proposed an interesting system in which producing FOS from sucrose using an enzyme commercial preparation Pectinex Ultra SP-L (Novozymes A/S, Denmark) after sugar mixture was processed using *Leuconostoc mesenteroides* B-512 FM dextransucrase to convert all remaining unreacted sugars to isomaltooligosaccharides which also increase the activity of bifidobacteria[47] used a preparation of commercial glucose oxidase to convert glucose generated to gluconic acid and was then precipitated as calcium gluconate using calcium carbonate for pH control of the reaction. This proved to be effective and the system occurred more than 90% (w/w) of FOS dry basis, the remainder being glucose, sucrose, and a small amount of calcium gluconate.

Yoshikawa et al.[67] produced FOS from sucrose with various enzyme preparations of FFase obtaining a yield of 62% with preparation FFaseI, then the reaction was just using glucose isomerase (GI) added in a ratio of activity 1:2 and FFase:GI obtaining a maximum yield of 69% FOS.

Various studies have used strains of *Saccharomyces cerevisiae*, *Zimmomonas mobilis*, and *Pichia heimii* to remove glucose and fructose accumulated during fermentation[7,68–70] where these were completely fermented and produced ethanol, carbon dioxide and a small amount of sorbitol as a by-product. A high content of FOS (98%) was obtained in the mixture after efficient removal of the released glucose and unreacted sucrose in the medium.[54]

Whereby the microbial treatment proved to be a good alternative to increase the percentage of FOS in the reaction mixture by removal of mono- and disaccharides , this being suitable process for the enzymatic production of FOS, but otherwise the implementation of this methodology is relatively new and still need to be developed to get good yields in addition to using this type of process would involve a further step in the purification to remove biomass and other metabolites formed during fermentation to obtain FOS with less contaminants which would increase the cost of production.

Other techniques used to remove sugars from the medium are nanofiltration and microfiltration systems which have been proposed in different jobs to remove low molecular weight carbohydrates of the oligosaccharide mixture.[71–74] These techniques have reached high blends of FOS above 90% gFOS/gSucrose[71] and above 80%.[74] These systems have demonstrated good production yields of FOS but unfortunately these make the process cost rises.

Different studies on the production of FOS have reported the use of enzymes immobilized in calcium alginate beads, methacrylamide polymeric beads, epoxy acrylic-activated beads (Eupergit C) , epoxy-activated poly-methacrylate (Sepabeads EC-EP5), glass porous ion exchange resin (Amberlite IRA 900 CI), and various polymeric and ceramic filter membrane.[11,75–80] Where it is recognized that the immobilization of enzymes has proved to be an effective tool to retain enzymes in the reactors as well as providing greater stability to the enzyme in pH and temperature changes, allowing continuous operation system. Among the disadvantages of these systems are microbial contamination, can occur adsorption of feed components and hard pipe columns.

Also other techniques have been implemented as enzyme engineering[22] discloses a process for the production of 6-kestose in which uses a modified invertase expressed by Saccharomyces which exhibits improved activity of transfructosylation where in the 6-kestose was produced with high specificity representing 95% total FOS which could make interesting use of genetic engineering as a tool to enhance the activity of the enzymes involved in the synthesis of FOS and improve their production.

The market of prebiotics is increasing, whereby to meet a growing world demand is necessary the implementation of different techniques and bioprocess strategies which opens new possibilities and trends for production of FOS.

11.3.4 FOS MARKET

There is an increasing trend toward the production of prebiotic ingredient-based food products to provide innovative solutions to the consumers. Consumer demands were changing and highly influenced by the increasing consumption of health products.[81]

Manufacturers are focusing on the development of new products to provide a wide range of end applications to answer the new demands of the consumer for new food products with nutritional and health added value.[81] As seen previous in this review FOS belong to the prebiotics group and has been used as a sweetener over the past few years primarily in the food and beverage industry.

Transparency Market Research (TMR)[82] forecasts that the global prebiotic ingredients market will improve, it states that the global prebiotic ingredients market is anticipated to reach US$4.5 billion by the end of 2018, and according to study by Grand View Research Inc.,[83] it is projected to reach USD 5.75 billion by 2020. In terms of volume, global prebiotics market was 581.0 kilo tons in 2013 and is expected to reach 1084.7 kilo tons by 2020, growing at a compound annual growth rate (CAGR) of 9.3% from 2014 to 2020.[83]

Prebiotics major application sectors are in food products such as dairy, bakery, cereals, meat, and sports drinks are also driving the growth of prebiotic ingredients market.[81]

Food and beverage was the largest application segment with market volume of 488.1 kilo tons in 2013.[83] Other applications include animal feeding, and pharmaceuticals. Animal feed has not yet been fully developed and explored and hence the main focus of manufacturers in animal feed is in research and development of sustainable products. Prebiotic ingredient demand for animal feed is expected to reach USD 429.3 million in 2018 and is expected to have a positive impact on the FOS market.[82]

Increasing demand for low calorie or fat free food which provides added health benefits is expected to have a positive impact on the market.[84] The inulin and sucrose FOSs are the novel dietary fibers that fulfill these considerations.[84] United States, Germany, Japan, and China dominate the functional food market and expected to witness growth owing to increased demand for dietary supplements. In East Asia, North America, and Europe, the FOS products are primarily employed as dietary fibers, thus growth of dietary fiber sector will increase FOS demand.[84]

Development of symbiotic combining FOS with targeted probiotic strains coupled with introduction of new products such as drinking yogurts, low fat reduction creams, chocolates, and bakery products are expected to be key drivers for the industry over the forecast period. The product is expected to substitute numerous sweeteners in the food and beverage industry including aspartame, sucralose, and xylitol on account of their superior properties and cost effectiveness.[84]

Among the FOS products commercially available are Beneshine™ (P-type powder and liquid (>95% purity)) from sucrose by the enzyme fructosyltransferase is conducted by Shenzhen Victory Biology Engineering Co., Ltd., China.[84] Meioligo manufactured by a key player in the FOS industry Meiji Seika Kaisha Ltd. in Japan. FortiFeed® P-95, it is another product of which consists of a prebiotic soluble fiber in the mixture which contains about 95% short chain FOS dry basis (Corn Products International, Inc.). Actilight® FOS product (Beghin-Meiji Industries, France), is a highly bifidogenic product, produced from sucrose. Another commercially available product is NutraFlora® US GNC Nutrition. Orafti Active Food Ingredients in United States produces Raftilose, an inulin that contains FOSs in addition to polysaccharides. Among the FOS products available on the market from inulin, OLIFRUCTINE-SP® manufactured by Company Nutriagaves of Mexico S.A. de C.V., a mixture of fructan extracted from the juice of the agave plant Tequilana Weber blue variety. Other manufacturers include Jarrow Formulas, Cheil Foods, and Chemicals Inc.

The following Table 11.2 summarizes the companies that handle products of FOS and the brand name of these products is done.

In addition to the marketing of food grade FOS, FOS are available in analytical grade market with purities of 80–99%. Companies that include more supply of FOS 1-kestose (GF2), nystose (GF3), and 1F-fructofuranosyl nystose β-(GF4) are Sigma-Aldrich, Megazyme, and Wako Chemicals GmbH.[6,85]

TABLE 11.2   Commercially Available Food Grade FOS.

Substrate	Manufacturer	Trademark
Sucrose	Beghin-Meiji industries, Francia	Actilight®
	Cheil Foods and Chemicals Inc., Korea	Oligo-Sugar
	GTC Nutrition, EU	NutraFlora®
	Meiji Seika Kaisha Ltd., Japan	Meioligo®
	Victory Biology Engineering Co., Ltd., China	Beneshine™ P-type
Inulin	Beneo-Orafti, Bélgica	Orafti®
	Cosucra Groupe Warcoing, Bélgica	Fibrulose®
	Sensus, Holanda	Frutalose®
	Nutriagaves de Mexico S.A. de C.V., Mexico	Olifructine-SP®

11.4   CONCLUSIONS

The prebiotics market is gaining strength and is growing greatly in recent years due to the increasing interest in people for these products because it has raised awareness for a healthier life. This document has been reviewed many positive effects on health and in reducing the risk of disease by FOS, in addition to its many physicochemical and physiological properties and its wide application in the food industry. We reviewed the various challenges that are in the production of FOS, noting that there has been a breakthrough in the techniques and technologies in bioprocesses to improve yields, quality and purity of the final product; however, due to the increasing demand evaluating more FOS producing strains with greater efficiency in shorter periods of time, the use of genetic engineering to create strains with better transfructosylation activity, and improve production processes where industrial and agro-industrial wastes are used in order to establish processes allow low-cost production with good yields of FOS.

KEYWORDS

- **fructooligosaccharides**
- **prebiotics**
- **nutraceutical**

- **functional foods**
- **solid-state fermentation**
- **fructosyltransferase**
- **β-fructofuranosidase**

REFERENCES

1. Bitzios, M.; Fraser, I.; Haddock Fraser, J. Functional Ingredients and Food Choice: Results from a Dual-mode Study Employing Means-end-chain Analysis and a Choice Experiment. *Food Policy*. **2011,** *36* (5), 715–725.
2. Kovács, Z.; Benjamins, E.; Grau, K.; Ur Rehman, A.; Ebrahimi, M.; Czermak, P. Recent Developments in Manufacturing Oligosaccharides with Prebiotic Functions. *Adv. Biochem. Eng. Biotechnol.* **2013,** *143,* 257–295. Retrieved from http://www.ncbi.nlm.nih.gov/pubmed/23942834
3. Howlett, J., Ed. *Functional Foods—from Science to Health and Claims;* ILSI Europe—Concise Monograph Series: Brussels, Belgium, 2008.
4. Szajewska, H. Probiotics and Prebiotics in Preterm Infants: Where are we? Where are We Going? *Early Hum. Dev.* **2010,** *86,* S81–S86.
5. De Preter, V.; Hamer, H. M.; Windey, K.; Verbeke, K. The Impact of Pre- and/or Probiotics on Human Colonicmetabolism: Does it Affect Human Health? *Mol. Nutr. Food Res.* **2011,** *55,* 46–57.
6. Dominguez, A. L.; Rodrigues, L. R.; Lima, N. M.; Teixeira, J. A. An Overview of the Recent Developments on Fructooligosaccharide Production and Applications. *Food Bioprocess. Technol.* **2013,** *7,* 1–14.
7. Mutanda, T.; Mokoena, M. P.; Olaniran, A. O.; Wilhelmi, B. S.; Whiteley, C. G. Microbial Enzymatic Production and Applications of Short-chain Fructooligosaccharides and Inulooligosaccharides: Recent Advances and Current Perspectives. *J. Ind. Microbiol. Biotechnol.* **2014,** *41,* 893–906.
8. Godshall, M. A. Future Directions for the Sugar Industry. *Int. Sugar. J.* **2001,** *103,* 378–384.
9. Biedrzycka, E.; Bielecka, M. Fructooligosaccharides Production from Sucrose by *Aspergillus* sp. N74 Immobilized in Calcium Alginate. *Trends Food. Sci. Technol.* **2004,** *15,* 170–175.
10. Yun, J. W. Fructooligosaccharides Occurrence, Preparation, and Application. *Enzyme Microb. Tech.* **1996,** *19* (2), 107–117.
11. Tanriseven, A.; Aslan, Y. Immobilization of Pectinex Ultra SP-L to Produce Fructooligosaccharides. *Enzyme Microb. Tech.* **2005,** *36,* 550–554.
12. Dominguez, A.; Nobre, C.; Rodrigues, L. R.; Peres, A. M.; Torres, D.; Rocha, I.; Lima, N. Y.; Teixeira, J. A. New Improved Method for Fructooligosaccharides Production by *Aureobasidium pullullans*. *Carbohydr. Polym.* **2012,** *89,* 1174–1179.

13. Quigley, M. E.; Hudson, G. J.; Englyst, H. N. Determination of Resistant Short-chain Carbohydrates (Non-digestible Oligosaccharides) Using Gas-liquid Chromatography. *Food Chem.* **1999**, 65, 381–390.

14. Al Sheraji, S. H.; Ismail, A.; Manap, M. Y.; Mustafa, S.; Yusof, R. M.; Hassan, F. A. Prebiotics as Functional Foods: A Review. *J. Funct. Foods.* **2013**, 5, 1542–1553.

15. Gibson, G. R.; Roberfroid, M. B. Dietary Modulation of the Human Colonic Microbiota. Introducing the Concept of Prebiotics. *J. Nutr.* **1995**, 125, 1401–1412.

16. Sako, T.; Matsumoto, K.; Tanaka, R. Recent Progress on Research and Applications of Non-digestible Galacto-oligosaccharides. *Int. Dairy J.* **1999**, 9, 69–80.

17. Mussatto, S. I.; Aguilar, C. N.; Rodrigues, L. R.; Teixeira, J. A. Colonization of *Aspergillus japonicus* on Synthetic Materials and Application to the Production of Fructooligosaccharides. *Carbohydr. Res.* **2009**, 344 (6), 795–800.

18a. Chen, S. C.; Sheu, D. C.; Duan, K. J. Production of Fructooligosaccharides Using β-fructofuranosidase Immobilized onto Chitosan-coated Magnetic Nanoparticles. *J. Taiwan Inst. Chem. Eng.* **2014**, 45 (4), 1105–1110.

18b. Scheid, M. M. A.; Moreno, Y. M. F.; Maróstica Junior, M. R.; Pastore, G. M. Effect of Prebiotics on the Health of the Elderly. *Food Res. Int.* **2013**, 53, 426–432.

19. Vitali, B.; Ndagijimana, M.; Maccaferri, S.; Biagi, E.; Guerzoni, M. E.; Brigidi, P. An in Vitro Evaluation of the Effect of Probiotics and Prebiotics on the Metabolic Profile of Human Microbiota. *Anaerobe.* **2012**, 18 (4), 386–391.

20. Duncan, S. H.; Louis, P.; Flint, H. J. Lactate-Uilizing Bacteria, Isolated from Human Feces, that Produce Butyrate as a Major Fermentation Product. *Appl. Environ. Microbiol.* **2004**, 70, 5810–5817.

21. Gomes, A. C.; Bueno, A. A.; De Souza, R. G. M.; Mota, J. F. Gut Microbiota, Probiotics and Diabetes. *Nutr. J.* **2014**, 13, 60. Retrieved from http://www.pubmedcentral.nih.gov/articlerender.fcgi?artid=4078018&tool=pmcentrez&rendertype=abstract

22. Marín Navarro, J.; Talons pearls, D.; Plain, J. One-pot Production of Fructooligosaccharides by a *Saccharomyces Cerevisiae* Strain Expressing an Engineered Invertase, *Appl. Microbiol. Biotechnol.* **2014**, 99, 2549–2555.

23. Serbian, D. E. Gastrointestinal Cancers: Influence of Gut Microbiota, Probiotics and Prebiotics. *Cancer Lett.* **2014**, 345, 258–270.

24. Domokos, M.; Jakus, J.; Szeker, K.; Csizinszky, R.; Csiko, G.; Neogrady, Z.; Csordas, A.; Galfi, P. Butyrate-induced Cell Death and Differentiation are Associated with Distinct Patterns of ROS in HT29-derived Human Colon Cancer Cells. *Dig. Dis. Sci.* **2010**, 55, 920–930.

25. Scharlau, D.; Borowicki, A.; Habermann, N.; Hofmann, T.; Klenow, S.; Miene, C.; Munjal, U.; Stein, K.; Glei, M. Mechanisms of Primary Cancer Prevention by Butyrate and Other Products Formed During Gut Flora-mediated Fermentation of Dietary Fiber. *Mutat. Res.* **2009**, 682, 39–53.

26. O'Hara, A. M.; Shanahan, F. The Gut Flora as a Forgotten Organ. *EMBO. Rep.* **2006**, 7, 688–693.

27. Dehghan, P.; Gargari, B. P.; Jafar abadi, M. A. Oligofructose-enriched Inulin Improves Some Inflammatory Markers and Metabolic Endotoxemia in Women with Type 2 Diabetes Mellitus: A Randomized Controlled Clinical Trial. *Nutrition.* **2013**, 30, 418–423.

28. Depeint, F.; Tzortzis, G.; Vulevic, J.; I'anson, K.; Gibson, G. R. Prebiotic Evaluation of a Novel Galactooligosaccharide Mixture Produced by the Enzymatic Activity of *Bifidobacterium Bifidum* NCIMB 41171, in Healthy Humans: A Randomized, Double-blind, Crossover, Placebo-controlled Intervention Study. *Am. J. Clin. Nutr.* **2008**, 87, 785–791.

29. Gibson, G. R.; Rastall, R. A. Recent Developments in Prebiotics to Selectively Impact Beneficial Microbes and Promote Intestinal Health. *Elsevier.* **2015,** *32,* 42–46. Elsevier Ltd. Retrieved from http://dx.doi.org/10.1016/j.copbio.2014.11.002

30. Levrat, M. A.; Favier, M. L.; Moundras, C.; Remesy, C.; Demigne, C.; Morand, C. Role of Dietary Propionic Acid and Bile Acid Excretion in the Hypocholesterolemic Effects of Oligosaccharides in Rats. *J. Nutr.* **1994,** *124,* 531–538.

31. Roberfroid, M. Functional Food Concept and its Application to Prebiotics. *Dig. Liver Dis.* **2002,** *34,* S105–S110.

32. Bornet, F. R. J.; Brouns, F.; Tashiro, Y.; Duviller, V. Nutritional Aspect of Short-chain Fructooligosaccharieds: Natural Occurrence, Chemistry, Physiology and Health Implications. *Dig. Liver Dis.* **2002,** *34,* S111–S120.

33. Moshfegh, A. J.; Friday, J. E.; Goldman, J. P.; Ahuja, J. K. Presence of Inulin, and Oligofructose in the Diets of Americans. *J. Nutr.* **1999,** *129,* 1407S–1411S.

34. Carabin, I. G.; Flamm, W. G. Evaluation of Safety of Inulin and Oligofructose as Dietary Fibre. *Regul. Toxicol. Pharm.* **1999,** *30,* 268–282.

35. Den Hond, E.; Geypens, B.; Ghoos, Y. Effect if High Performance Chicory Inulin on Constipation. *Nutr. Res.* **2000,** *20,* 731–736.

36. Król, B.; Grzelak, K. Qualitative and Quantitative Composition of Fructooligosaccharides in Bread. *Eur. Food Res. Technol.* **2006,** *223* (6), 755–758.

37. Maiorano, A. E.; Piccoli, R. M.; Da Silva, E. S.; De Andrade Rodrigues, M. F. Microbial Production of Fructosyltransferases for Synthesis of Pre-biotics. *Biotechnol. Lett.* **2008,** *30,* 1867–1877.

38. Sangeetha, P. T.; Ramesh, M. N.; Prapulla, S. G. Production of Fructosyl Transferase by *Aspergillus oryzae* CFR 202 in Solid-state Fermentation Using Agricultural By-products. *Appl. Microbiol. Biotechnol.* **2004,** *65* (5), 530–537.

39. Nobre, C. Teixeira, L. R. Y.; Rodrigues, L. R. Fructo-oligosaccharides Purificationfrom a Fermentative Broth Using an Activated Charcoal Column. *N. Biotechnol.* **2011,** *29,* 1871–6784.

40. Yun, J. W.; Kang, S. C.; Song, S. K. Continuous Production of Fructooligosaccharides from Sucrose by Immobilized Fructosyltransferase. *Biotechnol. Tech.* **1995,** *9,* 805–808.

41. Sangeetha, P. T.; Ramesh, M. N.; Prapulla, S. G. Production of Fructosyltransferase by *Aspergillus oryzae* CFR 202 in Solid-state Fermentation Using Agricultural by-products. *Appl. Microbiol. Biotechnol.* **2004,** *65,* 530–537.

42. Yoshikawa, J.; Amachi, S.; Shinoyama, H.; Fujii, T. Multiple *b*-fructofuranosidases by *Aureobasidium pullulans* DSM2404 and Their Roles in Fructooligosaccharide Production. *FEMS Microbiol. Lett.* **2006,** *265,* 159–163.

43. Jung, K. H.; Bang, S. H.; Oh, T. K.; Park, H. J. Industrial Production of Fructooligosaccharides by Immobilized Cells of *Aureobasidium pullulans* in a Packed Bed Reactor. *Biotechnol. Lett.* **2011,** *33,* 1621–1624.

44. Mussatto, S. I.; Ballesteros L. F.; Martins, S.; Maltos, D. A. F.; Aguilar, C. N. Teixeira, J. A. 2013. Maximization of Fructooligosaccharides and β-Fructofuranosidase Production by *Aspergillus japonicas* under Solid-State Fermentation Conditions. *Food Bioprocess. Technol..* **2013,** *6,* 2128–2134.11947-012-0873-y.

45. Hidaka, H.; Hirayama, M.; Sumi, N. A Fructooligosaccharide-producing Enzyme from *Aspergillus niger* ATCC 20611. *Agric. Biol. Chem.* **1998,** *52,* 1182–1187.

46. Nguyen, Q. D.; Mattes, F.; Hoschke, A.; Rezessy Szabo, J.; Bhat, M+. K. Production, Purification and Identification of Fructooligosaccharides Produced by B-fructofuranosidase from *Aspergillus niger* IMI 303386.*Biotechnol. Lett.* **1999,** *21* (3), 183–186.

47. Sheu, D. C.; Lio, P. J.; Chen, S. T.; Lin, C. T.; Duan, K. J. Production of Fructooligosaccharides in High Yield Using a Mixed Enzyme System of ??-Fructofuranosidase and Glucose Oxidase. *Biotechnol. Lett.* **2001,** *23* (18), 1499–1503.
48. Patel, V.; Saunders, G.; Bucke, C. Production of Fructool- igosaccharides by *Fusarium oxysporum. Biotechnol. Lett.* **1994,** *16* (11), 1139–1144.
49. Tanriseven, A.; Gokmen, F. Novel Method for the Production of a Mixture Containing Fructooligosaccharides and Isomaltooligosaccharides. *Biotechnol. Tech.* **1999,** *13* (3), 207–210.
50. Antosová, M.; Polakovic, M. Fructosyltrasnferases: The Enzymes Catalyzing Production of Fructooligosaccharides. *Chem. Pap.* **2001,** *55* (6), 350–358. Retrieved from http://www.chempap.org/?id=7&paper=422
51. Chen, W. C.; Liu, C. H. Production of b-fructofuranosidase by *Aspergillus japonicus. Enzyme. Microb. Technol.* **1996,** *18,* 153–160.
52. Dorta, C.; Cruz, R.; Neto, P. O.; Moura, D. J. C. Sugarcane Molasses and Yeast Powder Used in the Fructooligosac- charides Production by *Aspergillus japonicus*-FCL 119T and *Aspergillus niger* ATCC 20611. *J. Ind. Microbiol. Biotechnol.* **2006,** *33,* 1003–1009.
53. Park, J. P.; Oh, T. K.; Yun, J. W. Purification and Character- ization of a Novel Transfructosylating Enzyme from Bacillus Macerans EG-6. *Process. Biochem.* **2001,** *37,* 471–476.
54. Sangeetha, P. T.; Ramesh, M. N.; Prapulla, S. G. Fructooli- gosaccharide Production Using Fructosyl Transferase Obtained from Recycling Culture of *Aspergillus oryzae* CFR 202. *Process. Biochem.* **2005a,** *40,* 1085–1088.
55. Wang, X. D.; Rakshit, S. K. Improved Extracellular Trans-ferase Enzyme Production by *Aspergillus Foetidus* for Synthesis of Isooligosaccharides. *Bioprocess. Eng.* **1999,** *20,* 429–434.
56. L'Hocine, L.; Jiang, Z. W. B.; Xu, S. Purification and Partial Characterization of Fructosyltransferase and Invertase from *Aspergillus niger* AS0023. *J. Biotechnol.* **2000,** *81,* 73–84.
57. Da Silva, J. B.; Fai, A. E. C.; Dos Santos, R.; Basso, L. C.; Pastore, G. M. Parameters Evaluation of Fructooligosaccharides Production by Sucrose Biotransformation Using an Osmophilic *Aureobasium pullulans* Strain. *Procedia. Food Sci.* **2011,** *1,* 1547–1552.
58. Vandáková, M.; Platková, M.; Antošová, M.; Báleš, V.; Polakovič, M. Optimization of Cultivation Conditions for Production of Fructosyltransferase by *Aureobasidium pullulans. Chem. Pap.* **2004,** *58* (1), 15–22.
59. Sangeetha, P. T.; Ramesh, M. N.; Prapulla, S. G. Recent Trends in the Microbial Production, Analysis and Application of Fructooligosaccharides. *Trends. Food. Sci. Technol.* **2005b,** *16,* 442–457.
60. Shin, H. T.; Baig, S. Y.; Lee, S. W.; Suh, D. S.; Kwon, S. T.; Lin, Y. B.; Lee, J. H. Production of Fructo-oligosaccharides from Molasses by *Aureobasidium pullulans* Cells. *Bioresour. Technol.* **2004,** *93,* 59–62.
61. Wang, L. M.; Zhou, H. M.. Isolation and Identification of a Novel *A. Japonicus* JN19 Producing β-fructofuranosidase and Characterization of the Enzyme. *J. Food Biochem.* **2006,** *30,* 641–658.
62. Dhake, A. B.; Patil, M. B. Effect of Substrate Feeding on Production of Fructosyltransferase by *Penicillium purpurogenum. Braz. J. Microbiol.* **2007,** 38, 194–199.
63. Mussatto, S. I.; Teixeira, J. A. Increase in the Fructo-oligosaccharides Yield and Productivity by Solid-state fermentation with *Aspergillus japonicus* Using Agro-industrial Residues as Support and Nutrient Source. *Biochem. Eng. J.* **2010,** *53,* 154–157.

64. Couto, S. R.; Sanromán, M. A. Application of Solid-state Fermentation to Food Industry–A Review. *J. Food Eng.* **2006,** *76* (3), 291–302.

65. Orzua, M. C.; Mussatto, S. I.; Contreras Esquivel, J. C.; Rodriguez, R.;De La Garza, H.; Teixeira, J. A.; Aguilar, C. N. Exploitation of Agro Industrial Wastes as Immobilization Carrier for Solid-state Aermentation. *Ind. Crop. Prod.* **2009,** *30* (1), 24–27.

66. Mussatto, S. I.; Ballesteros, L. F.; Martins, S.; Maltos, D. A. F.; Aguilar, C. N.; Teixeira, J. A. Maximization of Fructooligosaccharides and β-Fructofuranosidase Production by *Aspergillus japonicus* under Solid-State Fermentation Conditions. *Food Bioprocess Technol.* **2013,** *6* (8), 2128–2134.

67. Yoshikawa, J.; Amachi, S.; Shinoyama, H.; Fujii, T. Production of Fructooligosaccharides by Crude Enzyme Preparations of Beta-fructofuranosidase from *Aureobasidium pullulans*. *Biotechnol. Lett.* **2008,** *30* (3), 535–539.

68. Crittenden, R. J.; Playne, M. J. Purification of Food Grade Oli- Gosaccharides Using Immobilised Cells of *Zymomonas mobilis*. *Appl. Microbiol. Biotechnol.* **2002,** *58,* 297–302.

69. Yoon, E. J.; Yoo, S. H.; Chac, J.; Lee, H. G. Effect of Levan's Branching Structure on Antitumor Activity. *Int. J. Biol. Macromol.* **2004,** *34,* 191–194.

70. Sheu, D. C.; Chang, J. Y.; Wang, C. Y.; Wu, C. T.; Huang, C. J. Continuous Production of High-purity Fructooligosaccharides and Ethanol by Immobilized *Aspergillus japonicus* and *Pichia heimii*. *Bioprocess Biosyst. Eng.* **2013,** *36* (11), 1745–1751.

71. Nishizawa, K.; Nakajima, M.; Nabetani, H. Kinetic Study on Transfructosylation by β-Fructofuranosidase from *Aspergillus niger* ATCC 20611 and Availability of a Membrane Reactor for Fructooligosaccharide Production. *Food Sci. Technol. Res.* **2001,** *7,* 39–44.

72. Goulas, A.; Tzortzis, G.; Gibson, G. R. Development of a Process for the Production and Purification of α-and β-galactooligosaccharides from Bifobacterium Bifidum NCIMB 41171. *Int. Dairy J.* **2007,** *17,* 648–656.

73. Goulas, A. K.; Kapasakalidis, P. G.; Sinclair, H. R.; Rastall, R. A.; Grandison, A. S. *J. Membr. Sci.* **2002,** *209* (1), 321.

74. Sheu, D. C.; Duan, K. J.; Cheng, C. Y.; Bi, J. L.; Chen, J. Y. *Biotechnol. Prog.* **2002,** *18* (6), 1282.

75. Yun, J. W.; Jung, K. H.; Oh, J. W.; Lee, J. H. *Appl. Biochem. Biotechnol.* **1990,** *24* (25), 299–308.

76. Hayashi S, Nonoguchi M, Takasaki Y, Ueno H & Imada K **1991** Purification and Properties of b-fructofuranosidase from Aureobasidium sp. ATCC 20524. *J. Ind. Microbiol.* **1191,** *7,* 251–256.

77. Hayashi, S.; Tubouchi, M.; Takasaki, Y.; Imada, K. *Biotechnol. Lett.* **1994,** *16,* 227. doi: 10.1007/BF00134616. URL http://dx.doi.org/10.1007/BF00134616

78. Chiang, C. J.; Lee, W. C.; Sheu, D. C.; Duan, K. J. Immobilization of β-fructofuranosidases from Aspergillus on Methacrylamide-based Polymeric Beads for Production of Fructool-igosaccharides. *Biotechnol. Prog.* **1997,** *13,* 577–582.

79. Ghazi, I.; Fernández Arrojo, L.; Garcia Arellano, H.; Ferrer, M.; Ballesteros, A.; Plou, F. J. Purification and Kinetic Char- acterization of a Fructosyltransferase from *Aspergillus aculeatus*. *J. Biotechnol.* **2007,** *128,* 204–211.

80. Csanádi, Z.; Sisak, C. *Hung. J. Indus. Chem.* **2008,** *36* (1–2), 23.

81. Markets, Markets. Prebiotic Ingredients Market by Type (Oligosaccharides, Inulin, Polydextrose), Application (Food & Beverage, Dietary Supplements, Animal Feed), Source (Vegetables, Grains, Roots), Health Benefits (Heart, Bone, Immunity, Digestive,

Skin, Weight Management), & by Region - Global Trends & Forecast to 2020. Report Code: FB 3567. 2015. Retrieved from: http://www.marketsandmarkets.com/Market-Reports/prebiotics-ingredients-market-219677001.html

82. Transparency Market Research. Prebiotic Ingredients Market (FOS, GOS, MOS, Inulin) for Food & Beverage, Dietary Supplements & Animal Feed - Global Industry Analysis, Market Size, Share, Trends, and Forecast 2012–2018. 2013. Retrieved from: http://www.transparencymarketresearch.com/prebiotics-market.html

83. Grand View Research Inc. Prebiotics Market Analysis By Ingredients (FOS, Inulin, GOS, MOS), By Application (Food & Beverages, Animal Feed, Dietary Supplements) And Segment Forecasts To 2020. ISBN Code: 978-1-68038-089-7. 2014. Retrieved from: http://www.grandviewresearch.com/industry-analysis/prebiotics-market

84. Bali, V.; Panesar, P. S.; Bera, M. B.; Panesar, R. Fructo-oligosaccharides: Production, Purification and Potential Applications. *Crit. Rev. Food Sci. Nutr.* **2013,** *55, 1475–1490.*

85. Nobre, C.; Teixeira, J. A.; Rodrigues, L. R. New Trends and Technological Challenges in the Industrial Production and Purification of Fructo-oligosaccharides. *Crit. Rev. Food Sci. Nutr.* **2013,** *55,* 1444–1455. doi:10.1080/10408398.2012. 697082.

CHAPTER 12

THE RSM-CI EMERGING TECHNOLOGY FOR ENABLING BIOCHEMICAL PROCESS: ETHANOL PRODUCTION FROM PALM PLANTATION BIOMASS WASTE IN INDONESIA

TEUKU BEUNA BARDANT[1], HERU SUSANTO[1,2*], and INA WINARNI[3]

[1]*Research Center for Chemistry, The Indonesian Institute of Science, Jakarta, Indonesia*

[2]*Department of Information Management, College of Management, Tunghai University, Taichung, Taiwan*

[3]*Forest Products Research and Development Center, Ministry of Forestry, Jakarta, Indonesia*

Corresponding author. E-mail: heru.susanto@lipi.go.id; susanto.net@gmail.com

CONTENTS

ABSTRACT

The emergence of innovative chemistry informatics (cheminformatics) is threatening the sustainability of bioethanol production from palm oil empty fruit bunch (EFB) in Indonesia. With the friendliness and convenience offered by cheminformatics, many bioethanol production processes nowadays prefer to determining optimum condition and forecasting the results computationally through response surface methodology (RSM). RSM is one of cheminformatics (called by RSM-CI; Response Surface Methodology—Chemistry Informatics) that explores the relationships between several explanatory variables and one or more response variables. Indonesia had established bioethanol production pilot plant by using fully automatic computerized system supported by sufficient ICT emerging technology and database. Bioethanol production from cellulose-based material is one of the popular bioprocess engineering research fields in Indonesia in the last five years, since palm plantation biomass waste became the most concern potency for bioethanol raw material since Indonesia is the world leading producer in palm oil. The purpose of this chapter is to assess the impact of RSM as a practical cheminformatics to support bioprocess technology; bioethanol technology processes by optimizing the operations' variables and easily adopted in automatic computerized system ICT emerging technology.

12.1 INTRODUCTION

The term cheminformatics was born referring to the combination of chemistry and informatics. Initially, this field of chemistry did not have a name until 1998, where cheminformatics is a term first coined by Frank K. Brown. His definition is "...mixing of those information resources to transform data into information and information into knowledge for the intended purpose of making better decisions faster in the area of drug lead identification and optimization."[1–3] The recent development of cheminformatics makes it possible to be applied to data analysis for various industries like paper and pulp, dyes, and such allied industries.[1,3]

Response surface methodology (RSM) is one of cheminformatics that explores the relationships between several explanatory variables and one or more response variables. RSM was firstly introduced by George E.P. Box and K.B. Wilson with the main objective is to obtain optimal response.

Box and Wilson suggesting second order polynomial equation as the model template.[4] RSM had already familiar to researcher in bioprocess engineering which usually dealing with long-time microbial growth and required to consider many operations' variables. RSM will significantly increase the research efficiency for determining optimum condition and forecasting the results. Bioethanol production from cellulose-based material is one of the popular bioprocess engineering research fields in Indonesia in the last five years. Palm plantation biomass waste became the most concern potency for bioethanol raw material since Indonesia is the world-leading producer in palm oil.

Currently, Indonesia had established ethanol production pilot plant by using palm oil empty fruit bunch (EFB) as raw material in a fully automatic computerized system. Introducing cheminformatics approach by presenting optimum condition in mathematical equation will give significant advantage. The mathematical model will easily apply in the automatic computerized system which supported by sufficient database.

This study emphasizes the application of RSM as a practical cheminformatics to support bioprocess technology by optimizing the operations' variables in term that easily adopted in automatic computerized system, mathematical model. Ethanol production process from palm plantation biomass waste in Indonesia is an interesting case study of how this cheminformatics applied. In the next section, we provide a brief explanation about the potency of Indonesian palm oil plantation biomass waste for the bioethanol raw material production. In Section 12.3, brief explanation about bioethanol production process from cellulose-based material will be delivered. The crucial variables to be optimized were substrate loading, enzyme loading, and surfactant addition dose. The importance of surfactant additions, the third variable that make RSM application required, will be explained in Section 12.4. Practical examples of optimization using RSM in bioethanol production from palm oil EFB will be described in Section 12.5. The mathematical model modification from EFB enzymatic hydrolysis will be applicated to palm oil trunk enzymatic hydrolysis. The modification process and its application will be discussed in Section 12.6. Quoting George E.P. Box statement, "essentially, all models are wrong, but some are useful."[4] Thus, discussion will be closed by a summary about some current limitation of mathematical model obtained by using cheminformatics and future prospect of cheminformatics in dealing with limitations and the application possibilities in broader fields, other than drug discoveries and its related topics.

12.2 INDONESIAN POTENCY IN PALM PLANTATION BIOMASS

Today, Indonesia is the largest palm oil producers in the world. Palm oil plantations were developed since 1968 from only 79.209 Ha which is belong to Indonesian government and the other 40.451 Ha private companies' plantation. Small holder industrialist started to contribute in producing palm oil in 1979, mainly as the impact of transmigration program created by the government which trigging agro industries outside Java island. Along with the plantation expansion as shown in Figure 12.1, the production was kept increasing. In 2015, Indonesian total palm oil production estimated to be 30,948,931 tons which only 2.49% was came from government plantation.

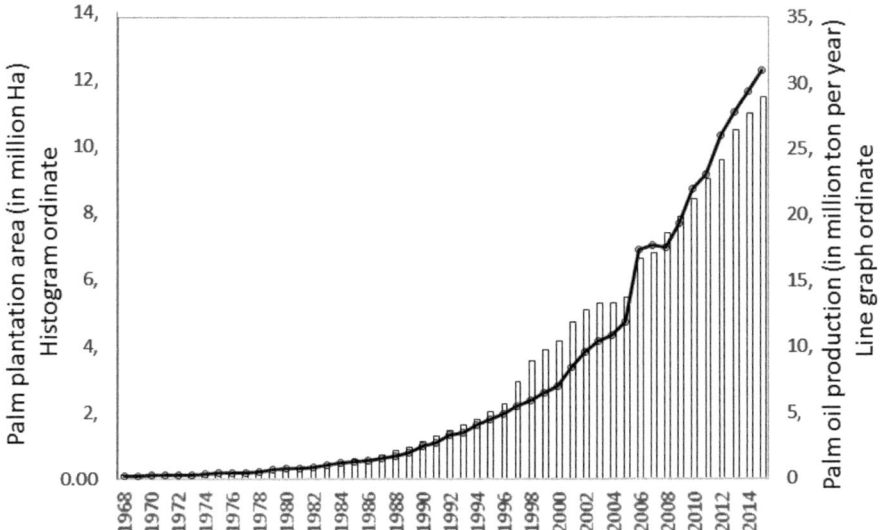

FIGURE 12.1   Indonesian palm oil production.

The EFB potency can be extrapolated from palm oil production since the EFB composition in fresh fruit bunch is similar with palm oil, 22–23%w.[5] This huge potency was waiting to be explored for the sake of Indonesian prosperity. The most efficient way to detach palm fruit from its bunch is by steam cooking fresh bunch in autoclave. This process caused the EFB was available in wet conditions which became the main challenge for utilizing it as construction material or as solid fuel by direct combustion just like palm kernel. Recent utilization of EFB was as in situ composting materials which consume transportation cost from extraction plant to the plantations.

The utilization was usually limited to the plantation that has the same owner with the extraction plant. Bioethanol production from EFB next to palm oil extraction plant gave a promising profitable solution. EFB as wet raw material will not be a problem since the pulping process will boil it in soda solution or other thermomechanical process. Energy produced from kernel and fiber combustion exceed the requirement of extraction plants which installing the latest technology. This excess energy can be used by bioethanol production in pulping and distillation unit. Recently, some extraction plants use the excess energy for generating power plant and supply electricity the adjacent dwellings. This utilization was limited by the wire infrastructure for distributions since the plant and plantation usually placed in relatively remote area. Converting the excess energy in form of bioethanol will give cheaper infestations in distribution infrastructure by using trucking system.

Other potential palm plantation biomass waste was the palm trunk. To keep the plantation productivity at its best, the palm tree cannot be older than 20 years old, thus, regeneration program need to be planned for the sake of sustainability. As can be seen in Figure 12.1, by assuming there will be no relocation happen in the existing plantation area, then theoretically there will be 224.528 Ha of palm oil plantation in regeneration program in all around Indonesia in 2016, similar with the number of plantation that developed in 1996. Palm oil plantation has 136–143 trees per Ha with the average weight of tree 330 kg.[5] Thus, the potency of palm trunk as biomass stock is 10–10.6 million ton in all over Indonesia in 2016. Since the expansion keep happening in 1997 and forward, the palm trunk potency was also climb up. Not to mention new generation of short-frond palm oil tree which gave possibility to plant more palm tree per hectare. The main obstacle of palm trunks utilization is its low mechanical strength that makes it not suitable for timber and furniture industries except particle board. Challenge for particle board industries in Indonesia is lack of resin availability in domestic market. Developing new resins factories in Indonesia still not feasible due to raw material availability and environmental law.

Utilization of palm plantation biomass waste as raw material for ethanol production is still feasibly competitive and have strong supported by Indonesian government. One of the significant supports is developing pilot plant in PP Kimia LIPI. Reports about the pilot plant performance had already delivered in previous study.[6] In the next section, the whole production process of bioethanol from cellulosic biomass as it was applied in the pilot plant will be briefly described.

12.3   ETHANOL PRODUCTION FROM CELLULOSIC BIOMASS WASTE: THE BASIC

Basically, the process consists of four steps, pretreatment which converts biomass waste to pulp, enzymatic hydrolysis which converts pulp to glucose, fermentation which converts glucose to ethanol, and the last step is distillation to purify the ethanol to meet fuel grade. Figure 12.2 gives brief description of how these four steps are connected. Converting biomass waste, in the most common material is wood-based materials, into pulp is an ancient technology that defines civilization. Since Tsai Lun invented paper to nowadays, the production technology has really evolved and established. This is a promising support for ethanol production from biomass. Room for improvement lies on increasing process efficiency by utilizing heat and chemicals as least as possible to meet next steps requirement.

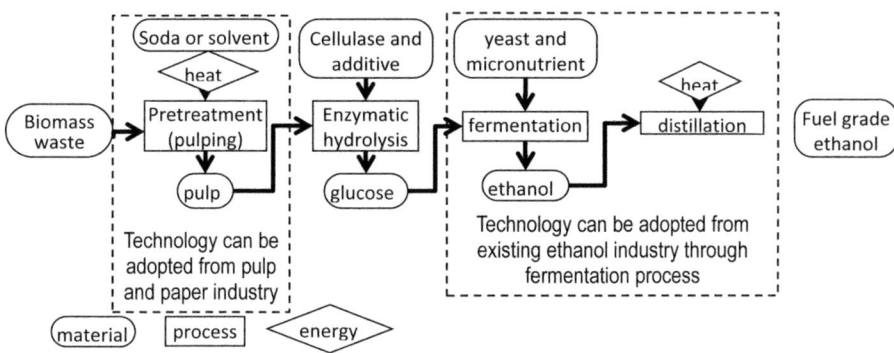

FIGURE 12.2   Flowchart of bioethanol production from biomass waste.

Fermentation and distillation technologies were also acknowledged since the history of human kind. Record and evidences of it can be found in every ancient civilization, that is, Egypt and China. This long lived technology had constantly improved so that nowadays, many industries offering fermentation and distillation licenses. In term to meet fuel grade requirement, ethanol–water mixture need to be separated above their azeotrope point. Mostly new developed technologies create improvement in this part of process to make the separation more efficient. One of the popular approaches was using mineral-based adsorbent to withdraw water.

Thus, enzymatic hydrolysis is the unique and crucial steps in the whole process which determining the economic feasibility. The hydrolysis rate decreases during process which leads to decreased yields and prolonged times which leads to higher production cost.[7,8] In 1998, the portion of enzyme cost in ethanol production from starch in Brazil was up to 40% of the total cost, higher than raw material, utilities, and other expenses which were 20, 28, and 12%, respectively. Through continuous research, the portion of enzyme cost became <8%. Although, there were contributions of increasing raw material prices and fuel which make their portion 32 and 35%, respectively. Research in developing high activity of amylase had reached the efficiency 1 g of enzyme (protein equivalent) for one gallon of ethanol. Recently, cellulase still lies in 100 g for a gallon of ethanol.[9]

Novozyme became the leading industry in the race for providing reliable cellulase in commercial scale. They claimed that their cellulase can be sold in US$ 2 per gallon which make ethanol production from corncob can be as feasible as if it was produced from cassava.[10] Danisco Inc. as the competitor was also introduced their latest cellulase in similar periode of time. As quoted by Reuters, Danisco spokesman, Rene Tronborg, claimed that by using their cellulase, bioethanol production from hayes has feasibilities comparable with the production process that using corn. This statement was announced in *Renewable Fuels Association's 15th Annual National Ethanol Conference* in Orlando, Florida. This event is a follow up action of Barack Obama Statement to reduce dependency on fossil fuel at the early of 2010.[11]

Other approach can be used in developing effective cellulose-to-ethanol process for liquid fuel application. To improve yields and reduce reduction time, and yet reducing the production cost, several technical approaches were studied. Conducting simultaneous hydrolysis and fermentation process for bioethanol production had already done by using pulp of palm oil EFB.[12–14] Kinetic study of enzymatic hydrolysis process had already conducted to synchronize the hydrolysis reaction rate with the fermentation rate.[15] This study had intensively developed to a pilot scale unit.[6] Conducting hydrolysis and fermentation simultaneously gave significant advantages of compared to hydrolysis and fermentation in series.[16]

The sense of chemical engineering that designing economically feasible process need to be involved in each different approach, which seldom overlooked by chemist and biologist whom developing enzyme. There is huge opportunity to adopt existing pulping, fermentation, and distillation technology, as shown in Figure 12.2, which will contribute to lower investment cost and maintenance. Thus, easier to calculate depreciation which leads to lower overall production cost. To adopt existing-commercially-reliable distillation unit for ethanol purification, hydrolysis, and fermentation need to be designed for producing fermentation broth with suitable specification. One of Indonesian well-known ethanol producer, P.T. Madubaru Jogjakarta, producing ethanol from molasses and setting 8–9%v ethanol concentration in its feed specification for the distillation unit operating normally. The distillation unit itself was designed to purify fermentation broth which contain 5%v of ethanol as its lowest. Feed with less ethanol concentration cannot be purified by using this similar unit. That is why in this study, 8%v was chosen as a benchmark in model validation as will be described later. The most logical approach for increasing ethanol concentration in fermentation broth was by loading more substrates in the hydrolysis process. The theoretical conversion of glucose to ethanol through fermentation using *Saccharomyces cerevisiae* is 51%. Thus, by also including cellulose content in pulp into account, the pulp loading in hydrolysis process need to be over 20%w of reaction system.

Performing cellulose hydrolysis experiments with high loading substrate is very difficult by only relying on usual laboratory apparatus such as orbital shaker or magnetic stirrer as reported by previous study.[17] Despite of this technical challenge, high loading substrate also needs to consider substrate inhibitions. This is some unique phenomena in reaction system that assisted by enzyme which was more reactant does not always mean more product. In case of cellulose hydrolysis, the enzyme can be trapped within cellulose fibers if the fibers are not dispersed adequately, which happened in reaction with high loading cellulose substrate. Not to mention the more negative impact caused by lignin impurities since more lignin in reaction system.

12.4   SURFACTANT ADDITION FOR ENHANCING CELLULOSE HYDROLYSIS PERFORMANCE; THE THIRD VARIABLE

Performing enzymatic hydrolysis of cellulosic biomass with high loading substrate implicated complex variables considerations since viscosity and substrate inhibitions will affect the end results. High enzyme concentrations are needed to reach high cellulose conversion, and enzyme recycling is difficult due to adsorption of enzymes to residual lignocelluloses. Enhancement of cellulose hydrolysis performance by adding surfactants to the hydrolysis mixture has been reported by several authors.[18–22] Increased hydrolysis yield by addition of surfactants has been reported for delignified steam-exploded wood[29] bagasse and corn stover.[23] Appearance of lignin impurities also affect the process since lignin can deactivate cellulase, thus reduce the performance of overall enzymatic hydrolysis. Surfactant addition to hydrolysis process can reduce the destructive effect of lignin impurities. The effect of surfactant on enzymatic hydrolysis of steam pretreated spruce (SPS), which relatively higher lignin impurities, was the object of intense research for use in an ethanol producing process.[25]

Surfactant addition in the hydrolysis process with high loading substrates using two naturally different raw materials, water hyacinth pulp and EFB pulp, were observed.[26] As can be seen in Figure 12.3 for the cellulose conversion of water hyacinth pulp, addition TWEEN 20 and SPAN 85 gave significant boost for cellulase. In cellulose hydrolysis conducted by 15 FPU/g-pulp cellulose, water hyacinth pulp produced by only boiling it in 1 M NaOH in atmospheric pressure gave 45.48%w cellulose conversion which can be increased to 52.29%w by producing the pulp in autoclave (2 atm pressure). Better results can be obtained by adding 1%v of TWEEN 20 to the water hyacinth boiled-pulp which gave 56.39%w. TWEEN 20 not only saving production cost by reducing energy for pulping process, but also by reducing the enzyme loading. Addition of 1%v of TWEEN 20 and SPAN 85 to water hyacinth boiled-pulp hydrolysis process gave 51 and 53.79%w cellulose conversion, respectively, by only using 10 FPU/g-pulp. Thus, all studies agreed that surfactant additions will improve the performance of hydrolysis step. The next issues then will be how the surfactants effect in the next step, fermentation process.

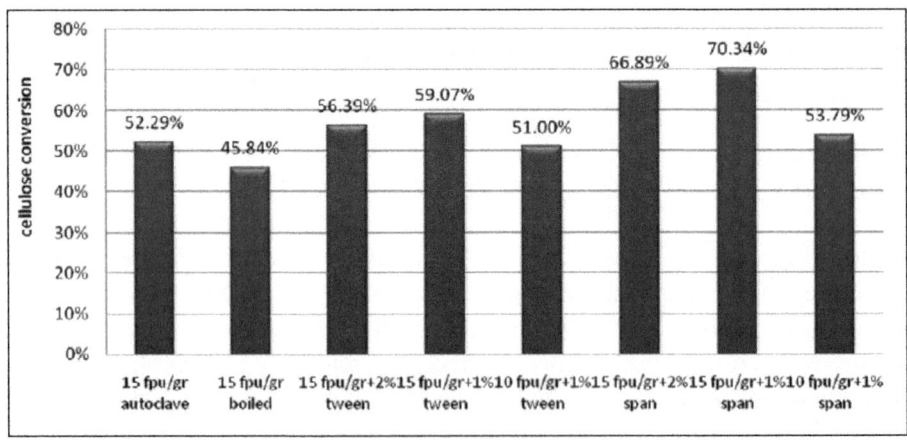

FIGURE 12.3 Comparison of TWEEN 20 and SPAN 85 addition in increasing cellulose conversion of enzymatic hydrolysis reaction to the reaction that using pulp produced by autoclave. Surfactant additions not only saving production cost by reducing energy for pulping process, but also by reducing the enzyme loading.

Surfactants were commonly introduced in fermentation systems to avoid foam generations in the systems which can interfere the monitoring and measurement tools, and assisting contamination by plugging the ventilation system. There are three types of surfactants, anionic, cationic, and nonionic surfactants. Most of the microbe membrane surfaces have electronegative charged and will be affected by anions and cationic surfactants. The toxicity of surfactant to microbial cell is attributed to an extraction of lipids in the cell membrane.[27] That is why these surfactants are also usually used as bactericides, such as quaternary ammonium compounds. There was a rule of thumb proposed to predict surfactant toxicity based on its h*ydrophile–lipophile balance* (HLB). Generally, increasing HLB tends to increase surfactant toxicity. As shown in previous study in fermentation of butanol production, surfactants with HLB value higher than 12 were toxic to microbes while surfactant with HLB in range 1–7 was not toxic.[27] Nonionic surfactants, such as linear alcohol ethoxylates, are biodegradable and show low toxicity in an aquatic environment.[28] On the other hand, a fortunate coincidence has been proven that the most effective surfactants that used as addition in cellulose enzymatic hydrolysis are nonionic surfactant.[18] Thus, addition of nonionic surfactant for enhancing lignocellulose enzymatic hydrolysis followed by fermentation in a system with no prior sterilization is preferable. The surfactant effect is higher at low cellulase concentration.[29]

The mechanism of surfactant effects in lignocellulose hydrolysis was extensively studied. It was concluded that nonionic surfactants can enhance the cellulase performance by overcoming the negative effect of cellulose crystalline.[19–22] Ooshima et al. compared amorphous cellulose with different types of crystalline celluloses (Avicel, tissue paper, and reclaimed paper). They showed that higher the crystallinity of the substrate, the more positive the effect of surfactant addition. Different mechanisms have been proposed for explaining positive effect of surfactant addition to an enzymatic hydrolysis of cellulose. The surfactant could change the nature of the substrate, for example, by increasing the available cellulose surface. Surfactant effects on enzyme–substrate interaction have been proposed by preventing it from inactivation enzymes that adsorbed into cellulose fiber. Surfactant addition facilitates desorption of enzymes from substrate.[23] Even these previous studies showed that surfactant could overcome the effect of cellulose crystalline; it should be noted that all the previous results were obtained from pure cellulose.

The effect of lignin impurities in pulp still beyond these explanations which having important role in utilizing pulp as raw material for bioethanol production process. Considering the effect of lignin impurities takes the effectiveness of pretreatment into account. Lignin-free pulp required high energy and/or chemical loading in the pretreatment, thus, high production cost. Defining optimum surfactant addition in the hydrolysis process can reduce the requirement in pretreatment process and lead to reduce production cost.

From another previous work in SPS pulp hydrolysis, it has been clarified that lignin has important role in preventing efficient conversion of lignocellulose.[24] Erickson had proposed explanation for surfactant effect on lignocellulose hydrolysis. It was postulated that the hydrophobic part of the surfactant binds, through hydrophobic interactions, with lignin on the lignocellulose fibers and the hydrophilic head group of the surfactant prevents unproductive binding of cellulase to lignin. Eriksson proposed mechanism gave more suitable explanation to this work. Then, surfactant effect in increasing the available cellulose surface or assisting enzymes desorption can be obtained by adding surfactant more than it required doses to overcome lignin problem.

Substrate loading and surfactant additions have their similarities. Adding too much substrate will cause substrate inhibition in hydrolysis process and thus lower the conversion but adding too few will cause higher energy required in the purification. Adding too much surfactant will give lethal effect for the yeast in the next process but adding too few cannot counter the negative impact of lignin impurities. Thus, defining optimum condition of surfactant doses will be as important as defining feasible substrate loading. Surfactant doses become the third variable to be considered besides substrate loading and enzyme loading.

12.5   RSM-CI APPLICATION IN DEFINING OPTIMUM CONDITION OF EFB ENZYMATIC HYDROLYSIS PROCESS

The appearance of surfactant doses as the third variable makes classical method in determining optimum condition no longer efficient. The classical method of optimization involves varying one parameter at a time and keeping the other constant. In two variables system, whether or not there were interrelations between variables, classical method will give same optimum condition, regardless of which variable was first set as constant. In system with three variables or more, classical method becomes inefficient and fails to explain relationships between the variables.

RSM was used in several optimizations in enzymatic reaction.[30–32] This methodology was recommended due to its ability to consider multivariable problem. In this study, ester sorbitan (mixture of 1,4-anhydrosorbitol, 1,5-anhydrosorbitol, and 1,4,3,6-dianhydrosorbitol) or also known as SPAN 85 was the chosen commercial surfactant based on one-point-test as shown in Figure 12.4. SPAN 85 has HLB value only 1.8. These results support the application of SPAN 85 as a nonionic surfactant in cellulase hydrolysis. The concerned hydrolysis process is the one that was followed by fermentation without any treatment for removing or reducing the effect of surfactant and cellulase prior fermentation.

In the hydrolysis system which using 10 FPU/g-substrates, addition 1%v of SPAN 85 could increase the cellulose conversion from 38.55% until up to 87.30% and clearly more effective than TWEEN 20. This result is comparable with the hydrolysis system which used 15 FPU/g-substrates. Thus, based on one-point-test, surfactant additions can save enzymes consumptions 33%. Since the enzyme cost is still become a big part of overall production cost, optimization of surfactant addition in cellulase hydrolysis reaction will be observed by using RSM method with the main objective to save more enzyme.

The application of RSM method was started by creating a three-level-three-factor central composite rotatable design (CCRD). The variable and their levels selected for the cellulase hydrolysis were SPAN 85 concentration (0–2%v), cellulase concentration (10–15 FPU/g substrate-dw), and substrate loading (20–30%dw). Substrate were palm oil EFB pulp made from EFB from Malimping, Indonesia by using kraft pulping with lignin impurities 9.34%w. Commercial cellulase Novozyme was used and SPAN 85 purchased from Merck. Fermentation processes were conducted in same condition, by adding 1%w of powdered yeast, 0.3%w of urea, and 0.1%w of nitrogen, phosphorus, and potash (NPK). Other affecting variables such as operation temperature was kept constant at Indonesian room temperature (25–28°C) and pH solution was set at 4.8 by using 50 mM sodium acetate buffer. All process conducted in a system with total volume 100 mL. Duration for hydrolysis and fermentation was set at 48 and 60 h, respectively. In this study, the observed variables were arranged as tabulated in Table 12.1, which using the least number of required experiments.

The data obtained from experiments were resulted reducing sugar, ethanol concentration, and the ethanol conversion. The experimental results, thus called actual value, for respective variables were also shown in the same table. Resulted reducing sugar is the measured reducing sugar concentration which obtained after the enzymatic hydrolysis process completed, as the feed for fermentation process. The resulted ethanol concentration is the ethanol concentration in the fermentation broth that measured after the fermentation process completed. The ethanol conversion was calculated as % weight, by comparing the resulted ethanol weight to the respective pulp weight as the process feed.

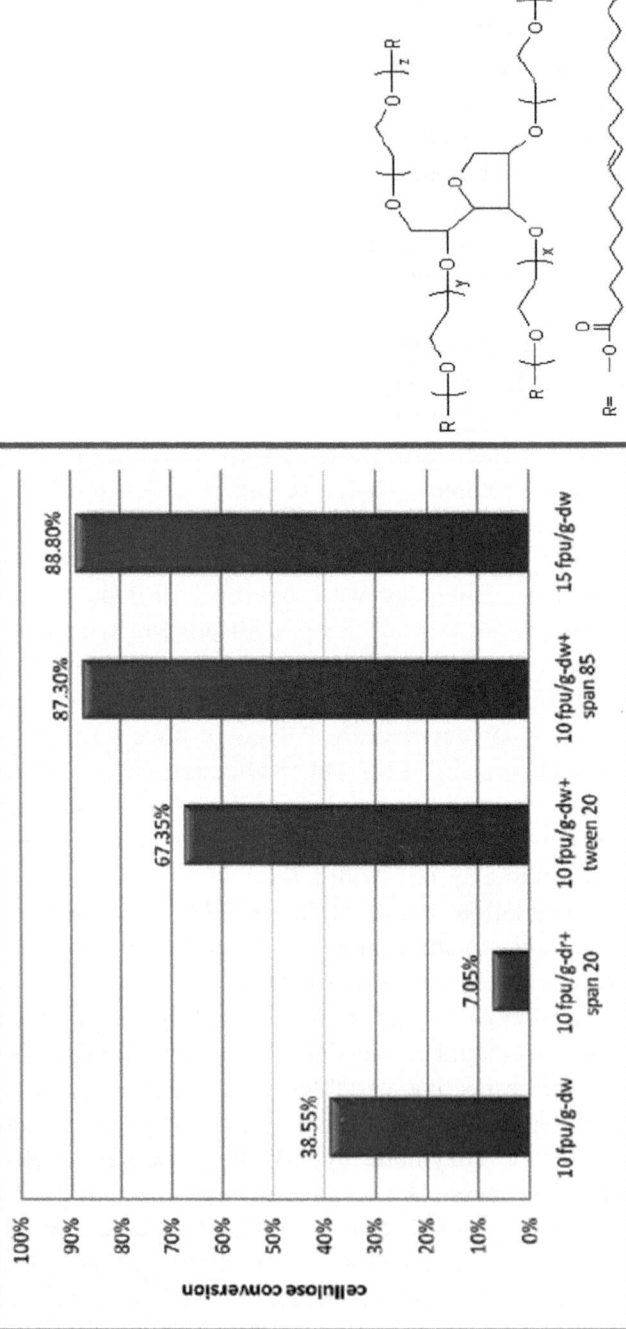

FIGURE 12.4 Left; the effect of TWEEN 20 and SPAN 85 addition in increasing cellulose conversion of enzymatic hydrolysis reaction. Right; the molecular structure of SPAN 85, surfactant that was used in this study.

TABLE 12.1 Central Composite Rotatable Quadratic Polynomial Model, Experimental Data, and Actual Predicted Values for Three-Level-Three-Factor Response Surface Analysis.

| Independent variables | | | Dependent variables | | | | | |
| Substrate loading (%dw) | enzyme (FPU/ gr substrate) | span 85 concentration (%vol) | Resulted reducing sugar (%dw) | | Resulted ethanol concentration (%v) | | Ethanol Conversion (%) | |
x1	x2	x3	actual	predicted	actual	predicted	actual	predicted
20 (-1)	10 (-1)	0 (-1)	11.98	11.94	4.39	4.69	21.95	23.10
25 (0)	15 (1)	1 (0)	21.76	20.49	6.97	6.46	27.88	26.18
30 (1)	12.5 (0)	2 (1)	22.74	25.17	5.6	6.10	18.67	20.05
20 (-1)	15 (1)	2 (1)	10.65	10.65	5.59	5.59	27.95	27.95
30 (1)	15 (1)	0 (-1)	23.53	25.93	7.1	7.90	23.67	26.20
25 (0)	10 (-1)	2 (1)	24.82	22.39	6.93	6.05	27.72	26.34
25 (0)	12.5 (0)	0 (-1)	22.93	20.57	7.16	6.43	28.64	24.96
30 (1)	10 (-1)	1 (0)	20.3	23.48	4.89	5.13	16.30	16.68
20 (-1)	12.5 (0)	1 (0)	8.92	9.08	6.34	6.67	31.70	33.16
Coefficient of determination (R^2)			0.9906		0.9623		0.9839	

All of these dependent variables were gone to model-fitting process to a second-order polynomial equation by using software for statistic SPSS 14. The basic principle of model fitting process was creating a mathematical model then uses it to recalculate all dependent variables that were tested in experiments. Dependent variables data obtained from recalculations, thus called predicted value, will be compared to the dependent variables data obtained from experiments by using a parameter called, coefficient of determination. The accepted mathematical model was the one that gave coefficient of determination above 0.95, in the range 0–1.

The accepted mathematical models were shown in term of dependent variables, Y_1 for resulted reducing sugar, Y_2 for resulted ethanol concentration, and Y_3 for ethanol conversion. The independent variables were stated as X_1 for substrate loading, X_2 for enzyme loading, and X_3 for the SPAN 85 doses. Based on their coefficient of determinations, all value was satisfyingly above 0.95. It can be concluded these statistical model will give good predictions for ethanol production process by using EFB pulp. By displaying the model prediction in 3D graph as shown in Figure 12.5, optimum condition was easier to be determined. These graphs were created by using CAD software and become strong statements for the existing cheminformatics. This is wonderful evidence that information technology gives strong support for chemistry and chemical engineering works.

$$Y_1 = -55.3491 + 7.5018X_1 - 5.6530X_2 - 2.0763X_3 - 0.1443X_1^2 + 0.1185X_2^2$$
$$+1.1450X_3^2 + 0.0983X_1X_2 - 0.0519X_1X_3 + 0.1327X_2X_3$$

$$Y_2 = 4.1471 + 0.0434X_1 - 0.4240X_2 + 10.8375X_3 - 0.04732X_1^2 - 0.2798X_2^2$$
$$-1.5456X_3^2 + 0.2541X_1X_2 - 0.6038X_1X_3 + 0.7078X_2X_3 \quad (12.1)$$

$$Y_3 = 9.4757 - 0.6878X_1 + 3.2328X_2 + 44.9815X_3 - 0.1651X_1^2 - 1.1526X_2^2$$
$$-6.0872X_3^2 + 0.8819X_1X_2 - 2.2216X_1X_3 + 2.2607X_2X_3$$

However, additional analyzes were required for these models to make sure they are aligned with existing biochemical theories. First analysis was focused on resulted reducing sugar that gave different trend compared to resulted ethanol concentration and ethanol conversion. Analyzing resulted reducing sugar was mainly about the enzyme performance and its kinetics. Based on basic theory of reaction kinetics, more substrate concentration will give more products. This theory was suitable with the model prediction that gave minimum value, and the value increase aligned with substrate loading.

Even though there is substrate inhibition theory in enzymatic reactions, but it still not happening in our substrate loading scope, 20–30%dw.

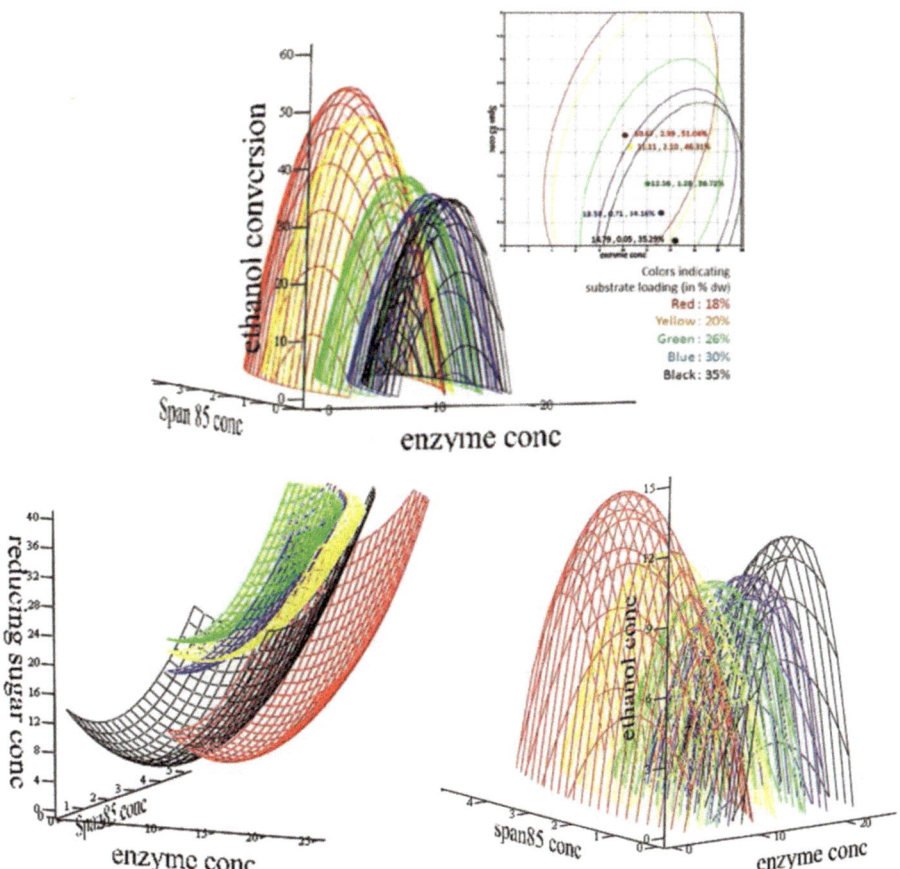

FIGURE 12.5   Profile of cellulose conversion, reducing sugar concentration, and ethanol concentration that predicted by the obtained mathematical model.

Surfactant addition was applied for reducing the negative effect of lignin impurities as aforementioned. Increasing substrate loading will require higher dose of surfactant addition. Thus, it is very reasonable the increasing in resulted reducing sugar will be obtained as surfactant doses increased. Small discrepancies were observed in the minimum value was not exactly at the point of origin (0,0).

In analyzing resulted ethanol concentration and ethanol conversion, *S. cerevisiae*, or yeast performance were included. The resulted mathematical model gave predictions that aligned with theory about surfactant effect to yeast by showing an optimum value. SPAN 85 doses lower than optimum value was indicating that the amount of surfactant could not overcome the negative effect of lignin impurities, thus lowering the cellulase performance. On the other hand, SPAN 85 doses higher than optimum values caused harmful effect to yeast by its increasing toxicity as previously explained. This effects cannot be recognized if the observed process is only hydrolysis step based on resulted reducing sugar. Of course, this is only happening if the process were conducted in similar duration and temperature with the one that used in this study, 48 h hydrolysis, 60 h fermentation in 25–28°C.

Interesting alignment between statistical conclusion with bioprocess theory and facts gave confidence for conducting model verification. In verification step, several random conditions that calculated to give ethanol concentration in fermentation broth over 8%v were used. This 8%v limit was adjusted to the lowest accepted ethanol concentration in fermentation broth to be directly sent to distillation unit in one of existing Indonesian ethanol production factory. The calculated results of depended variables were further called predicted value. List of tested random conditions and their respective predicted values were shown in Table 12.2. These random conditions were then sent to experimental test in laboratory to obtain the, so called, actual values. As can be observed in Table 12.2, the obtained actual values were very close with the predicted ones. Thus, the mathematical model had satis-fyingly proved as a reliable and useful prediction tools. As also can be seen in the series of tested variables, there are several conditions can be used for obtaining ethanol concentration over 8%v which adapting substrate loading and enzyme loading simultaneously. Suitable combination of substrate and enzyme loading was required and this can be fulfilled by RSM analysis. The mathematical model involving the term X_1X_2 in the equation which representing the mutual effect of substrate loading and enzyme loading. By involving this term, the sugar concentrations as feed for fermentation were adjusted to obtain ethanol concentration in fermentation broth over 8%v. As pulp substrate loading increased, the enzyme loading was adjusted at lowest possible concentration. This adaptation condition will be very useful in large scale/commercial application as a precaution if substrate or enzyme supply chain was interrupted. These results cannot be obtained from conventional analysis approach which observing the effect of one variable by keeping the other variables at constant value.

As can be seen in this case study, there will be a certain of doubt if the statistical results do not align with background facts and theories. In worse case, the model also cannot give satisfying predictions. In this problem,

TABLE 12.2 List Random Conditions and Their Respective Predicted Values that Tested in Laboratory Experiments. The Results from Laboratory were Called Actual Values.

Independent variables			Dependent variables							
Substrate loading (%dw)	enzyme (FPU/ gr substrate)	span 85 concentration (%vol)	Resulted reducing sugar (%dw)		Resulted ethanol concentration (%v)		Ethanol Conversion (%)			
x1	x2	x3	actual	predicted	actual	predicted	actual	predicted		
36.5	14.75	1	21.76	20.66	10.48	10.07	28.71	27.19		
23	12.5	1	15.63	16.50	8.23	8.41	35.78	36.20		
21.25	11	1	12.10	14.05	8.96	8.89	41.67	39.00		
20	10.25	1	10.51	11.01	7.77	8.91	38.85	40.12		
31	12	1	24.75	21.39	8.81	9.07	33.88	35.32		
26	13.25	1	21.98	23.37	8.25	8.24	26.61	27.46		
Coefficient of determination (R^2)				0.8880		0.9189		0.9163		

second order polynomial equation maybe not the suitable empirical model for the observed phenomena. The options for still using the same experimental data is very limited since no variables were set constant to be analyzed in conventional way. For avoiding waste of time and experimental resources, adequate reference studies of the similar or related field were required. In other problem, the optimum condition cannot be found. The most recommended way for searching it was to add more experiments, with broader variable range, which designed also in CCRD and repeat the RSM analysis sequences. Because usually the problem came from unsuitable selection of variable range which the optimum value was not lies in the range. Since RSM analyzing process is only dealing with empirical approach, it will not easy to create solid correlation between the results with scientific theory and fundamentals and thus, difficult to plan scientific improvement for the observed process.

12.6 MODIFICATION OF EFB MATHEMATICAL MODEL FOR TO BE USED IN PALM TRUNK ENZYMATIC HYDROLYSIS PROCESS

RSM had been demonstrated giving significant contribution for biochemical engineer in determining optimum conditions. However, obtaining optimum condition from laboratories experiments is not an easy task, especially in bioprocess like bioethanol from cellulose. It had been mentioned that for completing one experiment, 108 h were required and not including preparations. Possibility of using obtained equation for predicting similar or closely related raw material seems very worthy to be examined. This study continued by implementing the equation from hydrolysis and fermentation of palm oil EFB pulp to the process that use palm oil trunk.

As mentioned earlier, palm oil trunks become a potential raw material due to regeneration of palm oil plantation which actually happened in these recent years in Indonesia. Cross sectional profile of palm oil trunk can be seen in Figure 12.6. As part of monocotyledon plants, palm oil trunk does not have cambium with less structured arrangement of its fibers. By excluding the barks, palm oil trunk structure can be classified as parenchyma that lies inside the circle and vascular bundle at the outsides.

Vascular bundles have more structured fibers than parenchyma. Botanists believe this is due to its function for trunk outer protection. Parenchyma, on the other hand, is having softer structure and high water adsorption capacity which is very useful in transporting nutrient. It is also having a function as nutrient storage, thus small amount of starch and plant sap was found in parenchyma. Based on this structure and chemical composition, an attempt to direct utilization of parenchyma as feed for enzymatic hydrolysis process, without prior pulping, was conducted but gave unsatisfied results. In this study, palm oil trunk from Malingping, Indonesia were chopped and milled to 50 meshes. Some of the resulted material went to mechanical separation for obtaining vascular bundle and parenchyma part. The other part was not separated and both non-separated and vascular bundle part went through kraft pulping process that having similar condition with EFB. Pulp of whole trunks had 6.96%w lignin impurities and the vascular bundle had it 13.70%w. Later, both pulps were then separately used as raw material for bioethanol production process.

TABLE 12.3 Variables for Hydrolysis Whole Trunk Pulp and Vascular Bundle Pulp.

Optimum Condition Number	Substrate Loading (%dw)	Enzyme (FPU/g Substrate)	Span 85 Conc. (%v)
1	20	10.25	1
2	21.25	11	1
3	23	12	1
4	26	12.5	1
5	31	13.25	1
6	36.5	14.75	1

FIGURE 12.6 Left; Table 12.3 variables for hydrolysis whole trunk pulp and vascular bundle pulp. Right; cross sectional area of palm oil trunk. The part that inside the circle is parenchyma and the one outside the circle is vascular bundle.

Similar condition in hydrolysis and fermentation process for EFB pulp were applied to both of pulps, 48 h hydrolysis followed by 60 h fermentation in 25–28°C. Fermentation processes were conducted in same condition, by adding 1%w of powdered yeast, 0.3%w of urea, and 0.1%w of NPK. Observed variables were following the selected condition for EFB as shown in Table 12.3 with the similar goal, obtaining ethanol concentration in fermentation broth over 8%v. The resulted reducing sugar concentrations from hydrolysis of whole trunk pulps and its comparison with hydrolysis of EFB pulp were shown in Figure 12.7.

Data were performed in histogram with selected condition number as abscise. The numbers were referred to the similar optimum condition as mentioned in Table 12.3. The reducing sugar from palm oil empty bunch hydrolysis showed better results than resulted reducing sugar from palm oil trunks. The result of correlative test between trunks and EFB in high-loading substrate enzymatic hydrolysis based on their resulted reducing sugar was also shown in Figure 12.7.

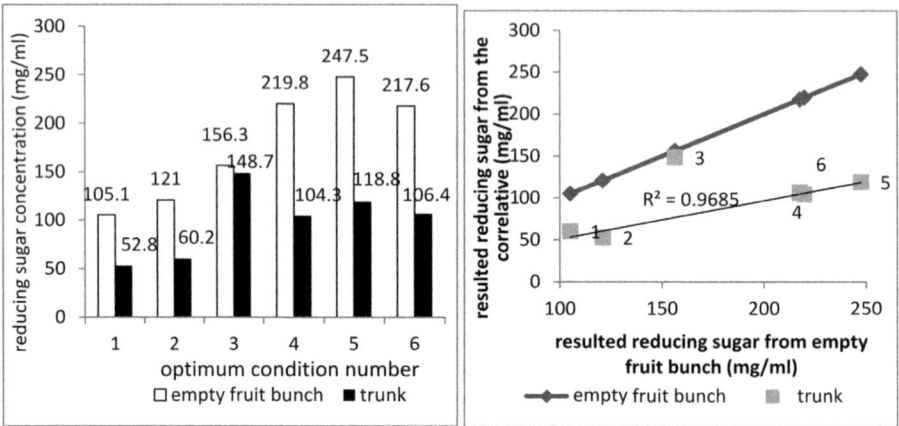

FIGURE 12.7   Resulted reducing sugar obtained from palm oil trunks hydrolysis.

Perfect correlation was simulated by correlating the EFB results to itself as showed in the graph as dashed line. Square dots were representing the resulted reducing sugar from palm oil trunks hydrolysis arranged correlatively to the ones obtained from palm oil EFB in the same optimum operation condition. By excluding data obtained from condition number 3, calculated correlative coefficient was 0.9993. This result supported the conclusion that optimum conditions for high-loading-substrate enzymatic hydrolysis of EFB were similar with optimum conditions for trunk pulp.

Reported data in Figure 12.8 were the comparison of resulted ethanol concentration between palm oil EFB and palm oil trunk. Since all operation condition of fermentation was set similar to all, then the difference solely caused by the difference of hydrolysis process and outside disturbances. No sterilization prior fermentation in order to reduce operational cost if this similar process applied in commercial scale. Unsterilized fermentations also allowed the enzymatic hydrolysis continued during the fermentation, which also called simultaneous saccharification fermentation (SSF). However, contamination possibilities need to be considered as outside disturbance.

It was shown in Figure 12.6 the resulted ethanol from trunks were lower than those obtained from EFB. This is aligned with the resulted reducing sugar reported above. In EFB pulp, 1%v of SPAN 85 could effectively reduce the negative effect of lignin residue up to 9.34%w. As previously described, the hydrophobic part of the surfactant binds through hydrophobic interactions to lignin on the lignocellulose fibers and the hydrophilic head group of the surfactant prevents unproductive binding of cellulase to lignin. The content of lignin residue in palm oil trunk pulp was only 6.96%w. Thus, lower cellulase performance in trunk pulp, which indicated by lower resulted ethanol concentration compare to EFB pulp, was not caused by lignin impurities. It could be concluded that the reason for lower resulted ethanol concentration was more dominantly caused by the higher crystalline cellulose content in whole trunk pulp. Later will be discussed there were two factors that affecting cellulase performance, lignin residue in pulp and cellulose crystalline in pulp.

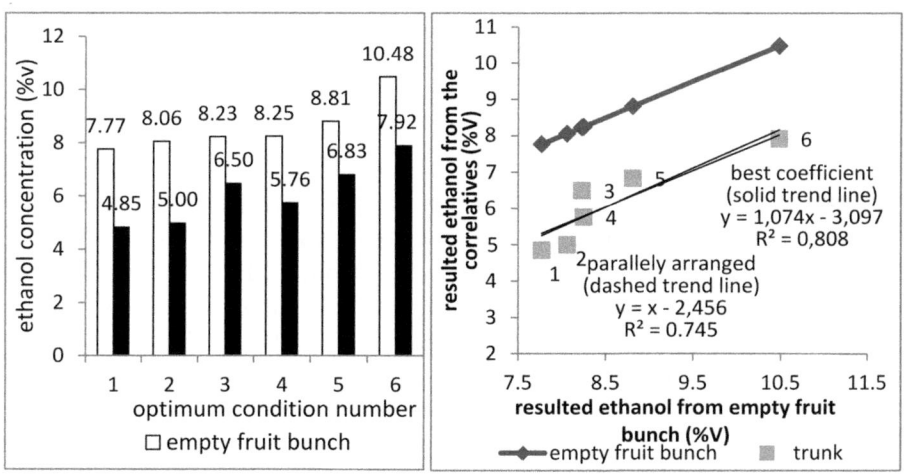

FIGURE 12.8 Ethanol obtained from palm oil trunks hydrolysis and continued by fermentation.

The result of correlative test between trunks and EFB in high-loading substrate enzymatic hydrolysis based on their resulted ethanol concentration was also shown in Figure 12.8. Unfortunately, none of tested conditions gave ethanol concentration over 8%v in fermentation broth. However, no data was excluded and the best correlative coefficient was calculated as 0.808. Aligning this conclusion with correlative test using sugar concentration, it is acceptable to assume that optimum conditions for high-loading-substrate enzymatic hydrolysis for EFB and those for trunk pulp were similar.

The trend line's slope for trunks was 1.074 which is very close to one. It means the trunks trend line can be considered as parallel to the dashed line for EFB. Arranged trend line to meet the slope equal to 1 reduced the correlative coefficient to 0.7451 and still considered acceptable. From the arranged trend line equation, there was a constant value obtained, −2.456%v then mathematical equation for predicting resulted ethanol concentration from palm oil trunks can be derived from eq 12.1.

$$Y_2 = (4.1471 - 2.456) + 0.0434X_1 - 0.4240X_2 + 10.8375X_3 - 0.04732X_1^2$$
$$-0.2798X_2^2 - 1.5456X_3^2 + 0.2541X_1X_2 - 0.6038X_1X_3 + 0.7078X_2X_3$$

$$Y_2 = 1.6904 + 0.0434X_1 - 0.4240X_2 + 10.8375X_3 - 0.04732X_1^2 - 0.2798X_2^2$$
$$-1.5456X_3^2 + 0.2541X_1X_2 - 0.6038X_1X_3 + 0.7078X_2X_3 \tag{12.2}$$

The predicting results of eq 12.2 were shown in Table 12.4. The coefficient of determination for eq 12.2 was much lower than prediction made by eq 12.1 for EFB but still usable in brief prediction. Thus, using RSM model that obtained from a certain substrate was very promising to be applied to similar substrates. Increasing more experimental number hopefully will give better correlations. Similar correlation test was conducted for vascular bundle pulp. Unfortunately, the correlation of EFB optimum conditions to vascular bundle was not as satisfying as the correlation between EFB and whole trunks. Resulted reducing sugar of vascular bundle pulp gave acceptable correlation with sugars obtained from EFB pulp from respective optimum conditions as shown in Figure 12.9. However, the correlation of resulted ethanol concentration between EFB and vascular bundle gave unacceptable results. The correlative coefficient only 0.219, as shown in

TABLE 12.4 Comparison Obtained Experimental Ethanol Concentration from Trunks Pulp and Empty Fruit Bunch Pulp to Their Correlative Prediction by Mathematical Model.

Optimum Condition Number	Substrate Loading (%dw)	Enzyme (FPU/g Substrate)	Resulted Ethanol Concentration from Trunks (%v)		Resulted Ethanol Concentration from Empty Fruit Bunch (%v)	
			Obtained	Predicted (eq 12.2)	Obtained	Predicted (eq 12.1)
1	20	10.25	4.85	5.55	7.77	8.01
2	21.25	11	5.00	5.63	8.06	8.09
3	23	12	6.50	5.65	8.23	8.11
4	26	12.5	5.76	5.78	8.25	8.24
5	31	13.25	6.83	6.61	8.81	9.07
6	36.5	14.75	7.92	7.61	10.48	10.07
Coefficient of determination				0.7451		0.9189

Figure 12.10, which is much lower than the previous study which using whole trunk, 0.7451.

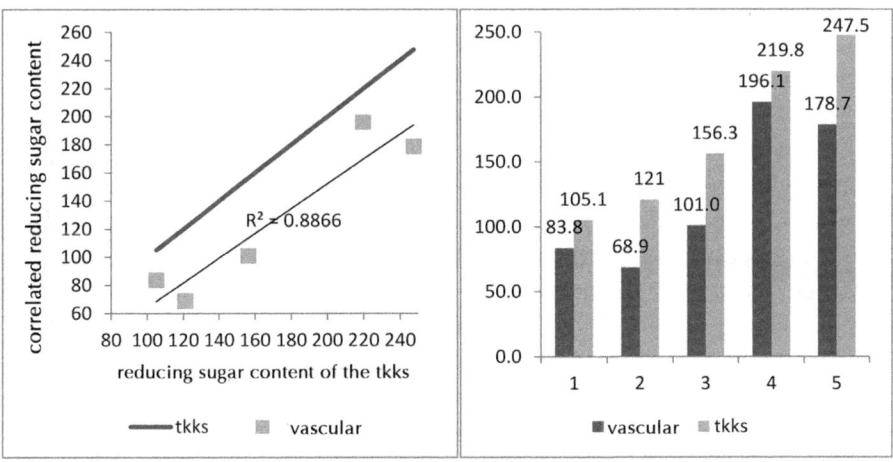

FIGURE 12.9 Ethanol obtained from palm oil vascular bundle hydrolysis and continued by fermentation.

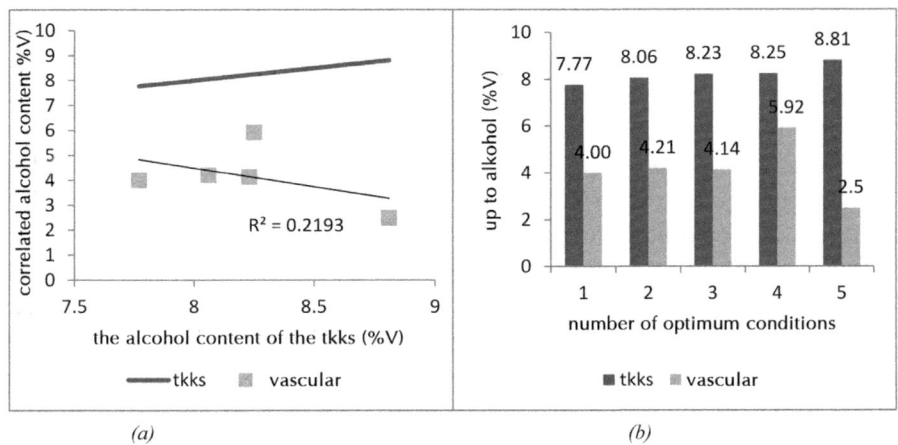

(a)   *(b)*

FIGURE 12.10   (a) The correlation curve and (b) alcohol content with tkks vascular bundle.

In a previous study, parts of palm oil trunks were separated mechanically to its vascular bundle and parenchyma then each part was measured its hemi-celluloses content and cellulose crystallinity index.[33] The results show that vascular bundle and parenchyma had hemicelluloses content 25.47%w and 35.57%w, respectively, and cellulose crystallinity index 76.78 and 69.70, respectively. Cellulose crystalline index of palm oil EFB was not found from previous studies. However, in another previous study in Miscantus, it was shown the correlation between hemicelluloses content and crystalline cellulose in its natural existence.[34] The higher hemicelluloses content in one observed part of plants, the lower cellulose crystallinity index. These previous studies in palm oil trunk and Miscantus were supporting a conclusion about correlation between hemicelluloses content and celluloses crystallinity index.

In another previous study, effects of non-cellulose sugar in natural plants to cellulose crystallinity were specified only to galactoglucuronoxylan.[35] Three monocotyledons which are Italian ryegrass, pineapple, and onion measured its galactoglucuronoxylan content and correlated the results to cellulose crystallinity index. The results showed that higher galactoglucuronoxylan content will lead to higher crystallinity index when the cellulose naturally biosynthesized. Glucuronogalactoxylan has similar properties with starch due to its water solubility and branched molecular structure. Starch was found in palm oil trunks but not in EFB. This fact could strengthen our previous conclusion that palm oil trunks have higher cellulose crystallinity index than its EFB. Not only because it had lower hemicelluloses contented but also because the existence of starch in palm oil trunks.

Hemicelluloses content in parenchyma, EFB, whole trunks, and vascular bundle were 35.57, 35.3, 31.8, and 25.47%w, respectively.[33,36–38] Cellulose crystallinity index for vascular bundle and parenchyma was 76.78 and 69.70, respectively. It will be reasonable to assume cellulose crystallinity index in EFB was close to parenchyma, 69.70, due to no significant difference between hemicelluloses content in EFB and parenchyma and the index for whole trunk would be between 69.70 and 76.70. Ordering from the highest cellulose crystallinity index to the lowest in its natural existence would be vascular bundle (76.78) > whole trunk > EFB = parenchyma (69.70). Since both raw materials were gone through kraft pulping process with similar operation conditions, then the cellulose crystallinity index in resulted pulp gave similar tendencies. These facts support the experimental finding that the order of enzymatic hydrolysis performance from the lowest was vascular bundle < whole trunk < EFB.

These results showed the limitation of RSM in searching for general model that can be applied to similar or closely related conditions. In case of bioethanol production from palm plantation biomass waste, general model cannot be obtained even the biomass came from one specific species. However, scientific and experimental facts were aligned with the RSM model which opens windows of opportunities for using RSM in this field.

One promising possibilities were establishing clear raw material characteristics, including lignin impurities and cellulose crystallinity index, as the term for the existing RSM model works satisfyingly.

12.7 CONCLUSION AND FUTURE RESEARCH

12.7.1 CONCLUSION REMARK

The RSM-CI is a medium for bioethanol processes to extend their activities beyond traditional boundaries that were characterized by space and time constraints, leading to accessible reach of processes directly anytime anywhere.

RSM-CI systems are one of cheminformatics that explores the relationships between several explanatory variables and one or more response variables that allow users to access bioethanol processing systems directly by clicks of a mouse or touches of a finger.

The adoption of RSM-CI systems by bioethanol production, enhancing cellulose hydrolysis, and optimum enzymatic condition of EFB enzymatic are high, threatening software as the systems allow users interact directly. Adoption of cheminformatics systems by users, especially during bioethanol production processes, in Indonesia is also high, threatening the competitiveness of determining optimum condition and forecasting the results.

However, in Indonesia, with the rapid advancement of technology, there is no doubt that traditional stages of bioethanol processes will face more challenges in the future. In general, RSM-CI promises more efficiency and effectiveness stages and forecasting to grab optimum and economical scalability of product rather than traditionally processes.

12.7.2 FUTURE RESEARCH DIRECTION

This study incorporates two main issues of RSM-CI features for enabling bioethanol production from palm oil of EFB in Indonesia; determining optimum condition and forecasting the result of production. With the promising features and user acceptance of RSM-CI performance, quality, reliability, and usability, it is highly recommended for bioethanol producer to consider adopting RSM-CI methodology as it has been valued as an effective, efficient tool to assist plant to get efficient-effective condition and helps to set a clear strategy to get optimum result confidently.

The study triggers a future research direction in bioethanol production strategy which would focus on enhancing bioethanol plant, forecasting cellulase performance, EFB enzymatic hydrolysis process, change management associated with the implementation of RSM-CI, RSM-CI software release management, ICT service continuity management, and configuration management to handle bioethanol production.

The future direction of this study can also accommodate and customize RSM-CI to fit with other bioprocesses, such as other material of bioethanol production, bioethanol supply chain management, and after production handling care. RSM-CI was designed as an adaptable framework that can be implemented and extended to process by customizing its controls.

KEYWORDS

- **chemistry informatics**
- **response surface methodology**
- **RSM-CI**
- **ICT emerging technology**
- **EFB**

REFERENCES

1. Susanto, H. Cheminformatics: The Promising Future: Managing Change of Approach Through ICT Emerging Technology. In: *Apllied Chemistry and Chemical Engineering Book;* Apple Academic Press–Francis and Taylor Publisher: Canada, 2016; Vol. 2, DOI: 10.13140/RG.2.1.4651.1604.
2. Almunawar, M. N.; Susanto, H.; Anshari, M. A Cultural Transferability on IT Business Application: iReservation System. *Journal of Hospitality and Tourism Technology*, **2016,** *4* (2), 155–176. *JHTT.* 2016, DOI: 10.1108/JHTT-05-2012-0015.
3. Susanto, H.; Chen, C. K.; Leu, F.; Y. Lesson Learn from IT as Enable of Business Process Re-Design. *J. Inform. System Technol.* **2016,** DOI: 10.2139/ssrn.2748586.
4. Pena, D. George Box: An Interview with the International Journal of Forecasting. *Int. J. Forecast.* **2001,** *17*, 1–9.
5. Pahan, I. *Panduan Lengkap Kelapa Sawit: Manajemen Agribisnis dari Hulu Sampai Hilir;* Niaga Swadaya: Bogor, Indonesia, 2006.
6. Sudiyani, Y.; Styarini, D.; Triwahyuni, E.; Sembiring, K. C.; Aristiawan, Y. Utilization of Biomass Waste Empty Fruit Bunch Fiber of Palm Oil for Bioethanol Production Using Pilot–Scale Unit. *Energy Procedia.* **2013,** *32,* 31–38.

7. Gregg, D.; Saddler, J. N. Factors Affecting Cellulose Hydrolysis and the Potential of Enzyme Recycle to Enhance the Efficiency of an Integrated Wood to Ethanol Process. *Biotechnol. Bioeng.* **1996,** *51,* 375–383.

8. Teuku, B. B.; Chihiro, O.; Masanobu, N.; Keiichi, K.; Yanni, S.; Tatsuhiko, Y.; Yasumitsu, U. Improvement of Saccharification of Empty Fruit Bunch and Japanese Cedar Pulps with an Amphiphilic Lignin Derivative. *Mokuzai Gakkaishi.* **2010,** *56,* 420–426.

9. Kim, Y. H. *The Current Status of Bioethanol Production Technology;* Departement of Chemical Engineering, Kwangwoon University: Seoul, Korea, 2007.

10. Novozyme press release. New Enzymes Turn Waste into Fuel. Feb 15, 2010, www.novozyme.com

11. Kinney, J. Reborn Qteros Rises Again with New Plan to Develop Technology to Make Ethanol. *The Republican,* Feb 1, 2012.

12. Triwahyuni, E.; Sudiyani, Y.; Abimanyu, H. The Effect of Substrate Loading on Simultaneous Saccharification and Fermentation Process for Bioethanol Production from Oil Palm Empty Fruit Bunches. *Energy Procedia.* **2015,** *68,* 138–146.

13. Triwahyuni, E.; Hariyanti, S.; Dahnum, D.; Nurdin, M.; Abimanyu, H. Optimization of Saccharification and Fermentation Process in Bioethanol Production from Oil Palm Fronds. *Procedia Chem.* **2015,** *16,* 141–148.

14. Triwahyuni, E.; Muryanto; Fitria, I.; Abimanyu, H. Pemanfaatan Limbah Tandan Kosong Kelapa Sawit untuk Produksi Bioetanol dengan Optimasi Proses Sakarifikasi dan Fermentasi. *Jurnal Energi dan Lingkungan (Enerlink).* **2014,** *10,* 27–32.

15. Barlianti, V.; Dahnum, D.; Triwahyuni, E.; Aristiawan, Y.; Sudiyani, Y. Enzymatic Hydrolysis of Oil Palm Empty Fruit Bunch to Produce Reducing Sugar and Its Kinetic. *Menara Perkebunan.* **2015,** *83* (1), 37–43.

16. Dahnum, D.; Tasum, S. O.; Triwahyuni, E.; Nurdin, M.; Abimanyu, H. Comparison of SHF and SSF Processes Using Enzyme and Dry Yeast for Optimization of Bioethanol Production from Empty Fruit Bunch. *Energy Procedia.* **2015,** *68,* 107–116.

17. Chu, Q.; Li, X.; Ma, B.; Xu, Y.; Ouyang, J.; Zhu, J.; Yu, S.; Yong, Q. Bioethanol Production: An Integrated Process of Low Substrate Loading Hydrolysis-High Sugars Liquid Fermentation and Solid State Fermentation of Enzymatic Hydrolysis Residue. *Bioresour. Technol.* **2012,** *123,* 699–702.

18. Ooshima, H.; Sakata, M.; Harano, Y. *Biotechnol. Bioeng.* **1986,** *28,*1727–1734.

19. Boussaid, A.; Saddler, J. N. Adsorption and Activity Profiles of Cellulases during the Hydrolysis of Two Douglas Fir Pulps. *Enzyme Microb. Technol.* **1999,** *24* (3–4), 138–143.

20. Mishima, D.; Kuniki, M.; Sei, K.; Soda, S.; Ike, M.; Fujita, M. Comparative Study on Chemical Pretreatments to Accelerate Enzymatic Hydrolysis of Aquatic Macrophyte Biomass Used in Water Purification Processes. *Bioresour. Technol.* **2008,** *99,* 2495–2500.

21. Sonderegger, M.; Jeppsson, M.; Larsson, C.; Gorwa-Grauslund, M. F.; Boles, E.; Olsson, L. *Biotechnol. Bioeng.* **2004,** *87,* 90–98.

22. Berlin, A.; Gilkes, N.; Kilburn, D.; Bura, R.; Markov, A.; Skomarovsky, A.; Okunev, O.; Gusakov, A.; Maximenko, V.; Gregg, D.; Sinitsyn, A.; Saddler, J. Evaluation of Novel Fungal Cellulase Preparations for Ability to Hydrolyze Softwood Substrates – Evidence for the Role of Accessory Enzymes. *Enzyme Microb. Technol.* **2005,** *37* (2), 175–184.

23. Kaar, W. E.; Holtzapple, M. Benefits from Tween during Enzymic Hydrolysis of Corn Stover. *Biotechnol. Bioeng.* **1998,** *59,* 419–427.

24. Eriksson, T.; Börjesson, J.; Tjerneld, F. Mechanism of Surfactant Effect in Enzymatic Hydrolysis of Lignocellulose. *Enzyme Microb. Technol.* **2002,** *31,* 353–364.

25. Börjesson, J.; Peterson, R.; Tjerneld, F. Enhanced Enzymatic Conversion of Softwood Lignocellulose by Poly(ethylene glycol) Addition. *Enzyme Microb. Technol.* **2007,** *40* (4), 754–762.

26. Bardant, T. B.; Haznan, A.; Nina, A. Effect of Pretreatment Technology on Enzyme Susceptibility in High Substrate Loading Enzymatic Hydrolysis of Palm Oil EFB and Water Hyacinth. Int. J. Environ. *Bioenergy.* **2012,** *3* (3), 193–200.

27. Wang, Z.; Dai, Z. Extractive Microbial Fermentation in Cloud Point System. *Enzyme Microb. Technol.* **2010,** *46* (6), 407–418.

28. Dorn, P.; Salanitrsoh, J. O.; Evans, A.; Kravetz, L. Assessing the Aquatic Hazard of Some Branched and Linear Nonionic Surfactants by Biodegradation and Toxicity. *Environ. Toxicol. Chem.* **1993,** *12,* 1751–1762.

29. Helle, S. S.; Duff, S. J. B.; Cooper, D. G. Effect of Surfactants on Cellulose Hydrolysis. *Biotechnol. Bioeng.* **1993,** *42,* 611–617.

30. Gunawan, E. R.; Basri, M.; Rahman, M. B. A.; Salleh, A. B.; Rahman, R. N. Z. A. Study on Response Surface Methodology (RSM) of Lipase-Catalyzed Synthesis of Palm-Based Wax Ester. *Enzyme Microb. Technol.* **2005,** *37,* 739–744.

31. Shaw, J. F.; Wu, H. Z.; Shieh, C. J. Optimized Enzymatic Synthesis of Propylene Glycol Monolaurate by Direct Esterification. *Food Chem.* **2003,** *81,* 91–96.

32. Faciolli, N. L.; Barrera-arelano, D. Optimisation of Enzymatic Esterification of Soybean Oil Deodoriser Distillate. *J. Sci. Food Agric.* **2001,** *81,* 1193–1198.

33. Lamaming, J.; Hashim, R.; Leh, C. P.; Sulaiman, O.; Sugimoto, T. Proceedings of the Second International Conference on Advances in Bio-Informatics, Bio-Technology and Environmental Engineering- ABBE 2014, ISBN: 978-1-63248-053-8, 2014.

34. Xu, N.; Zhang, W.; Ren, S.; Liu, F.; Zhao, C.; Liao, H.; Xu, Z.; Huang, J.; Li, Q.; Tu, Y.; Yu, B.; Wang, Y.; Jiang, J.; Qin, J.; Peng, L. *Biotechnol. Biofuels.* **2012,** *5,* 58.

35. Smith, B. G.; Harris, P. J.; Melton, L. D.; Newman, R. H. Crystalline Cellulose in Hydrated Primary Cell Walls of Three Monocotyledons and One Dicotyledon. *Plant Cell Physiol.* **1998,** *39* (7), 711–720.

36. Saka, S. Proceedings of JSPS-VCC Natural Resources & Energy Environment Seminar, Kyoto, Japan, 2005.

37. Singh, G.; Huan, L. K.; Leng, T.; Kow, D. L. Oil Palm and the Environment: A Malaysian Perspective; Malaysia Oil Palm Growers Council: Kuala Lumpur, 1999; pp 41–53.

38. Yang, H. P.; Yan, R.; Chen, H. P.; Lee, D. H.; Liang, D. T.; Zheng, C. G. *Fuel Process. Technol.* **2004,** *18,* 1814–1821.

CHAPTER 13

ASSESSMENT OF QUERCETIN ISOLATED FROM *ENICOSTEMMA LITTORALE* AGAINST FEW CANCER TARGETS: AN IN SILICO APPROACH

R. SATHISHKUMAR[*]

Department of Botany, PSG College of Arts and Science, Coimbatore, India

[*]*Corresponding author. E-mail:sathishbioinf@gmail.com*

CONTENTS

ABSTRACT

Cancer being a malignant disease, the cure and best treatment method has not been identified yet. Although, the drugs are reported to have narrow therapeutic index and are often palliative as well as unpredictable; in the last few decades, cancer chemotherapy has been developed as one of the major medical advancement in treating cancer types. To infer the existing difficulties and since the present era thrust in finding the alternatives from medicinal plants, quercetin, a compound isolated from *Enicostemma littorale*, was studied using molecular docking with its ability to predict the binding site and efficiency could be used in designing a drug. Few proteins involved in the cause of cancer are selected in this study, and their docking analysis was compared between the interactions of native ligands and quercetin. Among the ten proteins, quercetin was best interacting with Bcl-2 protein of G, with a score value −5.74 Kcal/mol and 9 number of interaction. Molecular dynamics analysis also indicated the stability of Bcl-2–quercetin complex at 74th sample among 100 samples generated during the simulation period of 1 ns. The reduction in potential energy was observed from −14658.9 ê to −14707.3 ê of 0–100th sample.

13.1 INTRODUCTION

13.1.1 IMPORTANCE OF CANCER SPECIFIC PROTEIN

13.1.1.1 ACUTE LYMPHOBLASTIC LEUKEMIAS

Acute lymphoblastic leukemias (ALL) is a form of leukemia or cancer of the white blood cells characterized by excess lymphoblasts. Malignant, immature white blood cells continuously multiply and overproduce in the bone marrow causing damage and death by crowding out normal cells in the bone marrow, and by spreading (infiltrating) to other organs. ALL is most common in childhood with a peak incidence at 2–5 years of age, and another peak in old age. The overall cure rate in children is about 80%, and about 45–60% of adults have long-term, disease-free survival.[69]

In general, cancer is caused by damage to DNA that leads to uncontrolled cellular growth and spread throughout the body, either by increasing chemical signals that cause growth, or interrupting chemical signals that control growth.[73] Damage can be caused through the formation of fusion genes, as well as the deregulation of a proto-oncogene via juxtaposition of the promoter of another gene, for example, the T-cell receptor gene. This damage may be caused by environmental factors such as chemicals, drugs, or radiation.

13.2.1 SYMPTOMS AND TREATMENT

The signs and symptoms of ALL are variable but follow from bone marrow replacement and/or organ infiltration. It causes generalized weakness and fatigue; anemia; frequent or unexplained fever and infection; weight loss and/or loss of appetite; excessive and unexplained bruising; bone pain, joint pain (caused by the spread of "blast" cells to the surface of the bone or into the joint from the marrow cavity); breathlessness; enlarged lymph nodes, liver and/or spleen; pitting edema (swelling) in the lower limbs and/or abdomen and petechia, which are tiny red spots or lines in the skin due to low platelet levels.

Prednisolone an active metabolite of prednisone which is a corticosteroid drug with predominant glucocorticoid receptor (GR), and low mineralocorticoid activity, making it useful for the treatment of a wide range of inflammatory and autoimmune conditions such as asthma ulcerative colitis, temporal arteritis and Crohn's disease, Bell's palsy, multiple sclerosis, cluster headaches, vasculitis, ALL and autoimmune hepatitis, systemic lupus erythematosus, and dermatomyositis.[16] With high affinity prednisolone irreversibly binds GR α and β. Virtually α-GR and β-GR are found in all tissues varies in numbers between 3000 and 10,000 per cell depending on the tissue involved.[39] Prednisolone can activate and influence biochemical behavior of most cells. The steroid/receptor complexes dimerize and interact with cellular DNA in the nucleus, binding to steroid-response elements and modifying gene transcription. They both induce synthesis of some genes and therefore some proteins and inhibit synthesis of others.[60]

Resistance or sensitivity to Glucocorticoids is considered to be of crucial importance for disease prognosis in childhood ALL. Prednisolone exerted a delayed biphasic effect on the resistant CCRF-CEM leukemic cell line, necrotic at low doses and apoptotic at higher doses. At low doses, prednisolone exerted a predominant mitogenic effect despite its induction on total cell death, while at higher doses, prednisolone's mitogenic, and cell death effects were counter balanced. Early gene microarray analysis revealed notable differences in 40 genes. The mitogenic/biphasic effects of prednisolone are of clinical importance in the case of resistant leukemic cells. This approach might lead to the identification of gene candidates for future molecular drug targets in combination therapy with glucocorticoids, along with early markers for glucocorticoid resistance.[39]

13.2.2   GLUCOCORTICOID RECEPTORS

The GR also known as NR3C1 (nuclear receptor subfamily 3, group C, and member 1) is a receptor that cortisol and other glucocorticoids bind to. The GR is expressed in almost every cell in the body which regulates genes controlling the development, metabolism, and immune response.[45] Because the receptor gene is expressed in several forms, it has many different (pleiotropic) effects in different parts of the body.[65] Like the other steroid receptors, the GR is modular in structure and contains the following domains (labeled A–F):

- A/B—N-terminal regulatory domain
- C—DNA-binding domain (DBD)
- D—hinge region
- E—ligand-binding domain (LBD)
- F—C-terminal domain

13.3   BREAST CANCER

Breast cancer also known as malignant breast neoplasm cancer originates from breast tissue, most commonly from the inner lining of milk ducts or the lobules that supply the ducts with milk. Cancers originating from ducts are known as ductal carcinomas whereas those originating from lobules are known as lobular carcinomas.[65] The size, stage, rate of growth and other characteristics of the tumor determine the kinds of treatment. Treatment may include surgery, drugs (hormonal therapy and chemotherapy), radiation, and/or immunotherapy. Surgical removal of the tumor provides the single largest benefit, with surgery alone being capable of producing a cure in many cases. To somewhat increase the likelihood of long-term, disease-free survival, several chemotherapy regimens are commonly given in addition to surgery.[21] Most forms of chemotherapy kill cells that are dividing rapidly anywhere in the body and as a result cause temporary hair loss and digestive disturbances. Radiation may be added to kill any cancer cells in the breast that were missed by the surgery which usually extends survival somewhat, although radiation exposure to the heart may cause heart failure in the future. Some breast cancers are sensitive to hormones such as estrogen and/or progesterone, which make it possible to treat them by blocking the effects of this hormones.[6]

Prognosis and survival rate varies and it greatly depends on the cancer type and staging. With best treatment and dependent on staging, 5-year relative survival varies from 98 to 23% with an overall survival rate of 85% (Cancer World Report 2008). Worldwide, breast cancer comprises of 22.9% of all non-melanoma skin cancers in women. In 2008, breast cancer caused 458,503 deaths worldwide, 13.7% of cancer deaths in women alone which was 100 times more common than men, although males tend to have poorer outcomes due to delays in diagnosis. The first noticeable symptom of breast cancer is typically a lump that feels different from the rest of the breast tissue. More than 80% of breast cancer cases are discovered when the woman feels a lump.[38] The earliest breast cancers are detected by a mammogram and lumps found in lymph nodes located in the armpits can also indicate breast cancer. Indications of breast cancer other than a lump may include changes in breast size or shape, skin dimpling, nipple inversion, or spontaneous single-nipple discharge.

Occasionally, breast cancer occur as metastatic disease, that is, cancer that has spread beyond the original organ which cause symptoms depending on the location of metastasis where common sites include bone, liver, lung, and brain. Unexplained weight loss can occasionally herald an occult breast cancer with symptoms of fevers or chills and sometimes with bone or joint, jaundice or neurological symptoms.[25] These symptoms are "nonspecific" as they can also be manifested in many other illnesses. Most symptoms of breast disorder do not turn out to represent underlying breast cancer. Benign breast diseases such as mastitis and fibroadenoma of the breast are more common causes of breast disorder symptoms.[82] The appearance of a new symptom should be taken seriously by both patients and their doctors, because of the possibility of an underlying breast cancer at almost any age.

13.3.1   HORMONE BLOCKING THERAPY

Some breast cancers require estrogen to continue growing. They can be identified by the presence of estrogen Receptors (ER) and progesterone Receptors (PR) on their surface (sometimes referred together as hormone receptors).[14] These ER cancers can be treated with drugs that either block the receptors, for example, tamoxifen, or alternatively block the production of estrogen.

13.3.2 ESTROGEN RECEPTOR

ER refers to a group of receptors that are activated by the hormone 17β-estradiol (estrogen). Two types of ER exist as ER, which is a member of the nuclear hormone family of intracellular receptors and as estrogen G protein-coupled receptor GPR30 (GPER) which is a G protein-coupled receptor. ERs are widely expressed in different tissue types; however, exhibits some notable differences in their expression patterns in which ERα is found in breast cancer cells respectively.[41] Tamoxifen is an antagonist in breast and is therefore used in breast cancer treatment.

13.3.3 ACTION OF TAMOXIFEN IN ESTROGEN RECEPTOR

Not all breast cancers responds to tamoxifen and many develop resistance despite initial benefit. Massarweh et al.[49] used an in vivo model of ER-positive breast cancer (MCF-7 xenografts) to investigate mechanisms of this resistance and developed strategies to circumvent it. Epidermal growth factor receptor (EGFR) and EGFR 2 (HER2), which were barely detected in controlling estrogen-treated tumors, increased slightly with tamoxifen and were markedly increased when tumors became resistant. Gefitinib, which inhibits EGFR/HER2, improved the antitumor effect of tamoxifen and delayed acquired resistance, but had no effect on estrogen-stimulated growth. Phosphorylated levels of *p42/44* and *p38* mitogen-activated protein kinases (both downstream of EGFR/HER2) were increased in the tamoxifen-resistant tumors and were suppressed by gefitinib. There was no apparent increase in phosphorylated AKT (also downstream of EGFR/HER2) in resistant tumors, but it was nonetheless suppressed by gefitinib. Phosphorylated insulin-like growth factor-IR (IGF-IR), which can interact with both EGFR and membrane ER, was elevated in the tamoxifen-resistant tumors compared with the sensitive group. However, ER-regulated gene products, including total IGF-IR itself and progesterone receptor remained suppressed even at the time of acquired resistance. Tamoxifen's antagonism of classic ER genomic function was retained in these resistant tumors and even in tumors that over expressed HER2 (MCF-7 HER2/18). In conclusion, EGFR/HER2 may mediate tamoxifen resistance in ER-positive breast cancer despite continued suppression of ER genomic function by tamoxifen. IGF-IR expression remains dependent on ER, but is activated in the tamoxifen-resistant tumors. Their study provides a rationale to combine HER inhibitors with tamoxifen in clinical studies, even in tumors that do not initially over express EGFR/HER2.[49]

13.4 CNS LYMPHOMA

A primary central nervous system lymphoma (PCNSL) also known as microglioma and primary brain lymphoma, is a primary intracranial tumor appearing mostly in patients with severe immunosuppression (typically patients with AIDS)[20]. PCNSLs represent around 20% of all cases of lymphomas in HIV infections (other types are Burkitt's lymphomas and immunoblastic lymphomas). Primary CNS lymphoma (PCNSLs) is highly associated with Epstein-Barr virus (EBV) infection (>90%) in immunodeficient patients (such as those with AIDS and those iatrogenically immunosuppressed) and does not have a predilection for any particular age group. In the immunocompetent population, PCNSLs typically appear in older patients in their 50s and 60s.[18] Importantly, the incidence of PCNSL in the immunocompetent population has been reported to have increased more than 10-fold from 2.5 cases to 30 cases per 10 million populations.[12]

Surgical resection is usually ineffective because of the depth of the tumor. Treatment with irradiation and corticosteroids often only produces a partial response and tumor recurs in more than 90% of patients. Median survival is 10–18 months in immunocompetent patients, and less in those with AIDS. The addition of IV methotrexate and folinic acid (leucovorin) may extend survival to a median of 3.5 years. If radiation is added to methotrexate, median survival time may increase beyond four years.[30] However, radiation is not recommended in conjunction with methotrexate because of an increased risk of leukoencephalopathy and dementia in patients older than 60. Dihydrofolate reductase (DHFR) is the intracellular protein involves in homocysteine remethylation in CNS through the synthesis of 5, 10-methyl-tetra-hydrofolate reductase which imparts both nucleic acid synthesis and homocysteine remthylation via 5, 10-methyltetrahydrofolate reductase and 5-methyltetrahydrofolate-homocysteine S-methyltransferase (MTR).[50]

13.4.1 DIHYDROFOLATE REDUCTASE

(DHFR) is an enzyme that reduces dihydrofolic acid to tetrahydrofolic acid using NADPH as electron donor which can be converted to the kinds of tetrahydrofolate cofactors used in 1-carbon transfer chemistry. In humans, the DHFR enzyme is encoded by the DHFR gene[10] found in all organisms, DHFR has a critical role in regulating the amount of tetrahydrofolate in the cell. Tetrahydrofolate and its derivatives are essential for purine and thymidylate synthesis which is important for cell proliferation and cell growth.[9]

DHFR plays a central role in the synthesis of nucleic acid precursors, and it has been shown that mutant cells that completely lack DHFR require glycine, a purine, and thymidine to grow.[23]

13.4.2 METHOTREXATE

Methotrexate (MTX) formerly known as amethopterin, is an antimetabolite and antifolate drug. It is used in the treatment of cancer, autoimmune diseases, ectopic pregnancy, and for the induction of medical abortions by inhibiting the metabolism of folic acid.[54] It competitively inhibits DHFR, an enzyme that participates in the tetrahydrofolate synthesis. The affinity of methotrexate for DHFR is about one thousand-fold that of folate.[67] DHFR catalyzes the conversion of dihydrofolate to the active tetrahydrofolate. Folic acid is needed for the de novo synthesis of the nucleoside thymidine, required for DNA synthesis. Also folate is needed for purine base synthesis therefore purine synthesis will be inhibited. Therefore, MTX inhibits the synthesis of DNA, RNA, thymidylates, and proteins.[58]

13.5 LUNG CANCER

Lung cancer is a disease that consists of uncontrolled cell growth in tissues of the lung. This growth may lead to metastasis which is the invasion of adjacent tissue and infiltration beyond the lungs.[52] The vast majority of primary lung cancers are carcinomas derived from epithelial cells and it is the most common cause of cancer-related death in men and women which is roughly estimated to cause about 1.3 million deaths worldwide.

Lung cancer may be seen on chest radiograph and computed tomography (CT scan). The diagnosis is confirmed through biopsy where treatment and prognosis depend on the histological type of cancer, the stage (degree of spread) and the patient's performance status.[77] Possible treatments include surgery, chemotherapy, and radiotherapy; however, survival depends on stage, overall health, and other factors, but overall only 14% of people diagnosed with lung cancer survive five years after the diagnosis. Symptoms include dyspnea (shortness of breath), hemoptysis (coughing up blood), chronic coughing or change in regular coughing pattern, wheezing, chest pain or pain in the abdomen, cachexia (weight loss), fatigue and loss of appetite, dysphonia (hoarse voice), clubbing of the fingernails (uncommon), and dysphagia (difficulty swallowing).[26]

Many of the symptoms of lung cancer (bone pain, fever, and weight loss) are nonspecific; in the elderly, these may be attributed to comorbid illness. In many patients, the cancer has already spread beyond the original site by the time they have symptoms and seek medical attention. Common sites of metastasis include the brain, bone, adrenal glands, contralateral (opposite) lung, liver, pericardium, and kidneys.[8] The Bcl-2 gene has been implicated in lung carcinomas as well as schizophrenia and autoimmunity. It is also thought to be involved in resistance to conventional cancer treatment. This supports a role for decreased apoptosis in the pathogenesis of cancer.

13.5.1   BCL-2

Bcl-2 (B-cell lymphoma 2) is the founding member of Bcl-2 family of apoptosis regulator proteins encoded by the BCL-2 gene. Bcl-2 derives its name from B-cell lymphoma 2 as it is the second member of a range of proteins initially described in chromosomal translocations involving chromosomes 14 and 18 in follicular lymphomas. BCL-2 protein located on outer membrane of mitochondria, consisting 239 amino acids which suppress apoptosis, regulates cell death by controlling the mitochondrial membrane permeability. It also inhibits caspase activity either by preventing the release of cytochrome c from the mitochondria and/or by binding to the apoptosis activating factor (APAF-1). Hypermethylation at the fifth carbon of the cytosine residues would lead to the origin of cancer.[66]

13.5.2   DOCETAXEL

Docetaxel (as generic or with trade name Taxotere) is a clinically well-established anti-mitotic chemotherapy medication (it interferes with cell division). It is used mainly for the treatment of breast, ovarian and non-small cell lung cancer.[11] Docetaxel has an FDA approved claim for treatment of patients who have locally advanced, or metastatic breast or non small-cell lung cancers that have undergone anthracycline-based chemotherapy and failed to stop cancer progression or relapsed (US-FDA, 2006). Docetaxel is of the chemotherapy drug class; taxane, and is a semi-synthetic analogue of paclitaxel (Taxol), an extract from the bark of the rare Pacific yew tree *Taxus brevifolia*.[11] Due to scarcity of paclitaxel, extensive research was carried out leading to the synthesis of docetaxel—an esterified product of 10-deacetyl baccatin III, which is extracted from the renewable and readily available European yew tree.

13.5.3 ACTION OF DOCETAXEL IN LUNG CANCER

Docetaxel affect microtubule polymerization and surprisingly differences in tumor sensitivity to the taxanes have also been observed. Docetaxel was superior to paclitaxel in inhibiting in vivo growth of human lung and prostate but not breast cancer models. They compared drug cytotoxicity effects on tubulin isoforms, markers of apoptosis and proteomic profiles in human prostate (LNCaP), lung (SK-MES, MV-522) and breast (MCF-7, MDA-231) cancer cell lines in vitro. Cytotoxicity was found in the order of SK-MES < MV-522 < LNCaP < MCF-7 < MDA-MB-231 in which docetaxel was more effective. Cytotoxicity was directly correlated with Bcl-2 expression in vitro whereas inversely correlated with docetaxel sensitivity in vivo. Proteomic profiling identified a protein expressed in lung and prostate cells, which was differentially regulated by docetaxel and paclitaxel in SK-MES. The superior activity of docetaxel in tumors with low Bcl-2 warrants further studies on biomarkers for drug sensitivity and investigation of docetaxel in combination with drugs that reduce Bcl-2 gene expression.[32]

13.6 ORAL CANCER

Oral cancer is a subtype of head and neck cancer said to be arising as a primary lesion originating in any of the oral tissues, by metastasis from a distant site of origin or by extension from a neighboring anatomic structure, such as the nasal cavity. or the It may originate in the tissues of the mouth and may be of varied histologic types such as teratoma, adenocarcinoma derived from a major or minor salivary gland, lymphoma from tonsillar or other lymphoid tissue, or melanoma from the pigment producing cells of the oral mucosa.[80] There are several types of oral cancers, but around 90% are squamous cell carcinomas (SCC) originating in the tissues that line the mouth and lips. Oral or mouth cancer most commonly involves the tongue and it may also occur on the floor of the mouth, cheek lining, gingiva (gums), lips or palate (roof of the mouth). Most oral cancers look very similar under the microscope and are called SCC.[64]

Skin lesion, lump, or ulcer on the tongue, lip or other mouth areas is usually small and most often look pale colored, dark or discolored where early sign may be a white patch (leukoplakia) or a red patch (erythroplakia) on the soft tissues of the mouth which are initially painless may develop with burning sensation or pain when the tumor is advanced. Additional symptoms that may be associated with this disease are tongue problems, swallowing difficulty, mouth sores. EGFR is overexpressed in the cells of certain types of human carcinomas—for example in oral and breast cancers, whereas high expression of EGFR is frequently observed in many solid tumor types including oral SCC. This study investigated whether treatment with gefitinib would inhibit the metastatic spread in OSCC cells. This was evaluated using orthotopic xenografts of highly metastatic OSCC. There were observed Metastasis in six of 13 gefitinib treated animals (46.2%), compared with all of 12 control animals (100%). After exposure to gefitinib, OSCC cells showed a marked reduction in cell adhesion ability to fibronectin and in the expression of integrin α3, α, β1, β 4, β5, and β6.[68]

13.6.1 EPIDERMAL GROWTH FACTOR

Epidermal growth factor (EGF) is a growth factor that plays an important role in the regulation of cell growth, proliferation, and differentiation by binding to its receptor EGFR. Human EGF is a 6045-Da protein with 53 amino acid residues and three intramolecular disulfide bonds.[5]

EGF is a low-molecular-weight polypeptide first purified from the mouse submandibular gland, but since then found in many human tissues including submandibular gland and parotid gland.[12] Salivary EGF is regulated by dietary inorganic iodine which plays an important physiological role in the maintenance of oro-esophageal and gastric tissue integrity[29]. The biological effects of salivary EGF includes healing of oral and gastroesophageal ulcers, inhibition of gastric acid secretion, stimulation of DNA synthesis as well as mucosal protection from intraluminal injurious factors such as gastric acid, bile acids, pepsin and trypsin and to physical, chemical and bacterial agents.[78]

13.6.2 GEFITINIB

Gefitinib is a drug used in the treatment of certain types of cancer and like erlotinib it is an EGFR inhibitor, which interrupts signaling through the EGFR in target cells. Gefitinib is the first selective inhibitor of EGFR's tyrosine kinase domain.[75] The target protein (EGFR) is also sometimes referred to as Her1 or ErbB-1 depending on the literature source. This leads to inappropriate activation of the anti-apoptotic Ras signaling cascade, eventually leading to uncontrolled cell proliferation.[56] Research on gefitinib-sensitive, non-small cell lung cancers has shown that a mutation in the EGFR tyrosine kinase domain is responsible for activating anti-apoptotic pathways. These mutations tend to confer increased sensitivity to tyrosine kinase inhibitors such as gefitinib and erlotinib.[70] Gefitinib inhibits EGFR tyrosine kinase by binding to the adenosine triphosphate (ATP)-binding site of the enzyme.[53] Thus, the function of the EGFR tyrosine kinase in activating the anti-apoptotic Ras signal transduction cascade is inhibited, and malignant cells are inhibited.

13.7　PROSTATE CANCER

Prostate cancer is a form of cancer that develops in the prostate gland of the male reproductive system. Most prostate cancers are slow growing; however, there are cases of aggressive prostate cancers. The cancer cells may metastasize (spread) from the prostate to other parts of the body, particularly the bones and lymph nodes. Prostate cancer may cause pain, difficulty in urinating, problems during sexual intercourse or erectile dysfunction. Other symptoms can potentially develop during later stages of the disease.

Rates of detection of prostate cancers vary widely across the world, with South and East Asia detecting less frequently than in Europe and especially the United States. Prostate cancer tends to develop in men over the age of fifty and although it is one of the most prevalent types of cancer in men, many never have symptoms, undergo no therapy, and eventually die of other causes.[76] This is because cancer of the prostate is, in most cases, slow-growing, symptom-free, and since men with the condition are older they often die of causes unrelated to the prostate cancer, such as heart/circulatory disease, pneumonia, other unconnected cancers, or old age. About 2/3 of cases are slow growing, the other third more aggressive and fast developing. Many factors, including genetics and diet, have been implicated in the development of prostate cancer. The presence of prostate cancer may be indicated by symptoms, physical examination, prostate-specific antigen (PSA), or biopsy.[17]

Early prostate cancer usually causes no symptoms. Often it is diagnosed during the workup for an elevated PSA noticed during a routine checkup. Its changes within the gland, therefore, directly affect urinary function. Because the "vas deferens" deposits seminal fluid into the prostatic urethra, and secretions from the prostate gland itself are included in semen content, prostate cancer may also cause problems with sexual function and performance, such as difficulty in achieving erection or painful ejaculation.[51] Advanced prostate cancer can spread to other parts of the body, possibly causing additional symptoms. The most common symptom is bone pain, often in the vertebrae (bones of the spine), pelvis, or ribs. Spread of cancer into other bones such as the femur is usually to the proximal part of the bone. Prostate cancer in the spine can also compress the spinal cord, causing leg weakness and urinary and fecal incontinence.

The androgen receptor (AR) helps prostate cancer cells to survive and is a target for many anti cancer research studies; so far, inhibiting the AR has only proven to be effective in mouse studies. Prostate specific membrane antigen (PSMA) stimulates the development of prostate cancer by increasing folate levels for the cancer cells to survive and grow; PSMA increases available folates for use by hydrolyzing glutamated folates.[83] Andarine is an orally active partial agonist for AR It has been shown to inhibit the development, progression, and metastasis as well in autochthonous transgenic adenocarcinoma of the mouse prostate (TRAMP) model, which spontaneously develops prostate cancer.[27]

13.7.1   ANDROGEN RECEPTOR

The AR also known as NR3C4 (nuclear receptor subfamily 3, group C, member 4) is a type of nuclear receptor that is activated by binding either of the androgenic hormones testosterone or dihydrotestosterone in the cytoplasm and then translocating into the nucleus. The AR is most closely related to the progesterone receptor and the progestin in higher dosages has the ability to block AR.[61] Androgens cause slow epiphysis or maturation of the bones, but more of the potent epiphysis effect comes from the estrogen produced by aromatization of androgens. Steroid users of teenage may find that their growth had been stunted by androgen and/or estrogen excess.[22]

13.7.2   ANDARINE

Andarine (GTx-007, S-4) is an investigational selective AR modulator (SARM) developed by GTX Inc. for treatment of conditions such as muscle wasting, osteoporosis and benign prostatic hypertrophy, using the non-steroidal androgen antagonist bicalutamide as a lead compound. Andarine is an orally active partial agonist for androgen receptors. It is less potent in both anabolic and androgenic effects than other SARMs. Perhaps, in an animal model of benign prostatic hypertrophy, andarine was shown to reduce prostate weight with similar efficacy to finasteride, but without producing any reduction in muscle mass or anti-androgenic side effects. This suggests that it is able to competitively block binding of dihydrotestosterone to its receptor targets in the prostate gland, but its partial agonist effects at AR prevent the side effects associated with the anti-androgenic drugs traditionally used for treatment of benign prostatic hyperplasis (BPH).

13.8 RECTAL CANCER

Colorectal cancer, less formally known as bowel cancer, is a cancer characterized by neoplasia in the colon, rectum, or vermiform appendix. Colorectal cancer is clinically distinct from anal cancer, which affects the anus. Colorectal cancers start in the lining of the bowel. If it left untreated, it could grow into the muscle layers underneath, and then through the bowel wall. Most begin as a small growth on the bowel wall: a colorectal polyp or adenoma. These mushroom-shaped growths are usually benign, but some develop into cancer over time. Localized bowel cancer is usually diagnosed through colonoscopy. Invasive cancers that are confined within the wall of the colon (Tetranitromethane (TNM) stages I and II) are often curable with surgery. Colorectal cancer is the third most commonly diagnosed cancer in the world, but it is more common in developed countries. More than half of the people who die of colorectal cancer live in a developed region of the world (http://globocan.iarc.fr/). GLOBOSCAN estimated that, in 2008 reported 1.23 million cases with colorectal cancer of which more than 600,000 people died.

13.8.1 SYMPTOMS

Symptoms include blood in the stool, narrower stools, a change in bowel habits and general stomach discomfort. As an anti-Rectal cancer, chemotherapy targets thymidylate synthase (TS) such as fluorinated pyrimidine fluorouracil or certain folate analogues to inhibit rectal cancer.

13.8.2   THYMIDYLATE SYNTHASE

TS is an enzyme used to generate thymidine monophosphate (dTMP), which is subsequently phosphorylated to thymidine triphosphate for use in DNA synthesis and repair. By means of reductive methylation, deoxyuridine monophosphate (dUMP) and N5, N10-methylene tetrahydrofolate together used formed dTMP, yielding dihydrofolate as a secondary product. This provided the sole de novo pathway for production of dTMP and therefore is the only enzyme in folate metabolism in which the 5, 10-methylenetetrahydrofolate is oxidized during one-carbon transfer.[28]

The enzyme is essential for regulating the balanced supply of the four DNA precursors in normal DNA replication. However, defects in the enzyme activity, affects the regulation process causing various biological and genetic abnormalities such as thymineless death.[35] is an important target for certain chemotherapeutic drugs. Therefore it is about 30–35 KDa in most species except in protozoan and plants where it exists as a bifunctional enzyme that includes a DHFR domain. A cysteine residue is involved in the catalytic mechanism (it covalently binds the 5, 6-dihydro-dUMP intermediate). The sequence around the active site of this enzyme is conserved from phages to vertebrates.

13.8.3   FOLINIC ACID

Folinic acid or leucovorin generally administered as calcium or sodium folinate (or leucovorin calcium/sodium) is an adjuvant used in cancer chemotherapy involving the drug methotrexate which works in synergy combination with the chemotherapy, 5-fluorouracil. Folinic acid is a 5-formyl derivative of tetrahydrofolic acid and is readily converted to other reduced folic acid derivatives (e.g., tetrahydrofolate) and therefore vitamin activity is equivalent to folic acid. Since it does not require the action of TS for its conversion, its function as a vitamin is unaffected by inhibition of this enzyme by drugs such as methotrexate. Therefore folinic acid allows purine/pyrimidine synthesis to occur in the presence of DHFR inhibition, so that some normal DNA replication and RNA transcription processes can proceed without interruption. Folate metabolism supports the synthesis of nucleotides as well as the transfer of methyl groups. Polymorphisms in folate-metabolizing enzymes have been shown to affect risk of colorectal neoplasia and other malignancies.

13.9   RENAL CANCER

Renal cell carcinoma (RCC, also known as hypernephroma) is a kidney cancer that originates in the lining of the proximal convoluted tubule, the very small tubes in the kidney that filter the blood and remove waste products. RCC is the most common type of kidney cancer in adults, responsible for approximately 80% of cases.[55] It is also known to be the most lethal of all the genitourinary tumors. Initial treatment is most commonly a radical or partial nephrectomy and remains the mainstay of curative treatment.[63] Where the tumor is confined to the renal parenchyma, 5-year survival rate is 60–70%, but this is lowered considerably where metastases have spread. It is resistant to radiation therapy and chemotherapy, although some cases respond to immunotherapy. Targeted cancer therapies such as sunitinib, temsirolimus, bevacizumab, interferon-alpha, and possibly sorafenib have improved the outlook for RCC (progression-free survival), although they have not yet demonstrated improved survival. It is reported that in a population of 58,240 (35,370 men and 22,870 women) around 13,040 men and women died of kidney and renal cancer.[33]

A wide range of symptoms have been encountered with renal carcinoma, depending on the areas of body it affects. The classic triad is hematuria (blood in the urine), flank pain, and an abdominal mass. This triad only occurs in 10–15% of cases, and is generally indicative of more advanced disease. Today, the majorities of renal tumors are asymptomatic and are detected incidentally on imaging, usually for an unrelated cause. Signs may include abnormal urine color (dark, rusty, or brown) due to blood in the urine (found in 60% of cases), joint pain (found in 40% of cases), Abdominal mass (25% of cases), Malaise, weight loss or anorexia (30% of cases), Polycythemia (5% of cases) and anemia resulting from depression of erythropoietin (30% of cases).[37] Also, there may be erythrocytosis (increased production of red blood cells) due to increased erythropoietin secretion.

The presenting symptom may be due to metastatic disease, such as a pathologic fracture of the hip due to a metastasis to the bone Varicocele, the enlargement of one testicle, usually on the left (2% of cases). This is due to blockage of the left testicular vein by tumor invasion of the left renal vein; this typically does not occur on the right as the right gonadal vein drains directly into the inferior vena cava. Vision abnormalities, pallor or plethora, Hirsutism—Excessive hair growth (females), constipation and hypertension (high blood pressure) resulting from secretion of renin by the tumor (30% of cases), elevated calcium levels (hypercalcemia), Stauffer syndrome—paraneoplastic, non-metastatic liver disease, night sweats and severe weight Loss.

13.9.1 mTOR KINASE

The mammalian target of rapamycin (mTOR) also known as mechanistic target of rapamycin or FK506 binding protein 12-rapamycin associated protein 1 (FRAP1) is a protein encoded by FRAP1 gene. mTOR is a serine/threonine protein kinase that regulates cell growth, cell proliferation, cell motility, cell survival, protein synthesis, and transcription. mTOR belongs to the phosphatidylinositol 3-kinase-related kinase protein family. mTOR integrates the input from upstream pathways, including insulin, growth factors (such as IGF-1 and IGF-2), and amino acids. mTOR also senses cellular nutrient and energy levels and redox status. The mTOR pathway is dysregulated in human diseases, especially in certain cancers. Rapamycin is a bacterial product that can inhibit mTOR by associating with its intracellular receptor FKBP12. The FKBP12–rapamycin complex binds directly to the FKBP12-rapamycin binding (FRB) domain of mTOR. mTOR is the catalytic subunit of two molecular complexes.[4]

13.9.2 TORISEL

Temsirolimus (CCI-779) is an intravenous drug for the treatment of RCC, developed by Wyeth Pharmaceuticals and approved by the U.S. Food and Drug Administration (FDA)[19] and was also approved by the European Medicines Agency (EMEA) during November 2007. It is a derivative of sirolimus and is sold as torisel. Temsirolimus is a specific inhibitor of mTOR and interferes with the synthesis of proteins that regulate proliferation, growth, and survival of tumor cells. Treatment with temsirolimus leads to cell cycle arrest in the G1 phase, and also inhibits tumor angiogenesis by reducing synthesis of vascular endothelial growth factor (VEGF).[79]

The potent inhibitor of the mTOR is temsirolimus which is comprised for cell cycle, angiogenesis, and proliferation and has been proven beneficial in the treatment of advanced RCC. Temsirolimus is officially approved for first line therapy in high-risk previously untreated mRCC patients.[24]

13.10 SQUAMOUS CELL CARCINOMA

SCC occasionally rendered as "squamous-cell carcinoma", is a histological form of cancer that arises from the uncontrolled multiplication of transformed malignant cells showing squamous differentiation and tissue architecture.

SCC is one of the most common histological forms of cancer, and frequently forms in a large number of body tissues and organs, including skin, lips, mouth, esophagus, urinary bladder, prostate, lung, vagina, and cervix, among others. Despite the common name, SCCs that present in different sites often show large differences in their presentation.[47] It is reported that 42,610 men and 31,400 women around 11,790 men and women died of cancer of the skin in US alone.[7]

Symptoms are highly variable depending on the involved organs. SCC of the skin begins as a small nodule and as it enlarges the center becomes necrotic and sloughs and the nodule turns into an ulcer. The lesion caused by SCC is often asymptomatic, ulcer or reddish skin plaque that is slow growing, intermittent bleeding from the tumor, especially on the lip. The clinical appearance is highly variable; usually the tumor presents an ulcerated lesion with hard, raised edges. The tumor may be in the form of a hard plaque or a papule, often with an opalescent quality, with tiny blood vessels. The tumor can lie below the level of the surrounding skin, and eventually ulcerates and invades the underlying tissue. The tumor grows relatively slowly, unlike basal cell carcinoma (BCC), SCC has a substantial risk of metastasis. Imiquinoid has been used with success for SCC in situ of the skin and the penis, but the morbidity and discomfort of the treatment is severe. An advantage is the cosmetic result: after treatment, the skin resembles normal skin without the usual scarring and morbidity associated with standard excision. Imiquinoid is not FDA-approved for any SCC.

13.10.1   TOLL-LIKE RECEPTOR 7

Toll-like receptor 7 also known as TLR7 is an immune gene possessed by humans and other mammals. The protein encoded by this gene is a member of the Toll-like receptor (TLR) family which plays a fundamental role in pathogen recognition and activation of innate immunity. TLRs are highly conserved from Drosophila to humans and share structural and functional similarities. They recognize pathogen-associated molecular patterns (PAMPs) that are expressed on SCC, and mediate the production of cytokines necessary for the development of effective immunity. The various TLRs exhibit different patterns of expression. This gene is predominantly expressed in lung, placenta and spleen, and lies in close proximity to another family member.[43]

13.10.2   IMIQUIMOD

Imiquinoid marketed by MEDA AB, Graceway Pharmaceuticals and iNova Pharmaceuticals under the trade names Aldara, Zyclara, and Mochida as Beselna activates immune cells through the toll-like receptor 7 (TLR7), commonly involved in pathogen recognition.[36]

   TLR7 activation by imiquimod has pleiotropic effects on innate immune cells, but its effects on T cells remain largely uncharacterized. Because tumor destruction and formation of immunological memory are ultimately T-cell mediated effects, SCC treated with imiquimod before excision contained dense T-cell infiltrates associated with tumor cell apoptosis and histologically evidenced tumor regression. Effector T cells from treated SCC produced more IFN-γ, granzyme, and perforin and less IL-10 and transforming growth factor-β (TGF-β) than T cells from untreated tumors. Treatment of normal human skin with imiquimod induced activation of resident T cells and reduced IL-10 production but had no effect on IFN-γ, perforin, or granzyme, suggesting that these latter effects arise from the recruitment of distinct populations of T cells into tumors. Thus, imiquimod stimulates tumor destruction by recruiting cutaneous effector T cells from blood and by inhibiting tonic anti-inflammatory signals within the tumor.[74]

13.11   THYROID CANCER

Thyroid cancer is a thyroid neoplasm that is malignant. It can be treated with radioactive iodine or chemotherapy. Most often the first symptom of thyroid cancer is a nodule in the thyroid region of the neck. However, many adults have small nodules in their thyroids, but typically fewer than 5% of these nodules are found to be malignant. Sometimes the first sign is an enlarged lymph node, followed by pain in the anterior region of the neck and changes in voice.[31] Thyroid nodules are of particular concern when they are found in those under the age of 20. The presentation of benign nodules at this age is less likely, and thus the potential for malignancy is far greater. It is reported that in 10,740 men and 33,930 women, around 1690 men and women were dead and diagnosed of cancer during 2010. But as the cancer grows, difficulty in swallowing or breathing, a benign goiter, hyperthyroidism, or hypothyroidism will be developed.

13.11.1 RAF PROTEIN

RAF proto-oncogene serine/threonine is a specific protein kinase also known as proto-oncogene c-RAF or simply c-Raf, is an enzyme that in humans is encoded by the RAF1 gene. Protein kinases are overactive in many of the molecular pathways that cause cells to become cancerous. These pathways include Raf kinase, PDGF (platelet-derived growth factor), VEGF receptor 2 and 3 kinases and c Kit the receptor for Stem cell factor. The c-Raf protein functions in the MAPK/ERK signal transduction pathway as part of a protein kinase cascade.[42] functions downstream of the Ras subfamily of membrane associated GTPases to which it binds directly. Once activated Raf-1 is phosphorylated and activate the dual specific protein kinases MEK1 and MEK2, which, in turn, phosphorylate to activate the serine/threonine-specific protein kinases ERK1 and ERK2. Activated ERKs are pleiotropic effectors of cell physiology and play an important role in the control of gene expression involved in the cell division cycle, apoptosis, cell differentiation, and cell migration.[71]

13.11.2 SORAFENIB

Sorafenib show promise for thyroid cancer and are being used for some patients who do not qualify for clinical trials. Numerous agents are in phase II clinical trials and XL184 has started a phase III trial. Sorafenib (a bi-aryl urea) is a small molecular inhibitor of several Tyrosine protein kinases (VEGFR and PDGFR) and Raf. Sorafenib is unique in targeting the Raf/Mek/Erk pathway (MAP Kinase pathway) by inhibiting some intracellular serine/threonine kinases (e.g., Raf-1, wild-type B-Raf and mutant B-Raf).[81].

However, the antiproliferative activity of sorafenib varies widely depending on the oncogenic signaling pathways driving proliferation. Sorafenib has also been shown recently to sequester Raf-1 and B-Raf in a stable inactive complex in treated tumor cell lines expressing wild-type B-Raf. This alteration of Raf-1 protein complexes by sorafenib may result in perturbation of other Raf-1 complexes with MST-1 and ASK-1, which are involved in thyroid cell survival signaling mechanisms.[81] In a netshield, over ten different types of cancers (ALL, breast cancer, CNS lymphoma, lung, oral, prostate, renal, rectal cancer, SCC and thyroid cancer) with their targets (GR, ER, dihydrofolate reductase (DHFR), Bcl-2 Protein, EGR, AR, mTOR, TS, TLR-7, RAF protein) and their drugs which are in market have been identified. Since, quercetin a naturally extracted and purified small molecule have been achieved and projected over the previous chapters have to be evaluated for its efficiency as an anti-cancer agent, it was further examined with the above objective and compared against the synthetic molecules (Drug in use) over in silico approaches.

Bioinformatics is the combination of biology and information technology. It records, annotates, stores, analyzes, and searches/retrieves the nucleic acid sequence, protein sequence and structures that provide new biological insights. Molecular docking plays an important role in analyzing the inhibitory action and interaction studies of small ligand molecules with protein structures[46] and it is a key tool in structural molecular biology and computer-assisted drug design. The goal of ligand–protein docking is to predict the predominant binding models of a ligand with a protein of known three-dimensional structure and to predict the affinity and activity of the small molecule in order to study whether it forms a stable complex. Therefore docking plays an important role in the rational drug designing. The result obtained from the docking study would be useful in both understanding the inhibitory mode as well as in rapidly and accurately predicting the activities of newly designed inhibitors on the basis of docking scores.[72] Hence, the models also provide some beneficial clues in structural modification for designing new inhibitors for the treatment of diseases such as cancer with much higher inhibitory activities against cancer inducing proteins. In our study, Molecular Interaction profile of the target proteins with the ligand obtained from the herb *E. littorale* was compared against the already available synthetic ligands. In fact, finally molecular dynamics (MD) of best G-scored complex was alone done for stabilization.

13.12 ADME-TOX PROPERTIES

ADME-Tox refers to absorption, distribution, metabolism, excretion, and toxicity properties of failures for candidate molecules in drug design. The early evaluation of these properties during drug design could save time and money because certain properties make a drug different from other compounds. An appropriate concentration of the drug must circulate in the body for a reasonable length of time to achieve a desired beneficial effect with minimum adverse effects. For this process, oral drugs have to dissolve or suspend in the gastrointestinal tract and be absorbed through the gut wall to reach the blood stream though the liver. From there, the drug will be distributed to various tissues and organs and finally binds to its molecular target and exert its desired action. The drug is then subjected to hepatic metabolism followed by its elimination as bile or *via* the kidneys. Several pharmacokinetics properties are involved in this mechanism.

Bioavailability depends on absorption and liver first-pass metabolism. The volume of its distribution, together with its clearance rate, determines the half-life of a drug and therefore its dosage. Poor biopharmaceutical properties, such as poor aqueous solubility and slow dissolution rate can lead to poor oral absorption and hence low oral bioavailability. The conversion of active compounds into qualified clinical candidates has proved to be a challenge. At the molecular level, a coordinated system of transporters, channels, receptors, and enzymes act as gatekeepers to foreign molecules affecting the ADME-Tox properties of a given molecule in very different ways. The optimal approach for the ADME-Tox support of discovery will be one that uses both in vitro and in silico experiments in a complementary way ensuring that ADME-Tox is used and considered at almost every stage of the discovery process, from hit identification to lead optimization.[85]

In the hit identification stage, the primary goal of the in silico ADME-Tox models is to identify compounds or series of compounds with at least acceptable drug-like properties that are then disregarded. Another goal is to identify potential weaknesses and liabilities in the selected series highlighting the issues that will be focused in the improvement/optimization efforts. In the lead identification stage, the objective is to identify a small number of chemical series with the activity, selectivity, and drug-like properties required for a potential candidate. The application of in silico ADME-Tox should focus on predictions of chemical modifications of compounds that will improve ADME-Tox properties. In vitro assays are used to measure the ADME-Tox properties of the newly synthesized compounds. This information is valuable for the refinement of the in silico ADME-Tox models. Similar to lead identification, the lead optimization on ADME-Tox properties (Table 13.1) consist of an iterative workflow, starting from in silico prediction, to chemical synthesis, to experimental testing and confirmation, and to model refinement.[2]

Lipinski's rule of five is a rule of thumb to evaluate drug likeness, or determine if a chemical compound with a certain pharmacological or biological activity has properties that would make it a likely orally active drug in humans. The rule was formulated by Christopher A. Lipinski in 1997, based on the observation that most medication drugs are relatively small and lipophilic molecules.[44] Lipinski's rule says that, in general, an orally active drug has no more than one violation of the following criteria:

✓ Not more than five hydrogen bond donors (nitrogen or oxygen atoms with one or more hydrogen atoms)

TABLE 13.1 ADME/Tox Properties Description by Qikprop.

Property or descriptor	Description	Range or recommended values
mol_MW	Molecular weight of the molecule	130.0–725.0
SASA	Total solvent accessible surface area (SASA) in square angstroms using a probe with a 1.4 Å radius	0.0–450.0
Volume	Total solvent-accessible volume in cubic angstroms using a probe with a 1.4 Å radius	500.0–2000.0
DonorHB	Estimated number of hydrogen bonds that would be donated by the solute to water molecules in an aqueous solution. Values are averages taken over a number of configurations, so they can be non-integer	0.0–6.0
AccptHB	Estimated number of hydrogen bonds that would be accepted by the solute from water molecules in an aqueous solution. Values are averages taken over a number of configurations, so they can be non-integer	2.0–20.0
QP logPC 16	Predicted hexadecane/gas partition coefficient	4.0–18.0
QP logP oct	Predicted octanol/gas partition coefficient	8.0–35.0
QP logP w	Predicted water/gas partition coefficient	4.0–45.0
QP logP o/w	Predicted octanol/water partition coefficient	–2.0–6.5
QP logS	Predicted aqueous solubility, log S. S in mol dm–3 is the concentration of the solute in a saturated solution that is in equilibrium with the crystalline solid	–6.5–0.5
Metab	Number of likely metabolic reactions	1–8
Human oral ab-sorption	Predicted qualitative human oral absorption: 1, 2, or 3 for low, medium, or high. The text version is reported in the output. The assessment uses a knowledge-based set of rules, including checking for suitable values of percent human oral absorption, number of metabolites, number of rotatable bonds, logP, solubility, and cell permeability	> 80% is high
Percent human oral absorption	Predicted human oral absorption on 0–100% scale. The pre-diction is based on a quantitative multiple linear regression model. This property usually correlates well with Human Oral Absorption, as both measures the same property	< 25% is poor
QP PCaco	Predicted apparent Caco-2 cell permeability in nm/sec. Caco2 cells are a model for the gut–blood barrier. QikProp predictions are for non-active transport	< 25 poor, > 500 great

TABLE 13.1   *(Continued)*

Property or descriptor	Description	Range or recommended values
QP log BB	Predicted brain/blood partition coefficient. Note: QikProp pre-dictions are for orally delivered drugs so, for example, dopamine and serotonin are CNS negative because they are too polar to cross the blood–brain barrier	−3.0–1.2
QP P MDCK	Predicted apparent MDCK cell permeability in nm/sec. MDCK cells are considered to be a good mimic for the blood–brain barrier. QikProp predictions are for non-active transport.	< 25 poor, > 500 great
Rule of five	The rules are: mol_MW < 500, QPlogPo/w < 5, donorHB ≤ 5, accptHB ≤ 10. Compounds that satisfy these rules are considered drug-like. (The "five" refers to the limits, which are multiples of 5.)	maximum is 3
Rule of three	The three rules are: QP logS > −5.7, QP PCaco > 22 nm/s, # Primary Metabolites < 7. Compounds with fewer (and preferably no) violations of these rules are more likely to be orally available	maximum is 4

✓ Not more than ten hydrogen bond acceptors (nitrogen or oxygen atoms)
✓ A molecular mass not greater than 500 daltons
✓ An octanol-water partition coefficient[40]log P not greater than five.

Rule of three explained by Jorgensen are for the oral intake of the drug, compounds violating the rule were not considered to be administered orally. The rule says that

✓ QP logS > −5.7 represents the aqueous solubility of the drug molecule,
✓ QP PCaco > 22 nm/s represents the permeability of drug molecule at the gut/brain barrier (Caco-2 cells),
✓ primary metabolites < 7.

13.13   MOLECULAR DYNAMICS

MD is a computer simulation of physical movements of atoms and molecules where they interacts each other for a period of time. Therefore the movement of molecules and atoms are determined numerically by Newton's equations of motion whereas the forces between the particles and potential energy are defined by molecular mechanics force fields (MMFFs). MD is also termed as "statistical mechanics by numbers" and "Laplace's vision of Newtonian mechanics" and the results are used to determine macroscopic thermodynamic properties. MD simulations have provided detailed information on the fluctuations and conformational changes of proteins and nucleic acids. These methods are now routinely used to investigate the structure dynamics and thermodynamics of biological molecules and their complexes. They are also used in the determination of structures from x-ray crystallography and from NMR experiments. Simply MD refers the understanding properties of assemblies of molecules in terms of their structure and the microscopic interactions between them. The two main families of simulation technique are MD and Monte Carlo (MC). Simulation act as a bridge between theory and experiment.[1] MD simulations are used in the field of kinetics and irreversible processes, equilibrium ensemble sampling and modeling tools. Moreover, it is an important tool for understanding the physical basis of the structure and function of biological macromolecules. The early view of proteins as relatively rigid structures has been replaced by a dynamic model in which the internal motions and resulting conformational changes play an essential role in their function.[48]

13.14 MATERIALS AND METHODS

13.14.1 DATABASES

The databases protein data bank (PDB) (http://www.pdb.org) was used to retrieve the three dimensional structures of the cancer proteins such as GR, ER, DHFR, BCL-2 protein, AR, TS, mTOR, toll-like receptors (TLR) 7, RAF protein and EGFR for ALL's, breast cancer, CNS lymphoma, lung cancer, prostate cancer, rectal cancer, renal cancer, SCC, thyroid cancer, and oral cancer respectively were retrieved with higher resolution > 3.00 Å respectively and stored in pdb file formats (Table 13.2). The synthetic ligands noted form the literatures such as prednisolone (ALL), tamoxifen (breast cancer), methotrexate (CNS lymphoma), docetaxel (lung cancer), andarine (prostate cancer), folinic acid (renal cancer), torisel (rectal cancer), imiquinoid (SCC), sorafenib (thyroid cancer), gefitinib (oral cancer), and

TABLE 13.2 List of Cancer with Specific Receptors Chosen as a Target for Investigation.

S. No.	Name of the cancer	Targets with PDB ID	Amino acid length	Ribbon representation of cancer specific protein viewed using PyMol
1	Acute lymphoblastic leukemia	Glucocorticoid receptor (1R4O)	92	
2	Breast cancer	Estrogen receptor (3ERD)	261	

TABLE 13.2 *(Continued)*

S. No.	Name of the cancer	Targets with PDB ID	Amino acid length	Ribbon representation of cancer specific protein viewed using PyMol
3	CNS lymphoma	Dihydrofolate receptor (1DRF)	186	
4	Lung cancer	BCL - 2 Protein (1GJH)	166	
5	Oral cancer	Epidermal growth factor (1IVO)	622	

TABLE 13.2 *(Continued)*

S. No.	Name of the cancer	Targets with PDB ID	Amino acid length	Ribbon representation of cancer specific protein viewed using PyMol
6	Prostate cancer	Androgen receptor (2AM9)	266	
7	Renal cancer	mTOR (1AUE)	100	
8	Rectal cancer	Thymidylate synthase (1HVY)	288	

TABLE 13.2 *(Continued)*

S. No.	Name of the cancer	Targets with PDB ID	Amino acid length	Ribbon representation of cancer specific protein viewed using PyMol
9	Squamous cell carcinoma	TLR 7 (1ZIW)	680	
10	Thyroid cancer	RAF protein (1C1Y)	167	

isolated compound quercetin were retrieved from the PubChem (http://pubchem.ncbi.nlm.nih.gov/) (Table 13.3).

13.14.2 PROTEIN AND LIGAND PROCESSING

Ligand binding site for the selected proteins were identified using Q-Site finder. Toxicity prediction was carried out using QikProp module by fast mode access for all ten ligand to compare with quercetin compound in order to predict for central nervous system activity, Metabolism activity, and logarithmic value of each atom predicted. Before docking process, the protein and ligand molecules are prepared. The preparation of a protein involves a number of steps, to include explicit hydrogens, refined and to hydrogenate the structures of ligand and the ligand–receptor complex to made it suitable for performing high accuracy docking in the Maestro graphical user interface.

Using LigPrep module the structures of prednisolone, tamoxifen, meth-otrexate, docetaxel, andarine, folinic acid, torisel, imiquinoid, sorafenib, gefitinib, and quercetin were prepared and refined by adding hydrogen atoms and minimized with OPLS_2005 force field. In both the case of protein and ligand, the resulting structure was energy minimized and used for docking studies. The grid was generated for which the retrieved struc-tures might adopt more than one conformation on binding and to ensure that possible actives are not missed. Docking was carried out using glide module.

TABLE 13.3   List of Ligands in Use for Treatment.

Ligand Name with Pubchem ID	2D Structure	Molecular Formula	Molecular Weight [g/mol]
Prednisolone (5755)		$C_{21}H_{28}O_5$	360.44402
Tamoxifen (2733526)		$C_{26}H_{29}NO$	371.51456

TABLE 13.3 *(Continued)*

Ligand Name with Pubchem ID	2D Structure	Molecular Formula	Molecular Weight [g/mol]
Methotrexate (126941)		$C_{20}H_{22}N_8O_5$	454.43928
Docetaxel (148124)		$C_{43}H_{53}NO_{14}$	807.87922
Gefitinib (123631)		$C_{22}H_{24}C_lFN_4O_3$	446.902363

TABLE 13.3 *(Continued)*

Ligand Name with Pubchem ID	2D Structure	Molecular Formula	Molecular Weight [g/mol]
Andarine (9824562)		$C_{19}H_{18}F_3N_3O_6$	441.35793
Folinic acid (6006)		$C_{20}H_{23}N_7O_7$	473.43932
Torisel (23724530)		$C_{56}H_{87}NO_{16}$	1030.28708

TABLE 13.3   *(Continued)*

Ligand Name with Pubchem ID	2D Structure	Molecular Formula	Molecular Weight [g/mol]
Imiquinoid (57469)		$C_{14}H_{16}N_4$	240.30364
Sorafenib (216239)		$C_{21}H_{16}C_lF_3N_4O_3$	464.82495
Quercetin* (5280343)		$C_{15}H_{10}O_7$	302.2357

*Represents the isolated, purified and pubchem deposited natural substance (Quercetin) from *E. littorale* was used for interaction studied against the native ligands.

13.14.3   ACTIVE SITE PREDICTION

Prediction of active site for each cancer specific protein by Q-Site finder was tabulated as Table 13. 4.

TABLE 13.4   Active Site Residues of Each Targeted Proteins.

Protein name	Residues
Glucocorticoid receptor	CYS 450,HIS 451,TYR 452,GLY 453,LYS 461,LEU 507,ALA 509, ARG 510,LYS 511
Estrogen receptor	MET 315,VAL 316,SER 317,ALA 318,LEU 319,LEU 320,ASP 321, ALA 322,GLU 323,TRP 360,ALA 361,LYS 362,ARG 363,VAL 364,PRO 365,GLY 366,PHE 367,LEU 379,ILE 386,TRP 393,GLY 442,GLU 443,GLU 444,PHE 445,VAL 446,CYS 447,LEU 448,LYS 449,SER 450,LEU 453,LEU 454,VAL 478,ILE 482
Dihydrofolate reductase	ILE 7,VAL 8,ALA 9,VAL 10,ILE 16,GLY 17,LYS 18,GLY 20,ASP 21,LEU 22,TRP 24,PHE 34, LYS 55,THR 56,SER 59,VAL 115,GLY 116,GLY 117,SER 118,TYR 121
BCL-2 protein	ASP 10,ASN 11,ARG 12,GLU 13,ILE 14,VAL 15,MET 16,LYS 17,TYR 18,ILE 19,HIS 20,TYR 21,LYS 22,LEU 23,SER 24,GLN 25,ARG 26,GLY 27,TYR 28,TRP 30,VAL 93,HIS 94,THR 96,LEU 97,ARG 98,GLN 99,ALA 100,GLY 101,ASP 102,ASP 103,PHE 104,SER 105,TRP 144,GLY 145,ARG 146,ILE 147,VAL 148,ALA 149,PHE 150,PHE 151,GLU 152,GLY 154,GLY 155,VAL 156,CYS 158,VAL 159,GLU 160,SER 161,VAL 162,SER 167,PRO 168,LEU 169,VAL 170, ASP 171,ILE 173,ALA 174,LEU 175,MET 177,THR 178,GLU 179,LEU 181,TRP 195,PHE 198,TYR 202
Epidermal growth factor	GLN 47,ARG 48,ASN 49,TYR 50,ASP 51,LEU 52,SER 53,LYS 56,THR 71,VAL 72,GLU 73,ARG 74,ILE 75,PRO 76,GLU 78,ASP 102,THR 106,GLY 107
Androgen receptor	LEU 701,LEU 704,ASN 705,LEU 707,GLY 708,AGLN 711,TRP 741,MET 742,MET 745,VAL 746,MET 749,ARG 752,PHE 764, AMET 780,MET 787,LEU 873,HIS 874,PHE 876,THR 877,LEU 880,PHE 891,MET 895,ILE 899
Thymidylate synthase	ARG 50,PHE 80,GLU 87,ILE 108,TRP 109,ASN 112,TYR 135,LEU 192,PRO 194,CYS 195,HIS 196,GLN 214,ARG 215,SER 216,GLY 217,ASP 218,LEU 221,GLY 222,VAL 223,PHE 225,ASN 226,HIS 256,TYR 258,MET 311
mTOR	TRP A2024,TRP A2028,LEU A2055,HIS A2056,ALA A2057,MET A2059,GLU A2060,GLY A2062, PRO A2063,GLU A2068,THR A2069,SER A2070,PHE A2071,ASN A2072,GLN A2073,ALA A2074,TYR A2075,GLY A2076,ARG A2077,LEU A2079,MET A2080,GLU A2081,ALA A2082,GLN A2083,GLU A2084,TRP A2085,CYS A2086,ARG A2087,TYR A2105

TABLE 13.4   *(Continued)*

Protein name	Residues
TLR 7	TYR 556,PHE 557,LEU 558,LYS 559,GLY 560,GLU 580,VAL 581,PHE 582,LYS 583,ASP 584,LEU 585
RAF protein	SER 11,GLY 12,GLY 13,VAL 14,GLY 15,LYS 16,SER 17,ALA 18,VAL 29,GLU 30,LYS 31, TYR 32,ASP 33,THR 35,ASP 57,THR 58,ALA 59,GLY 60,THR 61,LYS 117

13.14.4 MOLECULAR DYNAMICS

MD calculations simulate molecular movement over time using Newton's equations of motion. Therefore, MC/SD simulation was carried out for the protein–ligand complex.

13.15 RESULTS

13.15.1 DETERMINATION OF ADME/TOX PROPERTIES

ADME/Tox properties were analyzed for the already available and marketed synthetic ligands and compared with that of the isolated compound quercetin using Qikprop module of Schrodinger (Table 13.5). Interestingly, the phyto-compound–quercetin was revealed to satisfy the LR5 (Lipinski rule of five) in an unbiased manner was comparable to certain other synthetic ligands like prednisolone, tamoxifen, andarine, Imiquimod, sorafenib and geftinib in almost all the parameters analyzed. Among the synthetic compounds, the molecular weight, solvent accessible surface area, and number of rotatable bonds were found to be satisfied for 8–9 compounds excepting docetaxel and torisel, which biased the specified range. However, the same parameters were predicted to lie in the preluded range in the phyto-ligand (quercetin) that showed 302.24 g/mol of molecular weight, 514.002 of SASA and five numbers of rotatable bonds (Table 13.5).

Similarly, the parameters such as lop P for either hexadecane/gas or octanol/gas, were also not fruitfully met by the synthetic ligands docetaxel and torisel, which were higher or beyond the limitations of 18 and 35, respectively. Perhaps, the phyto-ligand was predicted to uphold the limitation criteria. Interestingly, all the ligands under investigation were observe to lie in the prescribed range between 4–45 for water/gas log P. however,

TABLE 13.5 ADME-Tox Properties of the Synthetic Ligands of Specific Cancer Proteins Compared Against the Phytocompound Quercetin.

QikProp parameters analyzed ligands name	Normal range	Ligands evaluated										
		Prednisolone	Tamoxifen	Methotrexate	Docetaxel	Andarine	Folinic acid	Torisel	Imiquimod	Sorafenib	Gefitinib	Quercetin
Molecular weight	130.00–725.00	368.512	371.521	454.444	807.89	441.363	473.444	800.13	313.31	464.831	446.908	302.24
Total SASA	300.00–1000.00	588.329	730.6	755.46	945.183	734.198	755.505	1099.967	477.011	766.562	644.275	514.002
No. of rotatable bonds	0.0–15.00	7	9	11	13	8	10	35	3	5	8	5
Solute as donor—hydrogen bond	0.00–5.00	5	0	6	3	3	7	0	2	3	1	4
Solute as acceptor—hydrogen bond	0.00–10.00	7	2	11	16	7	14	20	3	6	7	5
Predictions for properties												
QP log P for hexadecane/gas	4.00–18.00	11.141	13.532	16.487	20.719	13.416	16.856	22.679	8.461	14.585	12.098	10.666
QP log P for octanol/gas	8.00–35.00	22.79	15.626	31.853	40.363	23.322	35.355	42.556	13.267	23.629	19.174	19.513
QP log P for water/gas	4.00–45.00	15.548	5.033	24.007	22.688	14.573	28.175	16.456	8.145	15.67	10.268	14.379
QP log P for octanol/water	<5	1.502	6.533	-1.841	4.176	3.13	-0.844	5.766	2.48	4.097	3.494	0.362
QP log S for aqueous solubility	-6.5–0.5	-3.134	-5.937	-4.008	-5.241	-5.89	-3.61	-0.525	-3.283	-7.019	-3.05	-2.83
QP log BB for brain/blood	-3.0–1.2	-1.346	0.364	-4.798	-2.367	-1.92	-5.208	-3.28	-0.485	-0.986	0.162	-2.352
No. of Primary metabolites	<7	5	3	5	10	4	4	14	0	2	5	5
Apparent caco-2 permeability (nm/sec)	<25 Poor, >500 great	212	2199	0	69	118	0	147	1067	312	662	19
Lipinski rule of 5 violations	4	0	1	2	2	0	2	3	0	0	0	0
Jorgensen rule of 3 violations	3	0	1	1	1	1	1	1	0	1	0	1
% Human oral absorption in GI (+-20%)	<25% Poor	77	100	58	58	82	0	61	96	96	100	52
Qual. model for human oral absorption	>8% is high	High	High	Low	Low	Low	Low	Low	High	Low	High	Medium

some of the parameters like the "solute as donor for hydrogen bonding" of "percentage of oral absorption by human" were not satisfied by methotrexate and folinic acid, providing no reason to be ascertained as either of it influences or controls each other in a vice-versa.

Whatever the case may be, whether the ligands be if synthetic or phyto-ligand, if satisfies few parameters under consideration, could still be chosen for interactions studies in order to evaluate its performance as a good interactor, since that determines the drug likability of the molecule as it is evidenced by many researchers, here again in this perception, all the synthetic ligands in addition to the phyto-ligand was chosen to extend research on docking studies.

13.15.2 DOCKING STUDY

The cancer specific proteins were docked with the synthetic ligands and with phytocompound quercetin and compared, in which the ligands and receptor were represented in orange and green colors, respectively whereas the bondings were represented in blue dotted lines.

13.15.2.1 ACUTE LUEKEMIAS LYMPHOMA

The GR was docked with ligand prednisolone and quercetin compound, respectively, and compared. The analysis of GR docking with both prednisolone and quercetin was observed to interact with each other by four hydrogen bondings however, there was a single common interaction found by LYS at 511 bonds by 2.5 and 2.6 Å. Other than this, prednisolone should get three more hydrogen interaction by a hydrophobic residues and with two uncharged residues THR and LYS at position 512 and 450, respectively by the hydrogen bond of 2.7 Å and 2.2 Å units (Fig. 13.1a). Interestingly, unlike that of synthetic ligand, the interaction mediated by hydrogen bonding in phyto-ligand involved the positively charged residues that are basic in nature (i.e., ARG and HIS at 510 and 451 position by 2.3 and 2.0 Å distances) (Fig. 13.1b). Although the synthetic ligand and quercetin binds with the active sites of the protein both binds with different residues. However, the G. score was observed to be higher (−4.36 Kcal/mol) in quercetin interacted complex than in prednisolone which showed the score of 3.77 Kcal/mol only (Table 13.6).

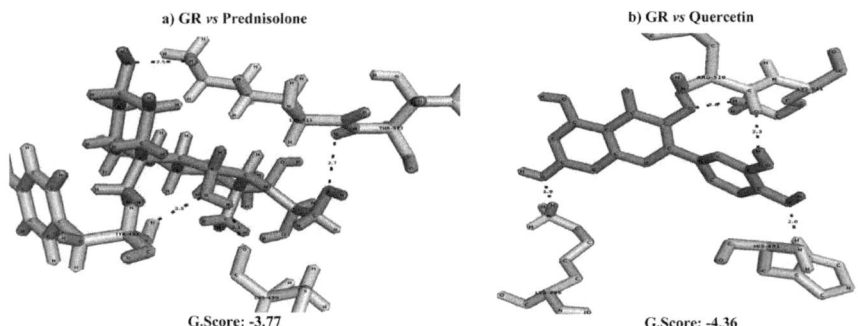

FIGURE 13.1   Snapshot of glucocorticoid receptor with prednisolone and quercetin docking for acute luekemias lymphoma.

13.15.2.2   BREAST CANCER

The ER was docked with synthetic ligand tamoxifen and quercetin compound and compared. The analysis of ER docking with tamoxifen was observed to interact with a negatively charged residue GLU 353 and a positively charged residue ARG 394 by O–H and H—O bond 2.1 and 2.2 Å distances (Fig. 13.2a). The same receptor with the purified substance quercetin interacted with a polar PRO 325, a negatively charged residue GLU 353 and hydrophobic residue TRP 393 by hydrogen bond interaction (Fig. 13.2b) with color index brown the bond length of 2.0, 1.7, and 2.1 Å, respectively. Here, both the ligands interacted with GLU at 353 but by exhibiting greater affinity with phyto-ligand by 1.7 Å distance only, which was evidenced by the higher G. score (−4.7Kcal/mol), unlike that of synthetic, which showed only −3.37 Kcal/mol (Table 13.6).

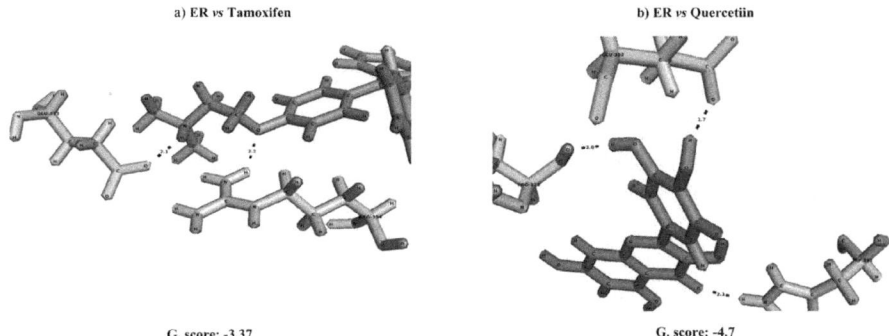

FIGURE 13.2   Snapshot of estrogen receptor with tamoxifen and quercetin docking for breast cancer.

13.15.2.3   CNS LYMPHOMA

The DHFR was docked with the synthetic ligand methotrexate and isolated pure compound quercetin, respectively. The interaction efficiency between them, when compared in terms of G. score, revealed quercetin to be higher −9.13 than methotrexate (−7.69 Kcal/mol) (Table 13.6) with three residues exhibiting good O–H interactions all being below 3 Å while one O–O inter-action mediated by a nonpolar residue VAL at 115 by extended bond length of 3.3 Å in addition to the O–H interaction. The interaction was observed with the positively charged residue ARG at 70 donating two hydrogen bonds of inter atomic distances 1.7, 1.8 Å. However the synthetic ligand interacted with negatively charged residue GLU 30 (2.0 Å) in addition to the hydro-phobic interaction which included the residues such as VAL 115 (2.5 Å), ILE 7 (1.9 Å) and also with uncharged amino acids GLN and ASN at position 35 and 64 forming O–H bonds of each 2.3 Å interatomic distances (Fig. 13.3a). Thus, altogether six different residues interacted by synthetic ligands, whereas only three different residues interact by phyto-ligand. Although the scoring seen to be high by phyto-ligand, could still not be considered favorable because of the fact that a specific O–O interaction especially by respected bond length was observed. The hydrogen boding was observed between O–H and O–O atoms with inter atomic distances of 2.1, 3.3, 2.9, 1.5, and 1.6 Å, respectively (Fig. 13.3b).

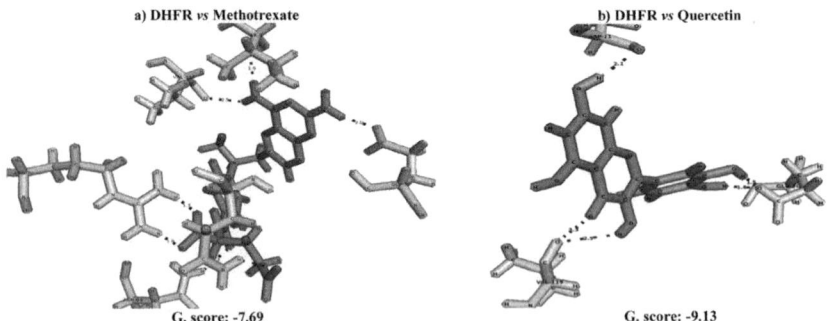

FIGURE 13.3   Snapshot of dihydrofolate receptor (DHFR) with methotrexate and quercetin docking for CNS lymphoma.

13.15.2.4   LUNG CANCER

The receptor BCL-2 protein docked with docetaxel and isolated pure compound quercetin with the receptor and compared. The docetaxel interacted with receptor by a single positively charged residue ARG by sharing four electrons one each from positions 106 and 109 2.1 Å inter atomic distance each and by two from position 26 by the distance of 2.2 Å inter atomic distances each (Fig. 13.4a and Table 13.6). This unfavorability could also be attributed to the special kind of interaction encountered by the phyto-ligand that has remarkably involved many different polar residues such as SER 105, TYR 108, ASP 102, and GLN 25 all being in favorable complementation (i.e., below 3 Å), in addition to the ALA hydrophobic residue to lie in the same platform. SER at 105, TYR at 108 (2 bonds), ASP at 102 (2 bonds) and GLN at 25 exhibited O–H interaction by 2.5 Å, 1.9 Å each by 2 different atomic co-ordinates, 2.4, 2.1, and 2.3 Å, respectively. Thus, the four polar residues alone imparted 6 O–H bonding. Moreover, a common interaction was also observed by ARG at position 26, but by longer distance (1.9 and 2.7 Å) through phyto-ligand interaction, where compound to synthetic ligand. On the whole, including hydrophobic residue, a total of nine interactions were observed making the complex a very stable or strong interaction, thereby interesting to extend research on confirming the stability of the complex through molecular simulation. In similar way, the interaction of quercetin with Bcl-2 was observed (Fig. 13.4 b).

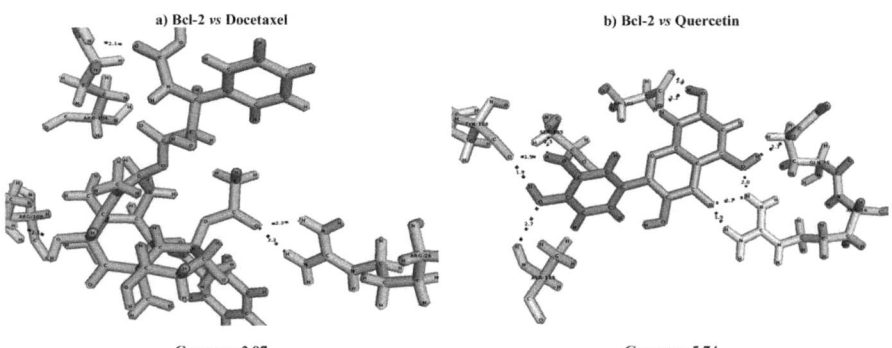

a) Bcl-2 *vs* Docetaxel   b) Bcl-2 *vs* Quercetin

G. score: -2.97   G. score: -5.74

FIGURE 13.4   Snapshot of Bcl-2 protein with docetaxel and quercetin docking for lung cancer.

13.15.2.5   ORAL CANCER

The EGFR was docked with synthetic ligand geftinib and quercetin compound and were compared each other. The docking of synthetic ligand

and phyto-ligand quercetin with EGFR where the receptor shares electron from the negatively charged residue GLU at 73 and ASN at 49 to both the ligands however, it shares with N atom in the case of synthetic ligand with 2.6 and 2.3 Å (Fig. 13.5a) whereas to O atom of quercetin with 2.3 and 2.1 Å inter atomic distances. Although these residues accepts electron from the H atom of the synthetic ligand forming totally four interactions, in terms of phyto-ligands forms two more interactions with polar positively charged ARG at 74, negatively charged ASP at 51 and hydrophobic LEU at 52 position of 1.8, 2.0, 2.6, and 2.5 Å where LEU at 52 position formed two hydrogen bonds between H and O atoms (Fig. 13.5b). However, the G. score was observed to be high (−5.21) in quercetin than in geftinib which showed the G. score of −4.3 only (Table 13.6).

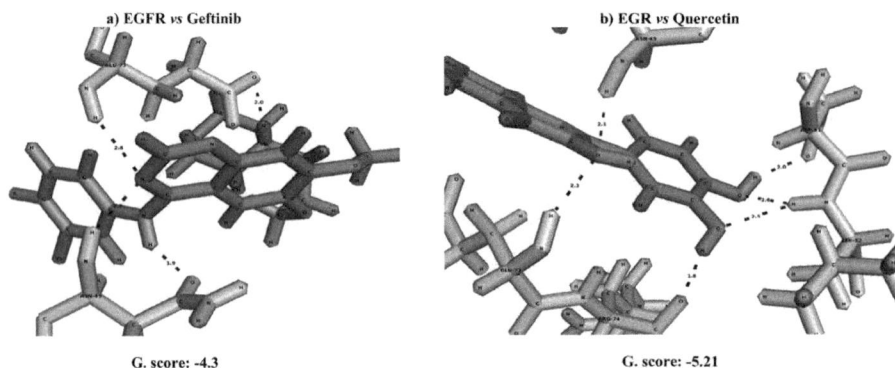

a) EGFR *vs* Geftinib   b) EGR *vs* Quercetin

G. score: -4.3   G. score: -5.21

FIGURE 13.5 Snapshot of epidermal growth factor (EGR) with gefitinib and quercetin docking for oral cancer.

13.15.2.6 PROSTATE CANCER

The AR was docked with synthetic ligand andarine and quercetin and was compared respectively. The receptor interacts with both the ligands at THR 877; however, the electron sharing pair differs, where in andarine O–O atoms interacts whereas in quercetin H and O interacts of 2.8 and 2.6 inter atomic distance. Apart from this interaction, andarine showed interaction with uncharged residue ASN at 705 with 2.0 Å distance (Fig. 13.6a) and quercetin interacts with negatively charged ASP at 51 (2.0 Å) and also with hydrophobic residue LEU at 704 position forming two hydrogen bonds between H and O atoms of 2.6 and 2.5 Å inter atomic distance (Fig. 13.6b). In spite of O–O interaction of residue THR 877 in case of andarine, it was observed that interaction between H and O in the case of quercetin eventually showed more stable interactions and higher G. score in quercetin (about −8.26 Kcal/mol) than andarine (−3.21 Kcal/mol).

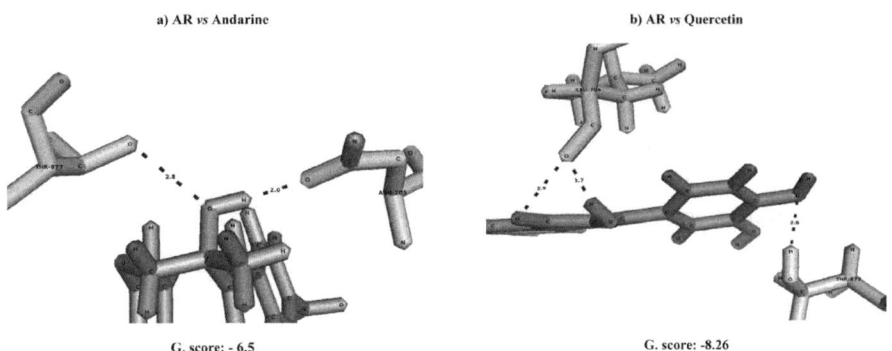

FIGURE 13.6 Snapshot of androgen receptor (AR) with andarine and quercetin docking for prostate cancer.

13.15.2.7 RECTAL CANCER

TS receptor was docked with synthetic ligand folinic acid and quercetin and interactions were compared respectively. The G. score obtained was found higher about −10.85 Kcal/mol in TS–folinic complex than −7.33 Kcal/mol in TS–quercetin complex (Table 13.6). Other four interactions with folinic acid were observed by the uncharged residue ASN at 112 and with hydrophobic residue MET at 309 of 1.9 Å and LYS at 77 position forming two hydrogen bondings of 1.9 and 2.4 Å distances (Fig. 13.7a).The TS shows six interactions with folinic acid and five with quercetin, however two hydrophobic residues PHE at position 80 and ILE at 108 interacts with both folinic acid and quercetin, where PHE 80 shows interaction between H and O atoms of the ligands and that ILE 108 shared electrons between O and H atoms in both the cases whereas in quercetin ILE forms one more O–O interaction of 3.3 Å distance which exceeds the beyond 3 Å (Fig. 13.7b). Therefore, apart from the three interactions with the common residues sharing bond with both the ligands quercetin forms two bondings with polar negatively charged residue GLU at 87 of 1.8 Å and hydrophobic residue LEU at 221 of 2.3 Å where the bond length except the O–O interaction by ILE not exceeds 3 Å distance.

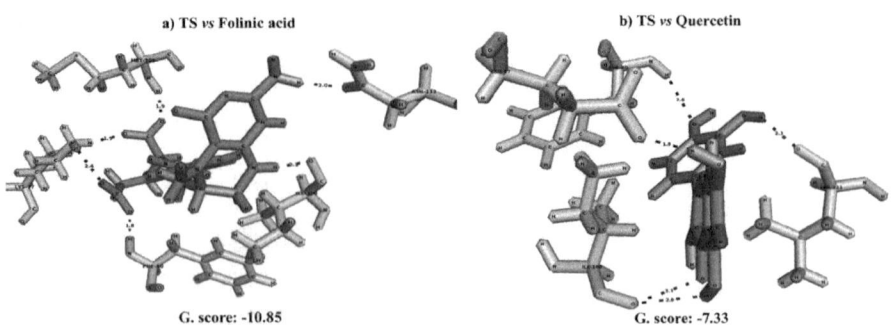

FIGURE 13.7   Snapshot of thymidylate synthase (TS) with folinic acid and quercetin docking for rectal cancer.

13.15.2.8 RENAL CANCER

The receptor mTOR was docked with synthetic ligand torisel and quercetin and was compared respectively. The docked complexes were observed with slightly varied G. score −3.1 Kcal/mol of mTOR-torisel and −3.19 Kcal/mol of mTOR-quercetin complex (Table 13.6). The interaction of torisel with mTOR showed the three electron donations of receptor by the hydrophobic residues such as VAL 2045 (2.8Å), LYS 2046 (2.3 Å), and ALA 2059 (1.8 Å) whereas the polar negatively charged residue GLU 2053 accepts electron from torisel and forms hydrogen bond of 2.0 Å inter atomic distance (Fig. 13.8a). The mTOR interactions with quercetin showed interactions between O and H atoms with polar uncharged residues such as ASN at 2072 (1.8 Å) and GLN at 2083 (1.9 Å) and also with polar positively charged residue HIS 2056 with 2.2 Å of inter atomic distances (Fig. 13.8b).

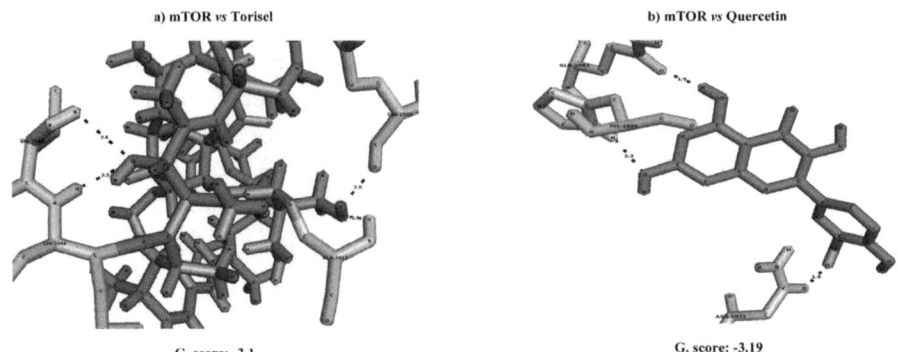

FIGURE 13.8   Snapshot of mTOR with torisel and quercetin docking for renal cancer.

13.15.2.9   SQUAMOUS CELL CARCINOMA

TLR-7 receptor was docked with synthetic ligand imiquimod and quercetin and the interactions were compared. The interaction of imiquimod and quercetin with TLR-7 with the same residue LYS at 559 with distance of 2.1 and 2.3 Å (Fig. 13.9a), apart from this imiquimod showed interaction with negatively charged residue ASP at 523 whereas quercetin interacts with receptor by residues LYS 583 and shares two bonding interaction with ASP 584 of O–O interaction and between O –H atoms of distance not exceeding 3 Å (Fig. 13.9b). The G. score was observed to be higher in TLR-7 quercetin about −3.39 Kcal/mol than TLR-7 imiquimod showed −2.17 only (Table 13.6).

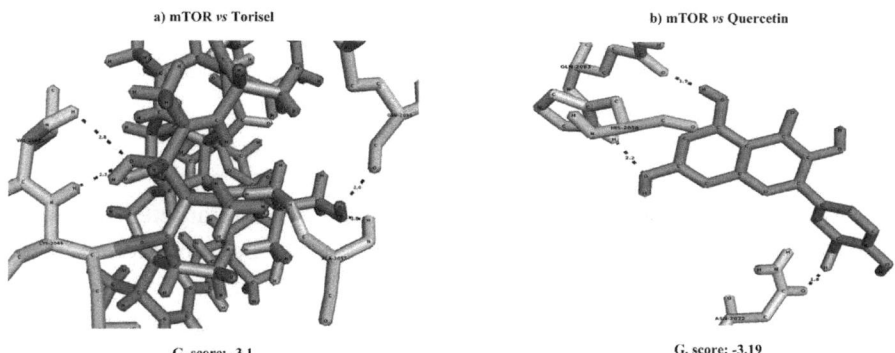

a) mTOR *vs* Torisel   b) mTOR *vs* Quercetin

G. score: -3.1   G. score: -3.19

FIGURE 13.9   Snapshot of TLR- 7 with Imiquinoid and quercetin docking for squamous cell carcinoma.

13.15.2.10   THYROID CANCER

The receptor RAF protein was docked with synthetic ligand sorafenib and quercetin and was compared respectively. The interactions of Sorafenib and quercetin with receptor by different residues where sorafenib shows hydrogen bonding interaction with polar uncharged ASN 116 and charged ASP 33 residues with 2.4, 1.8 and 2.2 Å, respectively (Fig. 13.10a) and that ASP 523 shares two with receptor to form two hydrogen bondings (Fig. 13.10a). When comparing the interaction of RAF with Sorafenib, quercetin showed interaction with hydrophobic residue LYS 31 (1.9 Å) and ALA 148 of 2.1 Å distance and also with polar uncharged residue ASP 119 (2.7 Å) shown in (Fig. 13.10b). The G. score obtained was comparatively high in receptor with quercetin (−7.82 Kcal/mol) than synthetic ligand sorafenib, which showed −2.54 only (Table 13.6).

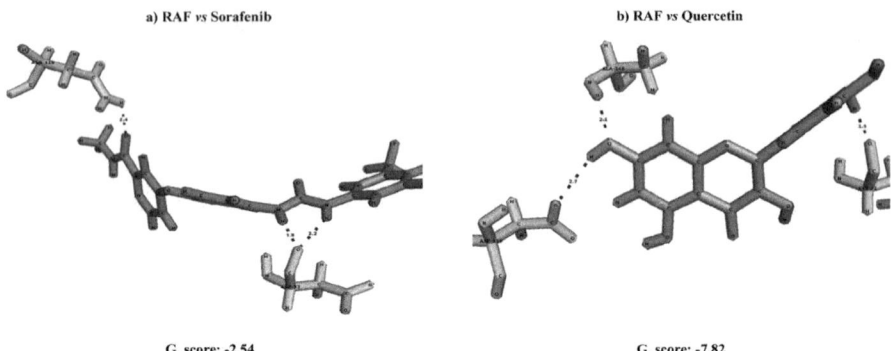

G. score: -2.54   G. score: -7.82

FIGURE 13.10   Snapshot of RAF protein with sorafenib and quercetin docking for thyroid cancer.

Thus, on the whole, among the ten proteins that were targeted, the level of interactions had been very excellent for the phyto-ligand (quercetin). In fact, the kind of interaction observed could be categorized into two: one which is showing more number of interactions by more or less the same residues (compared with synthetic and phto-ligand) as well with other important residues, thereby increasing the G. score by two fold or three fold; and the other could be attributable to the kind of residues (denotes the properties of the amino acid) showing additional interaction for instance O—O (e.g., In DHFR, VAL at 115th position also interacted by O—O in addition to O–H) or exhibited replaced interaction. In AR, THR at 877th position was exchanged by H–O instead of O–O which either effected or affected the docking energy G. score. For instance, the interaction between Bcl-2 protein and quercetin was observed with significant interactions (9 bonding) and best G. score (−5.74 Kcal/mol) followed by quercetin with EGF showing six number of interactions with docking energy of −5.21 Kcal/mol. However, the complex quercetin with RAF protein showed greater G. score of −7.82 Kcal/mol could still not be considered favorable as the hydrogen bond exhibits 3.0 Å inter atomic forces extended by this and perhaps with only few residues when compared to Bcl-2 interaction with phyto-ligand. Even though the interaction between folinic acid and TS showed 6 numbers of interactions with highest G. score about −10.85 Kcal/mol than quercetin still could not considered favorably as it violated the ADME-Tox properties, therefore even on the basis of toxicity prediction and binding affinity observation the Bcl-2 protein interaction with quercetin was found significant and further taken to the dynamics analysis.

Thus from the result obtained and analyzed so far the interaction could be considered as an ideal and most favorable irrespective of the number of parameter that are met with. Even though not cent percentage properties, are met through ADME-Tox, based on the kind of the amino acid interacting and based on the distances aiding in receptor-ligand complex formation that upholds the stabilized bond length (neither less than 1 nor greater than 3 Å) would enhance the chance of considering the interaction favorable for extending research or attempting either for stable conformation or for further pharmacokinetic and dynamic studies. Thus on these basis in the present study the phyto-ligand quercetin which complex with receptor Bcl-2 was considered suitable for continuing research on molecular simulation direction. Thus, it involves many different polar residues with more number of interactions in perhaps with good docking energy.

The interaction profile of cancer specific protein with their native ligands (Phase I) and quercetin compound (Phase II) were tabulated (Table 13.6) describing the bond length formed between the atoms of receptor residues and ligands with total number of interaction for each specific protein.

13.16 MOLECULAR DYNAMICS

The MD simulation was carried out for the docked complex Bcl-2 vs quercetin for Lung cancer which showed significant interaction and good G. score than the other proteins interaction with quercetin. Through the dynamic study it was decided to predict the variation undergone by the complex during the simulation process. The simulation was carried out for 1 ns and to retrieve 100 samples which were observed with reduction in potential energy from −14658.9 ê to −14707.3 ê of 0–100th sample.

The root mean square (RMS) fluctuation of each 100 sample obtained through 1 ns simulation showed stability attained by the Bcl-2 docked with quercetin complex was plotted as graph (Fig. 13.12) where the complex at 74th sample.

TABLE 13.6 A Comparative Interaction Profile of Cancer Specific Targets with Synthetic Ligands and Quercetin.

Receptor	Native ligand					Quercetin				
	Name of the ligand	Residues interacted	Bond length	No. of bond formed	G. score Kcal/mol	Name of the ligand	Residues interacted	Bond length	No. of bond formed	G. score Kcal/mol
Glucocorticoid Receptor	Prednisolone	TYR 452 (H–O)	2.5	4	–3.77	Quercetin	LYS 465 (H–O)	1.9	4	–4.36
		CYS 450 (O–H)	2.21				ARG 510 (O–H)	2.3		
		THR 512 (H–O)	2.7				**LYS 511 (H–O)**	**2.6**		
		LYS 511 (H–O)	**2.5**				HIS 451 (H–O)	2.0		
Estrogen Receptor	Tamoxifen	GLU 353 (O–H)	2.1	2	–3.37	Quercetin	PRO 325 (O–H)	2.0	3	–4.7
		ARG 394 (H–O)	2.2				GLU 353 (O–H)	1.7		
							TRP 393 (H–O)	2.1		
DHFR	Methotrexate	GLU 30 (O–H)	2.0	7	–7.69	Quercetin	ASP 21 (O–H)	2.1	5	–9.13
		ASN 64 (H–O)	2.3				**VAL 115 (O–O)**	**3.3**		
		ARG 70 (H–O)	1.7				**VAL 115 (O–H)**	**2.9**		
		ARG 70 (H–O)	1.8				GLU 30 (O–H)	1.5		
		VAL 115 (O–H)	**2.5**							
		ILE 7 (O–H)	1.9				GLU 30 (O–H)	1.6		
		GLN 35 (H–O)	2.3							
BCL - 2 Protein	Docetaxel	ARG 106 (H–O)	2.1	4	–2.97	Quercetin	SER 105(O–H)	2.5	9	–5.74
		ARG 26 (H–O)	**2.2**				TYR 108 (O–H)	1.9		
							TYR 108 (O–H)	1.9		
		ARG 26 (H–O)	2.2				ALA 113 (H–O)	2.5		

TABLE 13.6 *(Continued)*

Receptor	Native ligand					Quercetin				
	Name of the ligand	Residues interacted	Bond length	No. of bond formed	G. score Kcal/mol	Name of the ligand	Residues interacted	Bond length	No. of bond formed	G. score Kcal/ mol
Epidermal Growth Factor		ARG 109 (H–O)	2.1				ASP 102 (O–H)	2.4		
							ASP 102 (O–H)	2.1		
							ARG 26 (H–O)	**1.9**		
							ARG 26 (H–O)	**2.7**		
							GLN 25 (O–H)	2.3		
	Gefitinib	**GLU 73 (H–N)**	**2.6**	4	–4.3	Quercetin	**ASN 49 (H–O)**	**2.1**	6	**–5.21**
		GLU 73 (O–H)	**2.0**				**GLU 73 (H–O)**	**2.3**		
							ARG 74 (O–H)	1.8		
		ASN 49 (H–N)	**2.3**				ASP 51 (O–H)	2.0		
							LEU 52 (H–O)	2.6		
		ASN 49 (O–H)	**1.9**				LEU 52 (H–O)	2.5		
Androgen Receptor	Andarine	ASN 705 (O–H)	2.0	2	–3.21	Quercetin	THR 877 (H–O)	**2.6**	3	**–8.26**
		THR 877 (O–O)	**2.8**				LEU 704 (O–H)	1.7		
							LEU 704 (O–O)	2.9		
Thymidylate Synthase	Folinic acid	ASN 112 (O–H)	2.0	6	–10.85	Quercetin	PHE 80 (H–O)	2.6	5	**–7.33**
		ILE 108 (O–H)	**2.1**				GLU 87 (O–H)	1.8		
		PHE 80 (H–O)	**1.8**				ILE 108 (O–H)	2.6		
		MET 309 (H–O)	1.9				ILE 108 (O–O)	3.3		

TABLE 13.6 (Continued)

Receptor	Native ligand					Quercetin				
	Name of the ligand	Residues interacted	Bond length	No. of bond formed	G. score Kcal/mol	Name of the ligand	Residues interacted	Bond length	No. of bond formed	G. score Kcal/mol
		LYS 77 (H–O)	1.9				LEU 221 (O–H)	2.3		
		LYS 77 (H–O)	2.4							
mTOR	Torisel	VAL 2045 (H–O)	2.8	4	–3.1	Quercetin	GLN 2083 (O–H)	1.9	3	–7.27
		LYS 2046 (H–O)	2.3				HIS 2056 (H–O)	2.2		
		ALA 2059 (H–O)	1.8							
		GLU 2053 (O–H)	2.0				ASN 2072 (O–H)	1.8		
TLR 7	Imiquimod	LYS 559 (O–H)	**2.1**	2	–2.17	Quercetin	LYS 583 (H–O)	2.1	4	–3.39
							ASP 584 (O–O)	1.7		
							ASP 584 (O–H)	1.8		
		ASP 523 (O–H)	2.1				**LYS 559 (H–O)**	**2.3**		
RAF protein	Sorafenib	ASN 116 (H–O)	2.4	3	–2.54	Quercetin	ALA 148 (H–O)	2.1	3	–7.82
		ASP 33 (O–H)	1.8				LYS 31(O–H)	1.9		
		ASP 33 (O–H)	2.2				ASP 119 (O–H)	2.7		

Note: Residues mentioned in bold were involved in interactions with both the synthetic ligand and quercetin.

*LipophilicEvdW is Chemscore lipophilic pair term and fraction of the total protein –Ligand Van der Waal energy; HBond is H bond pair Term; Electro is Electrostatic rewards; HBPenal is Penalty for Ligand with large hydrophobic contact and low H bonds score; PhobicPenal is Penalty for exposed hydrophobic Ligand groups; RotPenal is Rotatable bond penalty

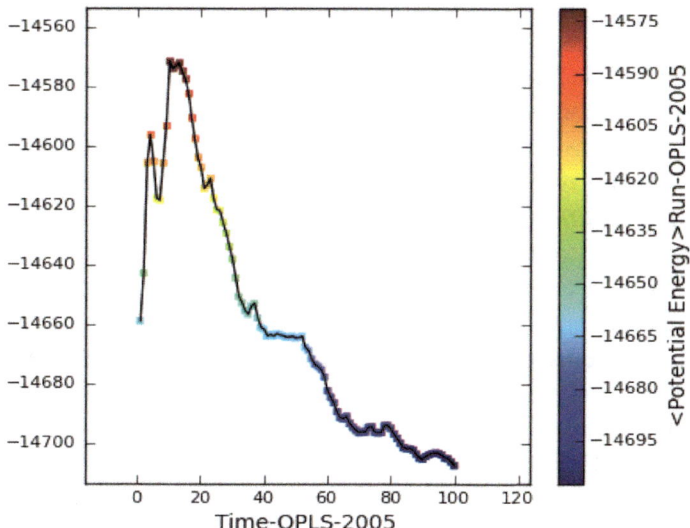

FIGURE 13.11   Dynamics plot against time vs potential energy OPLS 2005 in run platform upto 100 iteration.

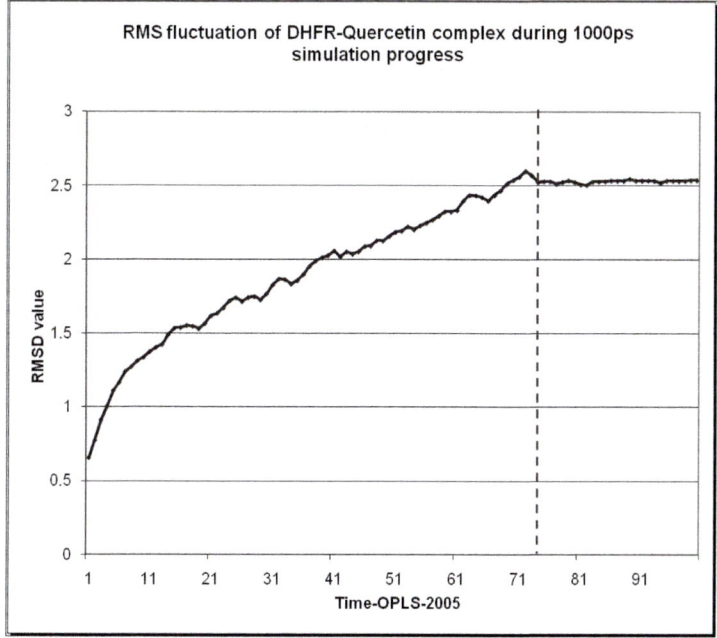

FIGURE 13.12   RMS fluctuation of Bcl-2 protein–quercetin complex.
The red line indicates the sample position attaining stability during the simulation.

13.17 DISCUSSION

In current *era* Molecular docking is one of the key tools in structural molecular biology and computer-assisted drug design. This study investigates the binding orientations and predicts binding affinities of native ligands which are currently used as synthetic drugs for cancers and also with isolated pure quercetin from *E. littorale* docking with the cancer specific proteins. Through the docking interaction analysis, the binding mode and G. score obtained with quercetin was compared with the native ligands for each cancer specific proteins the anti-cancer efficiency of quercetin was determined. The cancer and specific proteins selected for the study are ALL, breast cancer, CNS lymphoma, lung cancer, oral cancer, prostate cancer, renal cancer, rectal cancer, SCC and thyroid cancer and their specific proteins were glucocorticoid receptor (GR), ER, DHFR, B-cell lymphoma-2 (BCL-2) protein, EGFR, AR, mTOR protein, TS, toll-like receptor 7 (TLR-7) and RAF protein. The protein structures were retrieved from PDB database by concerning the resolution of the structure which should be < 2.5 Å and R-value should be nearly zero and R-free value should equal to the resolution (Table 13.1). Comparative study was carried out between the native ligands and quercetin hence to understand the important structural features required to enhance the inhibitory activities and further to help producing augmented inhibitory compounds than the existing synthetic drug molecules therefore to avoid the side effects produced by it. Thereby Arumugam et al.[3] investigated the inhibitory potency for diabetes of plant derived compound namely 1, 2 disubstituted idopyranose from *Vitex negundo* using GLIDE module of schrodinger. He carried out docking with four targets such as maltase glucoamylase, dipeptidyl peptidase, glycogen synthase kinase-3, and aldose reductase and compared docking of plant compound with the docking result obtained with the known ligand.

The ligand molecules were retrieved from the pubchem database in sdf format (Table 13.3) and prepared to optimize the structure before undergoing docking studies. The prepared native ligand molecules and isolated purified quercetin were subjected for ADME/Tox for predicting the absorption, distribution, mechanism, excretion, and toxicity, therefore, to analyze the pharmacokinetics and pharmacodynamics of the ligand by accessing the drug like properties. Through Qikprop the molecular weight, permeability through MDCK cells, log IC_{50} value for blockage of K^+ channels, gut–blood barrier, and violation of the lipinski's rule of five were analyzed. According to the Lipinski's rule of five, quercetin does not violate the rules and the other native ligands properties were discussed as follows.

The molecular weight and other properties predicted by Qikprop for each ligand molecules were tabulated (Table 13.4) in which docetaxel (807.89 Da) and torisel (800.13 Da) exceeded the normal range between 130 and 725 Daltons whereas quercetin's molecular weight is about 302.24 Da was lesser compared with the other ligand molecules. The hydrogen donors predicted showed that methotrexate (6.25) and andarine (7.25) possess more than 5 hydrogen donors and prednisolone posess exactly 5 and 4 by quercetin. Moreover, the hydrogen bond acceptors eliminated the drug molecules such as methotrexate (11.75), docetaxel (16.35), folinic acid (14.25), and torisel (20.7) where quercetin exerts 5.25 of hydrogen bond donors. A partition (P) or distribution coefficient (D) is the ratio of concentrations of a compound in the two phases of a mixture of two immiscible solvents at equilibrium. The last of Lipinski rule, that is, octanol/water accessibility was observed to be about 0.362 for quercetin whereas in other ligands such as tamoxifen (6.533) and torisel (5.766) exceeds 5. Thereby, quercetin was observed to obey the rules of Lipinski and also exerts significant values for other properties such as solvent accessible surface area, absorption in colorectal cells and in blood–brain barrier too suggesting that quercetin is orally administered at last the calculated percentage of oral absorption showed medium for quercetin.[59]

According to the Schrodinger tool Qikprop, Jorgensen's rule of three defines the properties of drug likeliness for oral administration. The three rules are, QP log S > −5.7, QP PCaco > 22 nm/s, and Primary metabolites should be <7. The QP logS value determines the solubility of the small molecule in aqueous solution which should be between −6.5 and 0.5, where all the native ligand molecules selected for the study obeys the aqueous solubility range except sorafenib about −7.019 whereas quercetin was observed with −2.83. Qikprop predicts the permeability through apparent colorectal adenocarcinoma cells (Caco) where every ligand molecules was observed above the range 25 despite quercetin has scored about 19. Thus none of the selected ligands was observed with acceptable Caco-2 cells permeability. Although the percentage of oral absorption in gastrointestinal tract (GI) was found to be about 52 for quercetin on the other hand methotrexate and folinic acid exhibited zero values. Tamoxifen and Gefitinib showed 100% of absorption in GI tract whereas imiquinoid and Sorafenib found with 96% and the other native ligands such as andarine, prednisolone, torisel, docetaxel observed with 82, 77, 61, and 58%.

The interaction study was carried out in two different phases where phase I includes the docking of cancer specific protein with synthetic ligands and phase II with quercetin and compared each other based on the interaction profile and (G. score) XP scored during the docking process. The G. score was calculated based on the LipophilicEvdW (Chemscore lipophilic pair term and fraction of the total protein–Ligand Van der Waal energy), HBond (H bond pair term), Electro (Electrostatic rewards), HBPenal (Penalty for Ligand with large hydrophobic contact and low H bonds score), PhobicPenal (Penalty for exposed hydrophobic Ligand groups), and RotPenal (Rotatable bond penalty). Comparing to the phase I study of synthetic ligands quercetin was observed with good binding affinity with the selected cancer specific proteins except in the case of rectal cancer where TS showed higher G. score about -10.85 Kcal/mol and more number of interactions with folinic acid than quercetin (-7.33 Kcal/mol) despite folinic acid has violated ADME-Tox properties. However, the synthetic ligands prednisolone, geftinib, andarine, and imiquimod undergo all the ADME-Tox properties these ligands were observed with minimum number of interactions compared to the phytoligand quercetin. The interactions of selected specific cancer proteins with quercetin based on their score and interactions number, was found significant with Bcl-2 protein and further carried for dynamics study. Quercetin interaction with all the ten selected proteins were observed to form hydrogen bonds between O and H, whereas bonds between O and O atoms were observed in case of synthetic ligands; geftinib alone showed electron sharing with N atoms of the receptor EPGR. In general, the interaction between two O atoms are said to enhance the stability of the bonding and the presence of such interaction in the active site region are due to the action of active site as a local storage site for small molecules resulting in increased effective concentration of the ligand[15]. And quercetin was found to have strong inhibitory effects on mammalian Thioredoxin reductase (TrxRs) with IC_{50} value of 0.97 μmol/L was shown to be dose and time-dependant and attack on the reduced COOH-terminal–Cys-Sec-Gly active site of TrxR, where TrxR is one of the thioredoxin systems which exerts a wide range of activities in cellular redox control, antioxidant function, cell viability and proliferation which also found overexpressed in many aggressive tumors.[34]

The best G. scored and effectively interacting Bcl-2 protein–quercetin was carried with MD studies to analyze the optimization of the complex so that the potential energy and to ability of the complex to attain stability were analyzed through MD study. At the initial stage of simulation (0th sample) the potential energy was observed to be -14658.9 ê which decreases gradually and at the end of the simulation about 1000nd pico second it was reduced to -14707.3 ê. The simulated protein at each 10th consecutive sample was observed for RMS deviation by and the graph plotted showed the stability attained by the protein–quercetin complex from the 74th sample (Fig. 13.12).

The Bcl-2 protein also seems to be the major inhibitor of cell death in acute myeloid leukemia where the protein was observed to be homology with

Bcl-x_L and the protein consists of seven α-helices and a long loop. The proteins solubility and the biological function was studied by Petros et al[57], which revealed that the deletion of long loop and C-termini of the protein does not interrupted the biological functions where it consists of two central, predominantly hydrophobic helices (helix 5 and 6) packed against four amphipathic α-helices. Helix 1 and 2 are oriented parallel to one another, crossing at an angle of about 45° and a long loop connects the helix 2 to helix 4 (residues 126–137) and helix 4, 5 (residues 144–163), and 6 (residues 167–192) are oriented in a nearly antiparallel fashion with a kink in helix 6 at histidine 184. An irregular turn composed of two glycine residues connects helix 6 to helix 7 (residues 195–202), which orients helix 7 orthogonal to helices 4, 5, and 6.

The protein consists of a hydrophobic groove on the surface on which the mutations in this region have been to abolish the antiapoptotic activity of Bcl-2 and block heterodimerization with other family members[84] and also reported that the hydrophobic groove includes the residues such as leucine, isoleucine, valine, tyrosine, phenylalanine, tryptophan, aspartate, glutamate, lysine, arginine, and histidine present in α helices 3, 4, 5, and 7) even though the exact position of the above residues were not discussed. During the simulation the protein was observed with slight structural movements, where every 10th sample of 100 samples was analyzed for its structural variation. The protein structure at 20th sample when superimposed showed the movement of the long loop that connects α1 and α2, loop connecting α3 and α4, and α4 helix. N and C-termini undergone greater fluctuation during the simulation progress, mainly the α3 and α4 helices were observed, therefore, it was already reported that half part of α3 contains the binding pocket of the protein. The BCL-2 protein binding interaction with quercetin was observed of involving the residues such as aspartic acid, serine, tyrosine, alanine, arginine, and glutamine in which three active residues Aspartic acid alone involved in the interaction in this study.

KEYWORDS

- **cancer targets**
- **drugs**
- **quercetin**
- **molecular docking**
- **dynamic simulation**

REFERENCES

1. Allen, M. P. *Introduction to Molecular Dynamics Simulation*; John von Neumann Institute for Computing: Julich, Germany, 2004, *23*, 1–28.
2. Ania, D. L. N.; Rolando, R. Current Methodology for the Assessment of ADME-Tox Properties on Drug Candidate Molecules. *Biotecnol. Apl.* **2008**, *25*, 97–110.
3. Arumugam, S.; Manikandan, R.; Chinnasamy, A. Molecular Docking Studies of 1,2 Disustituted Idopyranose from *Vitex negundo* with Anti-diabetic Activity of Type 2 Diabetes. *Int. J. Pharma. Bio. Sci.* **2011**, *2* (1), B68–B83.
4. Asnaghi, L.; Bruno, P.; Priulla, M.; Nicolin, A. mTOR: A Protein Kinase Switching between Life and Death. *Pharmacol. Res.* **2004**, *50* (6), 545–549.
5. Barnham, K. J.; Torres, A. M.; Alewood, D.; Alewood, P. F.; Domagala, T.; Nice, E. C.; Norton, R. S. Role of the 6–20 Disulfide Bridge in the Structure and Activity of Epidermal Growth Factor. *Protein Sci.* **1998**, 7 (8), 1738–1749.
6. Buchholz, T. A. Radiation Therapy for Early-stage Breast Cancer after Breast-Conserving Surgery. *N. Engl. J. Med.* **2009**, *360* (1), 63–70.
7. Edwards, K. B.; Ward, E.; Kohler, B. A.; Eheman, C.; Zauber, A. G.; Anderson, R. N.; Jemal, A.; Schymura, M. J.; Lansdorp-Vogelaar, I. Seeff, L. C.; Ballegooijen, M.; Goede, L.; Ries, L. A. G. Annual Report to the Nation on the Status of Cancer, 1975–2006, Featuring Colorectal Cancer Trends and Impact of Interventions (risk factors, screening, and treatment) to Reduce Future Rates. *Cancer*, **2009**, *116* (3), 544–573.
8. Carmona, R. H. *The Health Consequences of Involuntary Exposure to Tobacco Smoke*; a Report of the Surgeon General. U. S. Department of Health and Human Services, Rockville, MD, 2006.
9. Chen, M. J.; Shimada, T.; Moulton, A. D.; Cline, A.; Humphries, R. K.; Maizel, J.; Nienhuis, A. W. The Functional Human Dihydrofolate Reductase Gene. *J. Biol. Chem.* **1984**, *259* (6), 3933–3943.
10. Chen, M. J.; Shimada, T.; Moulton, A. D.; Harrison, M.; Nienhuis, A. W. Intronless Human Dihydrofolate Reductase Genes are Derived from Processed RNA Molecules. *Proc. Natl. Acad. Sci. U.S.A.* **1982**, *79* (23), 7435–7439.
11. Clarke, S. J.; Rivory, L. P. Clinical Pharmacokinetics of Docetaxel. *Clin. Pharmacokinet.* **1999**, *36* (2), 99–114.
12. Clauditz, T. S.; Reiff, M.; Gravert, L.; Gnoss, A.; Tsourlakis, M. C.; Sauter, G.; Bokemeyer, C.; Knecht, R.; Wilczak, W. Human Epidermal Growth Factor Receptor 2 (HER2) in Salivary Gland Carcinomas. *Pathology.* **2011**, *13*, 459–464.
13. Corn, B. W.; Marcus, S. M.; Topham, A.; Hauck, W.; Curran, W. J. Will Primary Central Nervous System Lymphoma be the most Frequent Brain Tumor Diagnosed in the Year 2000? *Cancer.* **1997**, *79* (12), 2409–2413.
14. Dahlman Wright, K.; Cavailles, V. Fuqua, S. A.; Jordan, V. C.; Katzenellenbogen, J. A.; Korach, K. S.; Maggi, A.; Muramatsu, M.; Parker, M. G.; Gustafsson, J. A. International Union of Pharmacology. LXIV. Estrogen Receptors. *Pharmacol. Rev.* **2006**, *58* (4), 773–781.
15. Daskalakis, V.; Varotsis, C. Binding and Docking Interactions of NO, CO and O_2 in Heme Proteins as Probed by Density Functional Theory. *Int. J. Mol. Sci.* **2009**, *10* (9), 4137–4156.
16. Davis, M.; Williams, R.; Chakraborty, J. Prednisone or Prednisolone for the Treatment of Chronic Active Hepatitis? A Comparison of Plasma Availability. *Br. J. Clin. Pharmacol.* **1978**, *5* (6), 501–505.

17. Djulbegovic, M.; Beyth, R. J.; Neuberger, M. M. Screening for Prostate Cancer: Systematic Review and Meta-analysis of Randomised Controlled Trials. *BMJ.* **2010,** *341*, c4543.
18. Eby, N. L.; Grufferman, S.; Flannelly, C. M.; Schold, S. C.; Vogel, F. S.; Burger, P. C. Increasing Incidence of Primary Brain Lymphoma in the US. *Cancer.* **1988,** *62* (11), 2461–2465.
19. FDA Approves New Drug for Advanced Kidney Cancer. 30 May 2007.
20. Fine, H. A.; Mayer, R. J. Primary Central Nervous System Lymphoma. *Ann. Intern. Med.* **1999,** *119* (11), 1093–1104.
21. Florescu, A.; Amir, E.; Bouganim, N.; Clemons, M. Immune Therapy for Breast Cancer in 2010—Hype or Hope? *Curr. Oncol.* **2011,** *18* (1), 9–18.
22. Frank, G. R. Role of Estrogen and Androgen in Pubertal Skeletal Physiology. *Med. Pediatr. Oncol.* **2003,** *41* (3), 217–221.
23. Funanage, V. L.; Myoda, T. T.; Moses, P. A.; Cowell, H. R. Assignment of the Human Dihydrofolate Reductase Gene to the q11----q22 Region of Chromosome 5. *Mol. Cell. Biol.* **1984,** *4* (10), 2010–2016.
24. Gerullis, H.; Ecke, T. H.; Imer, C.; Heuck, C. J.; Otto, T. mTOR-inhibition in Metastatic Renal Cell Carcinoma. Focus on Temsirolimus: A Review. *Minerva. Urol. Nefrol.* **2010,** *62* (4), 411–423.
25. Giordano, S. H.; Cohen, D. S.; Buzdar, A. U.; Perkins, G.; Hortobagyi, G. N. Breast Carcinoma in Men: A Population-based Study. *Cancer.* **2004,** *101* (1), 51–57.
26. Gorlova, O. Y.; Weng, S. F.; Zhang, Y. Aggregation of Cancer Among Relatives of Never-smoking Lung Cancer Patients. *Int. J. Cancer.* **2007,** *121* (1), 111–118.
27. Gupta, S.; Hastak, K.; Ahmad, N.; Lewin, J. S.; Mukhtar, H. Inhibition of Prostate Carcinogenesis in TRAMP Mice by Oral Infusion of Green Tea Polyphenols. *Proc. Natl. Acad. Sci. U.S.A.* **2001,** *98* (18), 10350–10355.
28. Hardy, L. W.; Stroud, R. M; Santi, D. V.; Montfort, W. R.; Jones, M. O.; Finer Moore, J. S. Atomic Structure of Thymidylate Synthase: Target for Rational Drug Design. *Science.* **1987,** *235* (4787), 448–455.
29. Herbst, R. S. Review of Epidermal Growth Factor Receptor Biology. *Int. J. Radiat. Oncol. Biol. Phys.* **2003,** *59* (2 Suppl), 21–26. doi:10.1016/j.ijrobp.2003.11.041. PMID 15142631.
30. Herrlinger, U.; Schabet, M.; Bitzer, M.; Petersen, D.; Krauseneck, P. Primary Central Nervous System Lymphoma: From Clinical Presentation to Diagnosis. *J. Neurosurg.* **2000,** *92,* 261–266.
31. Hu, M. I.; Vassilopoulou Sellin, R.; Lustig, R.; Lamont, J. P. Thyroid and Parathyroid Cancers. In Cancer Management: A Multidisciplinary Approach. 11th Edition; Pazdur, R., Wagman, L. D., Camphausen, K. A., Hoskins, W. J., Eds.; Cmp United Business Media: New York, 2008.
32. Izbicka, E.; Campos, D.; Carrizales, G.; Tolcher, A. Biomarkers for Sensitivity to Docetaxel and Paclitaxel in Human Tumor Cell Lines in vitro. *Cancer Genom. Proteom.* **2005,** *2*, 219–222.
33. Jemal, A.; Elizabeth, W. Cancer Statistics. *A Cancer J. Clin.* **2010,** *60* (5), 277–300.
34. Jun, L.; Laura, V. P.; Jianguo, F.; Salvador, R. N.; Boris, Z.; Arne, H. Inhibition of Mammalian Thioredoxin Reductase by Some Flavonoids: Implications for Myricetin and Quercetin Anticancer Activity. *Cancer Res.* **2006,** *66* (8), 4410–4418.
35. Kaneda, S.; Gotoh, O.; Shimizu, K.; Nalbantoglu, J.; Takeishi, K.; Seno, T.; Ayusawa, D. Structural and Functional Analysis of the Human Thymidylate Synthase Gene. *J. Biol. Chem.* **1990,** *265* (33), 20277–20284.

36. Konstantopoulou, M.; Lord, M. G.; Macfarlane, A. W. Treatment of Invasive Squamous Cell Carcinoma with 50-percent Imiquimod Cream. *Dermatol. Online J.* **2006,** *12* (3), 10.

37. Kumar, P.; Clark, M. *Clinical Medicine;* 6th ed.; London: Elsevier, 2005; pp 683–648.

38. Lacroix, M. Significance, Detection and Markers of Disseminated Breast Cancer Cells. *Endocr. Relat. Cancer.* **2006,** *13* (4), 1033–1067.

39. Lambrou, G. I.; Vlahopoulos, S.; Papathanasiou, C.; Papanikolaou, M.; Karpusas, M.; Zoumakis, E.; Tzortzatou Stathopoulou, F. Prednisolone Exerts Late Mitogenic and Biphasic Effects on Resistant Acute Lymphoblastic Leukemia Cells: Relation to Early Gene Expression. *Leuk Res.* **2009,** *33,* 1684–1695.

40. Leo, A.; Hansch, C.; Elkins, D. Partition Coefficients and Their Uses. *Chem. Rev.* **1971,** *71* (6), 525–616.

41. Levin, E. R. Integration of the Extranuclear and Nuclear Actions of Estrogen. *Mol. Endocrinol.* **2005,** *19* (8), 1951–1959.

42. Li, P.; Wood, K.; Mamon, H.; Haser, W.; Roberts, T. Raf-1: A Kinase Currently without a Cause but not Lacking in Effects. *Cell.* **1991,** *64* (3), 479–482.

43. Lien, E.; Ingalls, R. R. Toll-like Receptors. *Crit. Care. Med.* **2002,** *30* (1), 1–11.

44. Lipinski, C. A.; Lombardo, F.; Dominy, B. W.; Feeney, P. J. Experimental and Computational Approaches to Estimate Solubility and Permeability in Drug Discovery and Development Settings. *Adv. Drug Del. Rev.* **2001,** *46,* 3–26.

45. Lu, N. Z.; Wardell, S.E.; Burnstein, K. L.; Defranco, D.; Fuller, P. J.; Giguere, V.; Hochberg, R. B.; McKay, L.; Renoir, J. M.; Weigel, N. L.; Wilson, E. M.; McDonnell, D. P.; Cidlowski, J. A. International Union of Pharmacology. LXV. The Pharmacology and Classification of the Nuclear Receptor Superfamily: Glucocorticoid, Mineralocorticoid, Progesterone, and Androgen Receptors. *Pharmacol. Rev.* **2006,** *58* (4), 782–797.

46. Luscombe, N. M.; Greenbaum, D.; Gerstein, M. What is Bioinformatics? An Introduction and Overview. *Yearb. Med. Inform.* **2001,** *1,* 83–100.

47. Margaret, H. R.; Neil, A. F.; Leigh, A. S.; Frank, L. G. Histologic Variants of Squamous Cell Carcinoma of the Skin. *Cancer Control.* **2001,** *8* (4), 354–363.

48. Martin, K.; Andrew, J. M.; Molecular Dynamics Simulations of Biomolecules. *Nat. Struct. Biol.* **2002,** *9,* 646–652.

49. Massarweh, S.; Osborne, C. K.; Creighton, C. J.; Qin, L.; Tsimelzon, A.; Huang, S.; Weiss, H.; Rimawi, M.; Schiff, R. Tamoxifen Resistance in Breast Tumors is Driven by Growth Factor Receptor Signaling with Repression of Classic Estrogen Receptor Genomic Function. *Cancer Res.* **2008,** *68* (3), 826–833.

50. Michael, L.; Susanna, M.; Annika, J.; Matthias, S.; Alexander, S.; Katjana, O.; Axel, G.; Christopher, B.; Marlies, V.; Horst, U.; Ingo, G. H.; Hendrik, P.; Uwe, S. Association of Genetic Variants of Methionine Metabolism with Methotrexate-induced CNS White Matter Changes in Patients with Primary CNS Lymphoma. *Neuro. Oncol.* **2009,** *11,* 2–8.

51. Miller, D. C.; Hafez, K. S.; Stewart, A.; Montie, J. E.; Wei, J. T. Prostate Carcinoma Presentation, Diagnosis, and Staging: an Update from the National Cancer Data Base. *Cancer.* **2003,** *98* (6), 1169–1178.

52. Minna, J. D.; Schiller, J. H. *Harrison's Principles of Internal Medicine; 17th ed.;* McGraw-Hill: New York, 2008; pp 551–562.

53. Mok, T. S. Gefitinib or Carboplatin-paclitaxel in Pulmonary Adenocarcinoma. *N. Eng. J. Med.* **2009,** *361,* 947–957.

54. Mol, F.; Mol, B. W.; Ankum, W. M.; Van Der Veen, F.; Hajenius, P. J. Current Evidence on Surgery, Systemic Methotrexate and Expectant Management in the Treatment of

Tubal Ectopic Pregnancy: A Systematic Review and Meta-analysis. *Hum. Reprod. Update.* **2008,** *14* (4), 309–319.

55. Mulders, P. F.; Brouwers, A. H.; Hulsbergen van der Kaa, C. A.; van Lin, E. N.; Osanto, S.; de Mulder, P. H. Guideline 'Renal Cell Carcinoma' (in Dutch; Flemish). *Ned. Tijdschr. Geneeskd.* **2008,** *152* (7), 376–380.

56. Pao, W.; Miller. V.; Zakowski, M. EGF Receptor Gene Mutations are Common in Lung Cancers from "Never Smokers" and are Associated with Sensitivity of Tumors to Gefitinib and erlotinib". *Proc. Natl. Acad. Sci. U.S.A.* **2004,** *101* (36), 13306–13311.

57. Petros, A. M.; Medek, A.; David, G. N.; Daniel, H. K.; Yoon, H. S.; Kerry, S.; Edmund, D. M.; Tilman, O.; Stephen, W. F. Solution Structure of the Antiapoptotic Protein Bcl-2. *PNAS.* **2001,** *98* (6), 3012–3017.

58. Ravi Rajagopalan, P. T.; Zhang, Z.; McCourt, L.; Mary, D.; Benkovic Stephen, J.; Hammes Gordon, G. Interaction of Dihydrofolate Reductase with Methotrexate: Ensemble and Single-molecule Kinetics. *Proc. Nat. Acad. Sci.* **2002,** *99* (21), 13481–13486.

59. Ranelletti, F. O.; Maggiano, N.; Serra, F. G. Quercetin Inhibits p21-RAS Expression in Human Colon Cancer Cell Lines and in Primary Colorectal Tumors. *Int. J. Cancer.* **2000,** *85,* 438–445.

60. Rang, H. P.; Dale, M. M.; Ritter, J. M.; Moore, P. K. The Pituitary and the Adrenal Cortex. In *Pharmacology; 5th ed;* Hunter, l. Ed.; Churchill Livingstone: London, 2003; pp 413– 415.

61. Raudrant, D.; Rabe, T. Progestogens with Antiandrogenic Properties. *Drugs.* **2003,** *63* (5), 463–492.

62. Rhen, T.; Cidlowski, J. A. Antiinflammatory Action of Glucocorticoids-New Mechanisms for Old Drugs. *N. Engl. J. Med.* **2005,** *353* (16), 1711–1723.

63. Rini, B. I.; Rathmell, W. K.; Godley, P. Renal Cell Carcinoma. *Curr. Opin. Oncol.* **2008,** *20* (3), 300–306.

64. Saman, W.; Seppo, P.; Toru, N.; Victor, R. P.; Markku, P.; Heidi, K.; Onni, N. Demonstration of Ethanol-induced Protein Adducts in Oral Leukoplakia (pre-cancer) and Cancer. *J. Oral Pathol. Med.* **2008,** *37* (3), 157–165.

65. Sariego, J. Breast Cancer in the Young Patient. *Am. Surg.* **2010,** *76* (12), 1397–1400.

66. Saurabh, S.; Sanjaykumar, C.; Prashant, S.; Gomase, V. S. Computational Approach Towards the B-Cell Lymphoma-2 Protein: A Noticeable Target for Cancer Proteomics. *J. Pharmacol. Res.* **2010,** *1* (1), 1–8.

67. Scheinfeld, N. Three Cases of Toxic Skin Eruptions Associated with Methotrexate and a Compilation of Methotrexate-induced Skin Eruptions. *Dermatol. Online J.* **2006,** *12* (7), 15.

68. Shintani, S.; Li, C.; Mihara, M.; Nakashiro, K.; Hamakawa, H. Gefitinib ('Iressa'), an Epidermal Growth Factor Receptor Tyrosine Kinase Inhibitor, Mediates the Inhibition of Lymph Node Metastasis in Oral Cancer Cells. *Cancer Lett.* **2003,** *201* (2), 149–155.

69. Smith Malcolm, A. Secondary Leukemia or Myelodysplastic Syndrome after Treatment With Epipodophyllotoxins. *J. Clin. Oncol. (Am. Soc.Clin. Oncol.).* **2001,** *17* (2), 569–577.

70. Sordella, R.; Bell, D. W.; Haber, D. A.; Settleman, J. Gefitinib-sensitizing EGFR Mutations in Lung Cancer Activate Anti-apoptotic Pathways. *Science.* **2004,** *305* (5687), 1163–1167.

71. Sridhar, S. S.; Hedley, D.; Siu, L. L. Raf Kinase as a Target for Anticancer Therapeutics. *Mol. Cancer Ther.* **2005,** *4* (4), 677–685.

72. Srivastava, V.; Gupta, S. P.; Siddiqi, M. I.; Mishra, B. N. Molecular Docking Studies on Quinazoline Antifolate Derivatives as Human Thymidylate Synthase Inhibitors. *Bioinformation.* **2010,** *4* (8), 357–365.

73. Stams, W. A.; den Boer, M. L.; Beverloo, H. B. Expression Levels of TEL, AML1, and the Fusion Products TEL-AML1 and AML1-TEL versus Drug Sensitivity and Clinical Outcome in t(12; 21)-positive Pediatric Acute Lymphoblastic Leukemia. *Clin. Cancer Res.* **2005,** *11* (8), 2974–2980.

74. Susan, J. H.; Dirkjan, H.; George, F. M.; Thomas, S. K.; Adam, W. C.; Ilse, G. M.; Carl, F. S.; Danielle, M. M.; Chrysalyne, S.; Rachael, A. C. Imiquimod enhances IFN-γ Eproduction and Effector Function of T Cells Infiltrating Human Squamous Cell Carcinomas of the Skin. *J. Invest. Dermatol.* **2009,** *129* (11), 2676–2685.

75. Takimoto, C. H.; Calvo, E.; Pazdur, R.; Wagman, L. D.; Camphausen, K. A.; Hoskins, W. J. Principles of Oncologic Pharmacotherapy. In *Cancer Management: A Multidisciplinary Approach;* 11th ed.; 2008 pp 3–7.

76. Thompson, I. M.; Pauler, D. K.; Goodman, P. J. Prevalence of Prostate Cancer among Men with a Prostate-specific Antigen Level < or = 4.0 ng per Milliliter. *New Engl. J. Med.* **2004,** *350* (22), 2239–2246.

77. Thun, M. J.; Hannan, L. M.; Adams Campbell, L. L.; Hans Olov,.A. Lung Cancer Occurrence in Never-Smokers: An Analysis of 13 Cohorts and 22 Cancer Registry Studies. *PLoS Med.* **2008,** *5* (9), e185.

78. Venturi, S.; Venturi, M. Iodine in Evolution of Salivary Glands and in Oral Health. *Nutr. Health.* **2009,** *20* (2), 119–134.

79. Wan, X.; Shen, N.; Mendoza, A.; Khanna, C.; Helman, L. J. CCI-779 Inhibits Rhabdomyosarcoma Xenograft Growth by an Antiangiogenic Mechanism Linked to the Targeting of mTOR/HIF-1alpha/VEGF Signaling. *Neoplasia.* **2006,** *8* (5), 394–401.

80. Werning John, W. *Oral Cancer: Diagnosis, Management, and Rehabilitation;* Thieme Medical Publishers: New York, 2007, P 1, ISBN 978-1588903099.

81. Wilhelm, S. M.; Adnane, L.; Newell, P.; Villanueva, A.; Llovet, J. M.; Lynch, M. Preclinical Overview of Sorafenib, a Multikinase Inhibitor that Targets both Raf and VEGF and PDGF Receptor Tyrosine Kinase Signaling. *Mol. Cancer Ther.* **2008,** *7* (10), 3129–3140.

82. Yager, J. D.; Davidson, N. E. Estrogen Carcinogenesis in Breast Cancer. *New Engl. J. Med.* **2006,** *354* (3), 270–282.

83. Yao, V.; Berkman, C. E.; Choi, J. K.; O'Keefe, D. S; Bacich, D. J. Expression of Prostate-specific Membrane Antigen (PSMA), Increases Cell Folate Uptake and Proliferation and Suggests a Novel Role for PSMA in the Uptake of the Non-Polyglutamated Folate, Folic acid. *Prostate.* **2010,** *70* (3), 305–316.

84. Yin, X. M.; Oltvai, Z. N.; Korsmeyer, S. J. BH1 and BH2 Domains of Bcl–2 are Required for Inhibition of Apoptosis and Heterodimerization with Bax. *Nature.* **1994,** *369,* 321–323.

85. Yu, H.; Adedoyin, A. ADME-Tox Drug Discovery: Integration of Experimental and Computational Technologies. *Drug Discov. Today.* **2003,** *8,* 852–861.

INDEX